Build Your Own Transistor Radios

About the Author

Ronald Quan, a resident of Cupertino, California, has a BSEE degree from the University of California at Berkeley and is a member of SMPTE, IEEE, and AES. He has worked as a broadcast engineer for FM and AM radio stations, and for over 30 years he has worked for video and audio equipment companies (Ampex, Sony Corporation, Monster Cable, and Macrovision). Mr. Quan designed wide-band FM detectors for an HDTV tape recorder at Sony, and a twice color subcarrier frequency (7.16 MHz) differential phase measurement system for Macrovision, where he was a principal engineer. Also, at Hewlett Packard he developed a family of low-powered bar code readers that drained a fraction of the power consumed by conventional light pen readers.

Mr. Quan holds at least 70 U.S. patents in the areas of analog video processing, low-noise amplifier design, low-distortion voltage-controlled amplifiers, wide-band crystal voltage-controlled oscillators, video monitors, audio and video IQ modulation, in-band carrier audio single-sideband modulation and demodulation, audio and video scrambling, bar code reader products, and audio test equipment. In 2005 he was a guest speaker at the Stanford University Electrical Engineering Department's graduate seminar, talking on lower noise and distortion voltage-controlled amplifier topologies. In November 2010 he presented a paper on amplifier distortion to the Audio Engineering Society's conference in San Francisco's Moscone Center.

Build Your Own Transistor Radios

A Hobbyist's Guide to High-Performance and Low-Powered Radio Circuits

Ronald Quan

New York   Chicago   San Francisco   Lisbon
London   Madrid   Mexico City   Milan   New Delhi
San Juan   Seoul   Singapore   Sydney   Toronto

McGraw-Hill books are available at special quantity discounts to use as premiums and sales promotions, or for use in corporate training programs. To contact a representative, please e-mail us at bulksales@mcgraw-hill.com.

Build Your Own Transistor Radios

1 2 3 4 5 6 7 8 9 0   DOC/DOC   1 8 7 6 5 4 3 2

ISBN   978-0-07-179970-6
MHID   0-07-179970-2

This book is printed on acid-free paper.

Sponsoring Editor
Roger Stewart

Editorial Supervisor
Stephen M. Smith

Production Supervisor
Pamela A. Pelton

Acquisitions Coordinator
Molly T. Wyand

Project Manager
Patricia Wallenburg,
TypeWriting

Copy Editor
James K. Madru

Proofreader
Claire Splan

Indexer
Claire Splan

Art Director, Cover
Jeff Weeks

Composition
TypeWriting

Contents at a Glance

Contents

Preface

Ever since the invention of amplitude-modulated (AM) radio during the beginning of the twentieth century, there has been a unique interest on the part of hobbyists to build various types of receivers from the crystal radios of that era in the 1900s to the software-defined radios (SDRs) in the twenty-first century. In between the crystal receivers and SDRs are the tuned radio-frequency, regenerative, reflex, and conventional superheterodyne receivers. This book will cover these types of radios from Chapters 5 through 11. Chapter 12 will illustrate two types of front-end circuits for SDRs using analog and sampling methods to generate I and Q signals.

This book actually has two personalities. The first 12 chapters are organized as a do-it-yourself (DIY) book. Starting from Chapter 4, the reader can start building AM test generators for alignment and calibration of AM radios. Beginning from Chapter 5, low-powered tuned radio-frequency (TRF) or tuned radio designs are shown. By Chapter 11, a coil-less superheterodyne radio design is presented.

Throughout Chapters 4 to 12, there are circuit descriptions with some theory. Readers who have some electronics knowledge should find these technical descriptions of the radio circuits useful. Engineering students or engineers should relate even more to Chapters 4 through 11 based on their course work.

The last half of the book from Chapters 13 to 23 relates almost entirely to signals and circuits. These signals and circuits are explained in both an intuitive manner and with high school mathematics. Therefore, if the reader can still remember some basic algebra and trigonometry, the equations shown in these chapters should be understandable. Also, the mathematics behind Chapters 13 to 23 are presented in a step-by-step manner as if the equations or formulas are being written on a white board or chalk board during a lecture. Therefore, this book tries its best not to skip steps in the explanations.

Descriptions on how RF mixers and oscillators work, including large-signal behavior analysis, are generally taught in graduate engineering classes. However, even though modified Bessel functions are mentioned, the explanation of the modified Bessel functions is shown in an intuitive manner via tables and graphs.

For the engineer who has seen transistor amplifier analysis, this book will cover both small- and large-signal behavior, which also includes harmonic and intermodulation distortion.

Therefore, this book can be used as a complement to textbooks for circuits classes and lab courses. When this book was written, the goal was to design as many new or unconventional circuits as possible, along with some conventional examples. Any or some of these circuits can be used for teaching a laboratory electronics class in college and, to some degree, even in high school. For example, Chapter 23 shows how a circuit can be assembled on an index card without soldering any of the parts.

Ronald Quan

Acknowledgments

I would like to thank very much Roger Stewart for guiding me through the process of publishing my first book. Roger has been very supportive and helpful from the beginning, starting with the outline of the book. Also, I would like to give many thanks to Molly Wyand and the staff at McGraw-Hill. And, I wish to thank one of the editors, Patricia Wallenburg, who worked very hard in assembling this book.

And I really owe a great debt of gratitude to Paul Rako for posting two of my radios in *EDN Magazine*. Paul and the positive feedback from the readers of *EDN* encouraged me to write this book. Without Paul sending me to Roger Stewart at McGraw-Hill, the book would have never been written.

This project was very intense, and fellow engineer Andrew Mellows helped me immensely by proofreading all the material, including the manuscript and drawings, and by providing very helpful suggestions. Dr. Edison Fong also reviewed a couple of the more technical chapters and gave helpful suggestions.

I need to thank my mentors Robert G. Meyer, John Curl, John Ryan, and Barrett Guisinger, who all gave me a world-class education in electronics. And for all the crazy math in this book I thank my high school mathematics teacher, William K. Schwarze.

I also wish to thank the following people for their support and inspiration: Germano Belli, Alexis DiFirenzi Swale, Jo Acierto Spehar, and Jeri Ellsworth.

Finally, thank you to the members of my immediate family: Bill, George, Tom, and Frances. They all have taught me a great deal throughout the years.

And most of all, this book is dedicated to my parents, Nee and Lai.

Build Your Own Transistor Radios

Chapter 1

Introduction

This book will be a journey for both the hobbyist and the engineer on how radios are designed. The book starts off with simple designs such as an offshoot of crystal radios, tuned radio-frequency radios, to more complicated designs leading up to superheterodyne tuners and radios. Each chapter presents not only the circuits but also how each circuit was designed considering the tradeoffs in terms of performance, power consumption, availability of parts, and the number of parts.

In the engineering field, often there is no one best design to solve a problem. In some chapters, therefore, alternate designs will be presented.

Chapters 4 through 12 will walk the hobbyist through various radio projects. For those with an engineering background by practice and/or by academia, Chapters 13 through 23 will provide insights into the theory of the various circuits used in the projects, such as filter circuits, amplifiers, oscillators, and mixers.

For now, an overview of the various radios is given below.

Tuned Radio-Frequency (TRF) Radios

The simplest radio is the *tuned radio-frequency radio*, better known as the *TRF radio*. It consists mainly of a tunable filter, an amplifier, and a detector.

A tunable filter just means that the frequency of the filter can be varied. Very much like a violin string can be tuned to a specific frequency by varying the length of the string by using one's finger, a tunable filter can be varied by changing the values of the filter components.

Generally, a tuned filter consists of two components, a capacitor and an inductor. In a violin, the longer the string, the lower is the frequency that results. Similarly, in a tuned filter, the longer the wire used for making the inductor, the lower is the tuned frequency with the capacitor.

In TRF radios, there are usually two ways to vary the frequency of the tuned filter. One is to vary the capacitance by using a variable capacitor. This way is the most common method. Virtually all consumer amplitude-modulation (AM) radios use a variable capacitor, which may be a mechanical type such as air- or poly-insulated variable-capacitor type or an electronic variable capacitor. In the mechanical type of variable capacitor, turning a shaft varies the capacitance. In an electronic variable capacitor, known as a *varactor diode*, varying a voltage across the varactor diode varies its capacitance. This book will deal with the mechanical types of variable capacitors.

The second way to vary the frequency of a tunable filter is to vary the inductance of an inductor or coil via a tuning slug. This method is not used often in consumer radios because of cost. However, for very high-performance radios, variable inductors are used for tuning across the radio band. In this book, tunable or variable inductors will be used, but they will be adjusted once for calibration of the radio, and the main tuning will be done via a variable capacitor. Figure 1-1 shows a block diagram of a TRF radio.

Block Diagram of a TRF Radio

A TRF radio has a radio-frequency (RF) filter that is usually tunable, an RF amplifier for amplifying signals from radio stations, and a detector (see Figure 1-1). The detector converts the RF signal into an audio signal.

Circuit Description of a TRF Radio

For the AM radio band, the RF filter is tuned or adjusted to receive a particular radio station. Generally, an antenna is connected to the RF filter. But more commonly, a

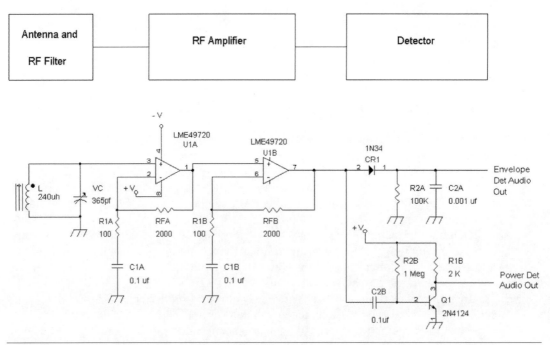

FIGURE 1-1   Block diagram and schematic of a TRF radio.

coil or an inductor serves as the "gatherer" of radio signals. The coil (L) may be a loop antenna (see Figure 3-1). A variable capacitor (VC) is used to tune from one station to another.

The output of the filter will provide RF signals on the order of about 100 microvolts to tens of millivolts depending on how strong a station is tuned to. Typically, the amplifier should have a minimum gain of 100. In this example, although the amplifier usually consists of a transistor, a dual op amp circuit (e.g., LME49720) is shown for simplicity. Each amplifier stage has gain of about 21, which yields a total gain of about 400 in terms of amplifying the RF signal.

The output of the amplifier is connected to a detector, usually a diode or a transistor, to convert the AM RF signal into an audio signal. A diode CR1 is used for recovering audio information from an AM signal. This type of diode circuit is commonly called an *envelope detector*.

Alternatively, a transistor amplifier (Q1, R1B, R2B, and C2B) also can be used for converting an AM signal into an audio signal by way of *power detection*. Using a transistor power detector is a way of demodulating or detecting an AM signal by the inherent distortion (nonlinear) characteristic of a transistor. Power detection is not quite the same as envelope detection, but it has the advantage of converting the AM signal to an audio signal and amplifying the audio signal as well.

Power-detection circuits are commonly used in regenerative radios and sometimes in superheterodyne radios.

It should be noted that in more complex TRF radios, multiple tuned filter circuits are used to provide better selectivity, or the ability to reduce interference from adjacent channels, and multiple amplifiers are used to increase sensitivity.

Regenerative Radio

This is probably the most efficient type of radio circuit ever invented. The principle behind such a radio is to recirculate or feed back some of the signal from the amplifier back to the RF filter section. This recirculation solves two problems in terms of providing better selectivity and higher gain. But there was another problem. Too much recirculation or regeneration caused the radio to oscillate, which caused a squealing effect on top of the program material (e.g., music or voice) (Figure 1-2).

Block Diagram of a Regenerative Radio

The regenerative radio in Figure 1-2 consists of a tunable filter that is connected to an RF amplifier. The RF amplifier serves two functions. First, it amplifies the signal from the tunable filter and sends back or recirculates a portion of that amplified RF signal to the tunable-filter section. This recirculation of the RF signal causes a positive-feedback effect that allows the gain of the amplifier to increase to larger than the original gain. For example, if the gain of the amplifier is 20, the recirculation technique will allow the amplifier to have a much higher gain, such as 100 or 1,000, until the amplifier oscillates. The second function of the amplifier is to provide

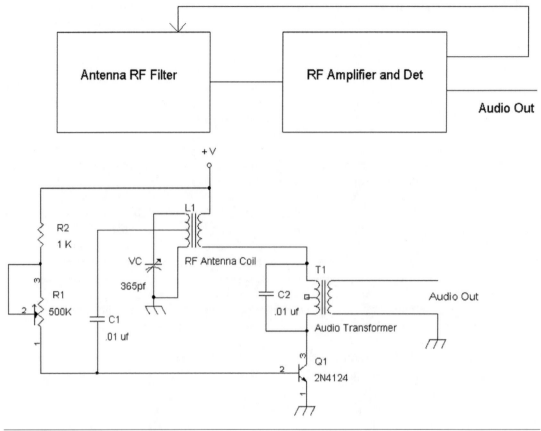

FIGURE 1-2   Block diagram and schematic of a regenerative radio.

power detection of the RF signal, which means that the amplifier also acts as an audio amplifier.

Circuit Description of a Regenerative Radio

In Figure 1-2, the tunable RF filter is formed by variable capacitor VC and antenna coil L1. Antenna coil L1 also has an extra winding, so this is more of an antenna coil-transformer. Also, because transistors have a finite load resistance versus the "infinite" input resistance of a vacuum tube or field-effect transistor, the base of the transistor is connected to a tap of antenna coil L to provide more efficient impedance matching.

 In a parallel-capacitor-coil resonant circuit (aka parallel-capacitance-inductance circuit) for a tunable RF filter, the quality factor, Q indicates that the higher the selectivity, the better is the separation of radio stations. A low Q in an antenna coil and variable-capacitor resonant circuit will allow unwanted adjacent stations to bleed into the tuned station. But a higher Q allows the RF signal of the desired tuned radio station to pass while attenuating RF signals from other stations. The Q in a parallel tank circuit is affected by the input resistance of the amplifier to which it is connected. The higher the input resistance, the higher the Q is maintained. So an amplifier with an input resistance on the order of at least 100 $k\Omega$ (e.g., typically 500 $k\Omega$ or more) allows for a high Q to be maintained. If an amplifier has a moderate input resistance (e.g., in the few thousands of ohms), tapping the coil with a stepped-down turns ratio allows the Q to be maintained, but at a tradeoff of lower signal output. For example, if an antenna coil has a 12:1 step-down ratio or 12:1 tap, the signal output will be $\frac{1}{12}$ in strength, but when connected to an amplifier of 3 $k\Omega$ of input resistance, the effective resistance across the whole coil and variable capacitor is $12 \times 12 \times 3$ $k\Omega$, or 432 $k\Omega$, which maintains a high Q.

Transistor Q1 serves a dual purpose as the RF amplifier and detector. The (collector) output signal of Q1 is connected to an audio transformer T1 that extracts audio signals from detector Q1, but Q1's collector also has amplified RF signals, which are fed back to coil L1 via the extra winding. By varying resistor R1, the gain of the Q1 amplifier is varied, and thus the amount of positive feedback is varied. The user tunes to a station and adjusts R1 to just below the verge of oscillation. Too much positive feedback causes the squealing effect. But when adjusted properly, the circuit provides very high gain and increased selectivity.

Reflex Radio

In a reflex radio, which also uses a recirculation technique, an amplifying circuit is used for purposes: (1) to amplify detected or demodulated RF signals and (2) to amplify RF signals as well. The demodulated RF signal, which is now an audio signal, is sent back to the amplifier to amplify audio signals along with the RF signal. So although reflex radios have similar characteristics as regenerative radios, they are not the same. Reflex radios do not recirculate RF signals back to the amplifier. And unlike regenerative radios, reflex radios do not have a regeneration control to increase the gain of the amplifier. A reflex radio would be essentially the same as a TRF radio but with use of a recirculation technique to amplify audio signals. Thus, in terms of sensitivity and selectivity, a reflex radio has the same performance as a TRF radio (Figure 1-3).

Block Diagram of a Reflex Radio

In Figure 1-3, the reflex radio consists of a tunable filter, an amplifier, and a detector. In essence, this reflex radio has the same components as the TRF radio in Figure 1-1. The difference, however, is that the output of the detector circuit (e.g., an envelope detector or diode), a low-level audio signal, is fed back and combined with the RF

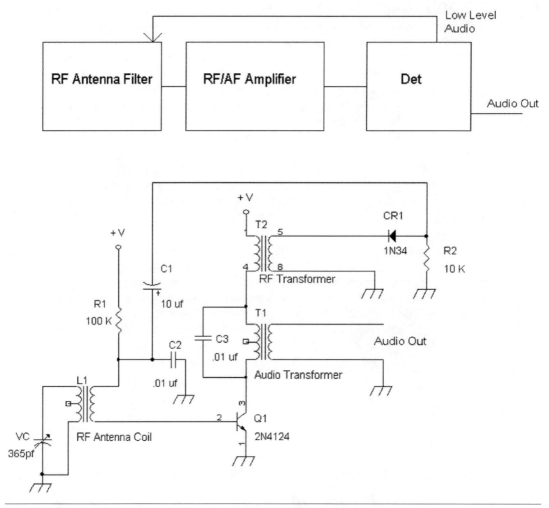

FIGURE 1-3   Block diagram and schematic of a reflex radio.

signal from the RF filter section. The audio output, which typically in other radios is taken from the output of the detector, is taken from the output of the radio-frequency/audio-frequency (RF/AF) amplifier instead.

Circuit Description of a Reflex Radio

The RF filter section is formed by variable capacitor VC and coil/inductor L1, which also has a (stepped-down) secondary winding connected to the base of transistor Q1. Note that the base of Q1 is an input for amplifier Q1. RF signals are amplified via Q1, and the RF signals are detected or demodulated by coupling through an RF transformer T2 to diode CR1 for envelope detection. At resistor R2 is a low-level audio signal that is connected to the input of Q1 via AF coupling capacitor C1 and the secondary winding of L1. RF coupling capacitor C2 is small in capacitance to direct RF signals to the emitter of transistor Q1 without attenuating the low-level audio signal. Audio transformer T1 is connected to the output of the amplifier at the collector of Q1. T1 thus extracts amplified audio signal for Q1.

Superheterodyne Radio

The superheterodyne radio overcomes shortfalls of the TRF, regenerative, and reflex radios in terms of sensitivity and selectivity. For example, the TRF and reflex radios generally have poor to fair selectivity and sensitivity. The regenerative radio can have high selectivity and sensitivity but requires the user to carefully tune each station and adjust the regeneration control so as to avoid oscillation or squealing.

A well-designed superheterodyne radio will provide very high sensitivity and selectivity without going into oscillation. However, this type of radio design requires quite a few extra components. These extra components are a multiple-section variable capacitor, a local oscillator, a mixer, and an intermediate-frequency (IF) filter/amplifier. In many designs, the local oscillator and mixer can be combined to form a converter circuit. Selectivity is defined mostly in the intermediate frequency filter (e.g., a 455-kHz IF) circuit. And it should be noted that an RF mixer usually denotes a circuit or system that translates or maps the frequency of an incoming RF signal to a new frequency. The mixer uses a local oscillator and the incoming RF signal to provide generally a difference frequency signal. Thus, for example, an incoming RF signal of 1,000 kHz is connected to an input of a mixer or converter circuit, and if the local oscillator is at 1,455 kHz, one of the output signals from the mixer will be 1,455 kHz minus 1,000 kHz, which equals 455 kHz.

One of the main characteristics of a superheterodyne radio is that it has a local oscillator that tracks the tuning for the incoming RF signal. So the tunable RF filter and the oscillator are tied in some relationship. Usually, this relationship ensures that no matter which station is tuned to in the oscillator, it changes accordingly such that the difference between the oscillator frequency and the tuned RF signal frequency is constant.

Thus, if the RF signal to be tuned is 540 kHz, the local oscillator is at 995 kHz, the RF signal to be tuned is at 1,600 kHz, and the local oscillator is at 2,055 kHz. In both cases, the difference between the oscillator frequency and the tuned RF frequency is 455 kHz.

Although the superheterodyne circuit is probably the most complicated system compared with other radios, it is the standard bearer of radios. Every television tuner, stereo receiver, or cell phone uses some kind of superheterodyne radio system, that is, a system that at least contains a local oscillator, a mixer, and an IF filter/amplifier.

Block Diagram of a Superheterodyne Radio

One of the main characteristics of a superheterodyne radio is that it has a local oscillator that tracks the tuning for the incoming RF signal (Figure 1-4).

The tunable RF filter is connected to an input of the converter oscillator circuit. The converter oscillator circuit provides an oscillation frequency that is always 455 kHz above the tuned RF frequency. Because the converter output has signals that are the sum and difference frequencies of the oscillator and the incoming tuned RF signal, it is the difference frequency (e.g., 455 kHz) that is passed through the IF filter and amplifier stage. So the output of the IF amplifier stage has an AM waveform whose carrier frequency has been shifted to 455 kHz.

To convert the 455-kHz AM waveform to an audio signal, the output of the IF amplifier/filter is connected to a detector such as a diode or transistor for demodulation.

Circuit Description of a Superheterodyne Radio

The tunable RF filter is provided by variable capacitor VC_RF and a 240-µH antenna coil ($L_{Primary}$) with a secondary winding ($L_{Secondary}$). The converter oscillator circuit includes transistor Q1, which is set up as an amplifier such that positive feedback for deliberate oscillation is determined by the inductance of Osc Transf 1 and variable capacitor VC OSC. In superheterodyne circuits both the VC RF and VC OSC variable capacitors share a common shaft to allow for tracking. At the base of Q1 there is the tuned RF signal, and at the emitter of Q1 there is the oscillator signal via a tapped winding from oscillator coil Osc Transf 1. The combination of the two signals at the base and emitter of Q1 results in a mixing action, and at the collector of Q1 is a signal whose frequency is the sum and difference of the tuned RF frequency and the oscillator frequency.

A first IF transformer (T1 IF) passes only the signal with a difference frequency, which is 455 kHz in this example. The secondary winding of T1 IF is connected to Q2's input (base) for further amplification of the IF signal. The output of Q2 is connected to a second IF transformer, T2 IF. The secondary winding of T2 IF is connected to the input (base) of the second-stage IF amplifier, Q3. It should be noted that in most higher-sensitivity superheterodyne radios, a second stage of amplification for the IF signal is desired. The output of Q3 is connected to a third IF transformer, T3 IF, whose output has sufficient amplitude for detector D2 to convert the AM 455-kHz signal into an audio signal.

Software-Defined Radio Front-End Circuits

A software-defined radio (SDR) is a superheterodyne radio in which there is a minimum of hardware components that allow a computer or dedicated digital logic chip to handle most of the functional blocks of the superheterodyne radio. So, in a typical SDR, the front-end circuits mix or translate the RF channels to a very low

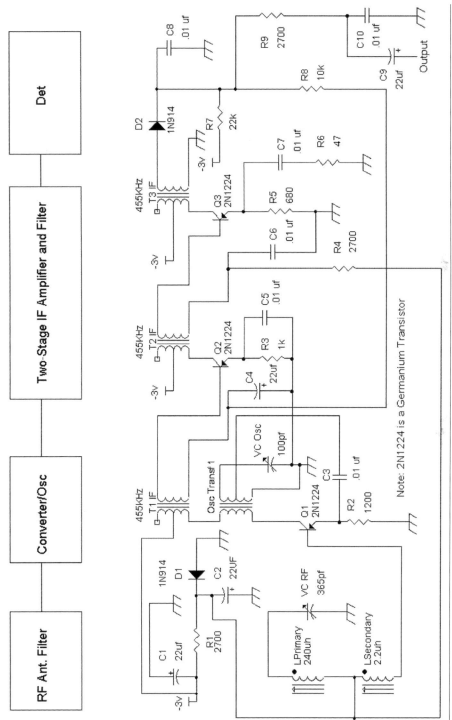

FIGURE 1-4   Block diagram and schematic of a superheterodyne radio.

9

IF (e.g., <455 kHz, such as 5 kHz to 20 kHz). This very low IF analog signal then is converted to digital signals via an analog-to-digital converter. The digital signal then is processed to amplify and detects not only AM signals but also frequency-modulated (FM) signals, single-sideband signals, and so on.

Fortunately, building a front-end circuit for a hobbyist's SDR is not too difficult. It involves a wide-band filter, a mixer, a local oscillator, and low-frequency amplifiers (e.g., bandwidths of 20 kHz to 100 kHz).

In the preceding description of a superheterodyne radio, a tuned filter preceded the converter oscillator or mixer. The tuned filter passes the station frequency that is desired and rejects signals from other stations to avoid interference. However, this tuned filter also rejects an "image" station that has a frequency twice the IF frequency away from the desired frequency to be received. Thus, for example, if the tuned station is 600 kHz, and if the tuned filter does not sufficiently attenuate an "image" station at 1,510 kHz (2×455 kHz + 600 kHz = 1,510 kHz), the image station will interfere with the 600-kHz station.

In SDRs, though, rarely is any variable tuned filter used at the front end. Instead, a wide-band filter is used. To address the image problem, the low-frequency IF signal is processed in a way to provide two channels of low-frequency IF. The two channels are 90 degrees out of phase with each other, forming an I channel and a Q channel. The I channel is defined as the 0-degree phase channel, and the Q channel is defined as the channel that is 90 degrees phase shifted from the I channel. It should be noted that having the I and Q channels allows for easy demodulation for AM signals via a Pythagorean process without the use of envelope detectors.

Block Diagram of a Software-Defined Radio Front End

Figure 1-5 shows a front-end system for a software-defined radio. An antenna is connected to a fixed (not variable) wide-band RF filter. The output of the wide-band RF filter then is connected to a two-phase quadrature mixer. This mixer generates two channels of low-frequency IF signals of 0 and 90 degrees via the oscillator, which also has 0- and 90-degree phase signals. The two channels (channel 1 and channel 2) form I and Q channels, which are amplified and sent to the stereo (audio) inputs to a computer or computing system. The computer then digitizes the I and Q channels, which contain a "block" of radio spectrum to be tuned to. For example, if the sound card in the computer samples at a 96-kHz rate, a bandwidth of 48 kHz of radio signals can be tuned into via the software-defined radio program in the computer.

A practical example would be listening into the 80-meter amateur radio band for a continuous-wave (CW) signal (Morse code), which spans from 3,675 kHz to 3,725 kHz (50 kHz of bandwidth). Most of this 50-kHz block of radio spectrum can be mixed down to about 100 Hz to 48 kHz. And the computer's software-defined radio program then can tune into each of the CW or Morse code carrier signals and demodulate them for the listener.

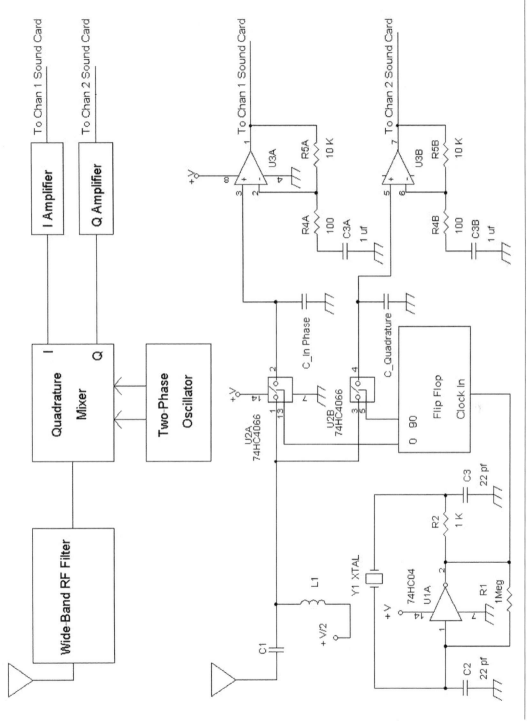

FIGURE 1-5 Block diagram and schematic for a front end of an SDR.

Description of Front-End Circuits for a Software-Defined Radio System

Figure 1-5 also shows an antenna (e.g., a long wire or whip antenna) connected to a fixed, nonvariable wide-band RF filter consisting of capacitor C1 and inductor L1. The output of this wide-band filter is connected to two analog switches that form a two-phase mixer via U2A and U2B. One switch, U2A, is toggled by a 0-degree phase signal from a flip-flop circuit, whereas the other switch, U2B, is toggled by a 90-degrees phase signal from another output terminal of the flip-flop circuit. The 0-degree switch U2A samples the RF signal from the RF filter and produces a low IF frequency. The sampling capacitor C_InPhase forms a low-pass filtering effect and thus provides a low-frequency IF signal to the I amplifier U3A. Similarly for U2B, the sampling capacitor C_Quadrature forms a low-pass filtering effect and also provides a low-frequency IF signal, which is amplified by the Q amplifier U3B. The output of the Q amplifier provides a low-frequency IF signal that is 90 degrees out of phase from the I channel amplifier's output.

Because the frequency of the IF signal is low, there is no need for special high-speed operational amplifiers (op amps). Moderate-bandwidth (e.g., 10 MHz to 50 MHz) op amps are sufficient to provide amplification.

The local oscillator circuit that provides the 0- and 90-degree signals for the quadrature mixer consists of a crystal oscillator and two flip-flop circuits. In a typical operation, the oscillator runs at four times the desired frequency for mixing, and the two flip-flop circuits provide the one times frequency for mixing while also generating 0- and 90-degree phase signals of the one times oscillator frequency.

The crystal oscillator consists of inverter gate U1A that serves as an amplifier bias via R1. Low-pass-filter circuit R2 and C3 along with crystal Y1 and C2 form a three-stage phase-shifting network to provide 180 degrees of phase shift at resonance or near resonance of the crystal, which allows for oscillation to occur at the crystal's frequency.

Comparison of the Types of Radios

Type	Sensitivity	Selectivity	Parts Count	Cost	Power
TRF	Low–medium	Low–medium	Low	Low	Low
Regenerative	High	High	Low	Low–medium	Low–medium
Reflex	Low	Low	Low–medium	Medium	Medium
Superhet	High	High	High	Medium–high	Medium
SDR without computer	Medium–high	High with computer	Low–medium	Medium	Medium

Chapter 2

Calibration Tools
and Generators for Testing

Many of the radios in this book will include adjustable inductors for intermediate-frequency transformers and oscillator coils. Therefore, special adjustment tools are required. This book will include not only radio projects but also circuits that require test generators to test and verify some of the electronics theory. For example, the generators will be helpful in testing both radio-frequency (RF) and audio-frequency amplifiers. Other test equipment, including a volt-ohm milliampere meter, an oscilloscope, a capacitance meter, and an inductance meter, will prove very useful for building and troubleshooting the various circuits.

Alignment Tools

The types of tools needed for the projects in this book are simple flat-blade alignment tools. These are used instead of flat-head screwdrivers so that there is no damage to adjustable inductors. Adjustable inductors use ferrite slugs to change the value of the inductor. Sometimes, when a regular screw driver is used, the ferrite slug can crack.

In addition, the alignment tools have thinner and wider blades than many screwdrivers, which relieve stress on the ferrite material. These alignment tools also have blades that are thin enough for adjustment to the trimmer or padder capacitors in poly-varicon variable capacitors (Figure 2-1).

The alignment tool in the middle of the figure below the board is suitable for adjusting the variable inductor and the poly-varicon's trimmer capacitors at the top right. The other alignment tool at the bottom of the figure is suitable for adjusting the inductor, but its blade is too thick for adjustment on the poly-varicon.

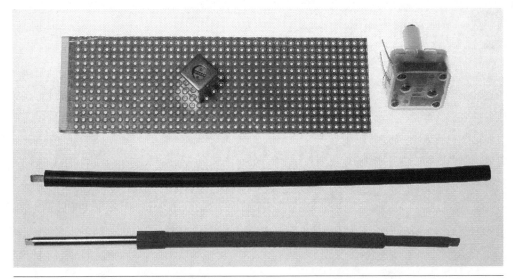

FIGURE 2-1   Alignment tools.

Test Generators

The radios shown in this book generally will not need test generators or test oscillators. Chapter 3 will show how to make two inexpensive test generators. However, buying a test generator or a test oscillator is always a good investment.

A function generator is a useful device because it will provide not just sine waves but usually triangle and square waves as well. Some will provide variable-duty-cycle pulses, and some also will provide an amplitude-modulated (AM) signal (Figure 2-2).

The function generator in the figure will produce waveforms from almost DC (direct current) to 2 MHz, which covers frequencies of the broadcast AM band plus the amateur radio 160-meter band. It also has an amplitude modulator. Thus, once a radio is built, this generator can be used for alignment and testing purposes.

Inductance Meter

An inductance meter is very handy to have around because some of the inductors for the radio projects in this book may not be available, and alternate inductors must be modified. An inductance meter allows hobbyists to wind their own coils and measure their inductances. For example, using normal hookup wire, one can wind an antenna coil to the correct inductance via an inductance meter. An inductance meter also can measure the value of unmarked coils, which then allows the hobbyist to determine the capacitance of a matching variable capacitor. Figure 2-3 shows an inductance meter measuring an inductor at 132.5 µH.

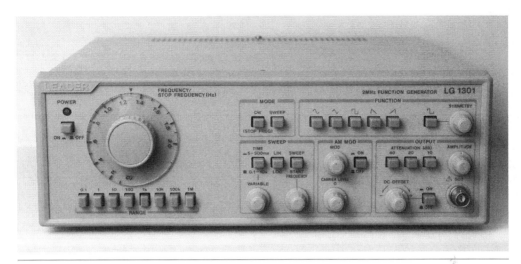

FIGURE 2-2   Function generator.

FIGURE 2-3   Inductance meter.

Capacitance Meter

The preceding inductance meter also measures the value of a capacitor. But there are also some digital volt-ohmeters that can measure capacitance. Variable capacitors come in a variety of values, such as maximum capacitances of 140 pF, 270 pF, 365 pF, and 500 pF. Often they are unmarked, and thus a capacitance meter is needed to determine their values (Figure 2-4).

The capacitance meter in the figure is measuring a 0.018-µF capacitor as 0.01744 µF. Note that it also will measure frequency (hertz or Hz). This particular EXTECH model measures up to 200 kHz. The company's newer unit, however, the EXTECH MN26T, will measure frequencies up to 10 MHz, which is suitable for measuring the frequency of oscillator circuits used in radio projects.

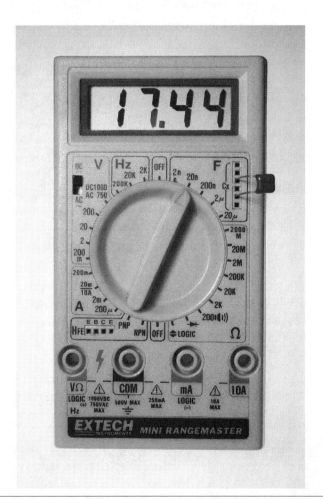

FIGURE 2-4   Capacitance meter.

Oscilloscopes

An oscilloscope is a voltage-measuring device that allows one to view voltages as a function of time. This instrument is useful in measuring signals from oscillators, amplifiers, and tuned RF circuits, as well as the AM signal.

But an oscilloscope is not really required for the projects in this book. However, having an oscilloscope allows the hobbyist to troubleshoot faster and understand radio and electronics better. The waveforms probed at particular parts of the radio reveal what is happening. A one-channel 10-MHz oscilloscope is a minimum requirement. Either an analog or a digital oscilloscope will suffice for the radio projects.

And often one can pick up a good used oscilloscope at an auction or on the Internet. *But beware of the sellers and make sure that there is a good return policy if the oscilloscope is defective.* Figure 2-5 is an example of a four-channel 200-MHz analog oscilloscope.

Radio Frequency (RF) Spectrum Analyzers

An RF spectrum analyzer allows one to view the frequency components of an RF signal. For example, one can view the spectral components of an AM signal, which includes the carrier and any of the AM signal's sidebands.

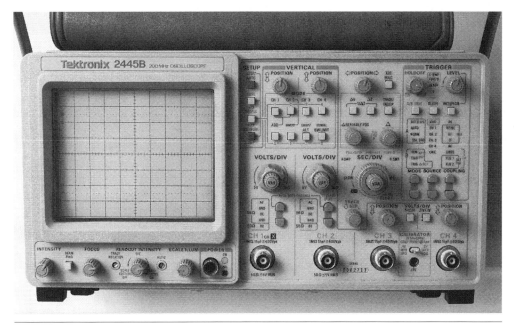

FIGURE 2-5   Oscilloscope.

Fortunately, this book will not require any RF spectrum analyzers. But one can download a program from the Internet to convert your computer into a low-frequency (e.g., 1 Hz to 22 kHz for a 44.1 kHz sampling rate, or 1 Hz to 48 kHz for a 96 kHz sampling rate) spectrum analyzer. For example, download the Spectran program from the web at http://digilander.libero.it/i2phd/spectran.html.

Where to Buy the Tools and Test Equipment

1. Digi-Key Corporation at www.digikey.com
2. Mouser Electronics at www.mouser.com
3. Frys Electronics at www.frys.com
4. MCM Electronics at www.mcmelectronics.com
5. Jameco Electronics at www.jameco.com

Chapter 3

Components and Hacking/ Modifying Parts for Radio Circuits

This chapter will present some of the basic components or parts needed for building radios. These components include variable capacitors, antenna coils, and transformers. Other parts that will be used in the projects include transistors, diodes, capacitors, and inductors.

Antenna Coils

Basically, the antenna coils that will be used in this book are the ferrite rod or ferrite bar types (Figure 3-1). These types of antenna coils are used commonly in all portable amplitude-modulated (AM) broadcast radios. They are small in size but receive radio-frequency (RF) signals equivalently in strength to the older, large air-core-loop antennas.

The antenna coil at the top of the figure is much longer than the other two, which allows for more sensitivity. That is, given the same RF signal, the longer rod antenna coil will yield more signal at its coil winding. This coil also has a secondary winding, which is "stepped" down by 10- to 20-fold to load into low-impedance transistor amplifiers. The primary winding of this antenna coil is normally connected to a tuning capacitor (variable capacitor). The primary winding inductance was measured at 430 µH, which matches with a variable capacitor of about 180 pF to 200 pF.

In the center of the figure is an antenna coil that is more miniaturized and will have less sensitivity to the antenna coil at the top of the figure. However, its primary winding inductance is actually higher at about 640 µH, which matches to a (more commonly available) 140-pF variable capacitor. This antenna coil also has a secondary winding that is stepped down.

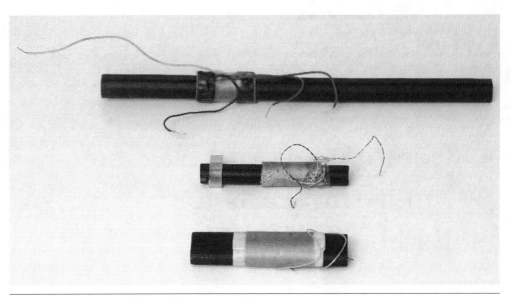

FIGURE 3-1 Ferrite-bar/rod antenna coils.

Finally, the bar antenna coil at the bottom of the figure has an inductance of about 740 μH. At 740 μH of inductance, this is a bit higher than needed, and some portion of the winding will have to be removed for use with standard 140-pF, 180-pF, 270-pF, or 365-pF variable capacitors.

It should be noted that all three antenna coils in Figure 3-1 allow changing the inductance further by sliding the coil to different locations on the ferrite rod or bar. For example, to increase inductance, slide the coil to the middle, and to decrease inductance, slide the coil toward either end of the rod or bar.

Ferrite antenna coils are readily available on the Web such as on eBay. An alternative to making an antenna coil is to buy ferrite rods or bars and wind your own coil. The ferrite material should be at least 2 inches long, and a paper insert of about 1.5 inches should be wrapped around the ferrite material such that the insert can slide. The magnet wire of about 30 American Wire Gauge (AWG) or No. 40 Litz wire is wound in a single layer over about 1.3 inches of the paper insert. With an inductance meter, measure the inductance when the insert is in the middle of the ferrite material and when it is toward the end of the ferrite material. If there is too much inductance, unwind some of the wire while measuring the inductance. If there is not enough inductance, splice the wire by soldering and wind in the same direction as the first single layer.

In most high-fidelity home stereo receivers today, the AM radio antenna is just an air dielectric loop (Figure 3-2).

The loop antenna in this figure has insufficient inductance to work with any of the standard variable capacitors (e.g., 140 pF to 365 pF). Therefore, this antenna is connected to a step-up RF transformer, and the RF transformer is matched with a

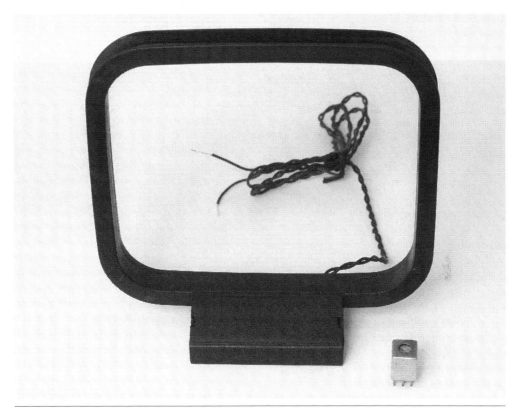

FIGURE 3-2   AM band loop antenna.

standard variable capacitor. In this book, oscillator coils and/or hacked intermediate-frequency (IF) transformers (see lower right-hand corner of Figure 3-2) will be used as the RF transformer for these types of loop antennas. It should be noted that these types of loop antennas are commonly available at MCM Electronics as replacement antennas for stereo receivers.

Variable Capacitors

These days, choosing variable capacitors for AM radios is limited to roughly two types of poly-varicon variable capacitors. Poly-varicon variable capacitors use polyester sheets between the plates as opposed to air-dielectric variable capacitors (Figure 3-3). A multiple gang variable capacitor such as a two, three, or four gang variable capacitor refers to the number of sections it has and all sections share a common tuning shaft. In general, a multiple gang variable capacitor is equivalent to a multiple section variable capacitor. However, some multiple section variable capacitors such as dual trimmer variable capacitors have two independent adjustments for varying the

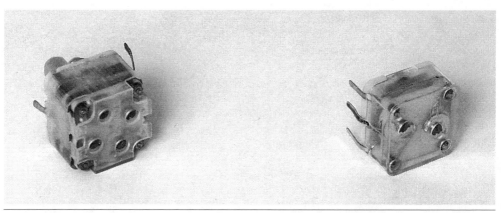

FIGURE 3-3   Variable capacitors using poly material for insulation between plates.

capacitance of each section. For this book, the main tuning capacitor is described as an "x" gang variable capacitor or equivalently, an "x" section variable capacitor.

The capacitor on the left in the figure is a twin-section variable capacitor, which commonly has 270 pF in each section. This type of variable capacitor is ideal for a one- or two-section tuned radio-frequency (TRF) radio. It should be noted that a 2.5-mm metric screw is used for adding an extended shaft (via a spacer).

For a superheterodyne radio, the first 270-pF section is matched with an antenna transformer or antenna coil of 330 μH and then with a series capacitor of about 300 pF to 330 pF with the second 270-pF section and a 180-μH coil to form an oscillator/ converter circuit.

Four trimmer capacitor shown, but only two are needed. This allows adding roughly up to 20 pF to the main sections of 270 pF. In some cases, the twin variable capacitor comes with 330 pF for each section instead of 270 pF.

On the right side of Figure 3-3 is another variable capacitor. This capacitor has two unequal sections. One section at 140 pF is dedicated to the antenna coil or antenna transformer, and the other section at 60 pF is used for an oscillator circuit. For identifying the various sections of the twin variable capacitor, see Figure 3-4.

For the twin gang/section variable capacitor, the trimmer capacitors' ground connection is internally connected to the ground tab, as seen in the figure. The trimmer-tab connections in general should be tied to each associated main section's connections (e.g., trimmer section A to main section A and trimmer section B to main section B). A detailed description of the second poly-varicon variable capacitor is provided in Figure 3-5.

The two trimmer capacitors for the variable capacitor in this figure are always internally tied to each of the antenna section and the oscillator section. Be careful to note the labeling of each tab or lead for "Ant." and "Osc." because they are not the same capacitance. If you look carefully, you can see that the oscillator section tab of a two-section variable capacitor leads to fewer plates than the upper tab, which is marked "Ant."

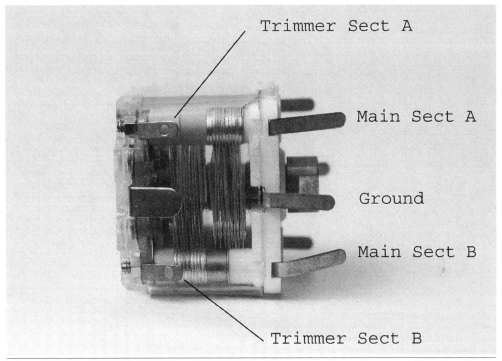

FIGURE 3-4   Twin gang/section variable capacitor.

FIGURE 3-5   Two gang variable capacitors for typical transistor radios.

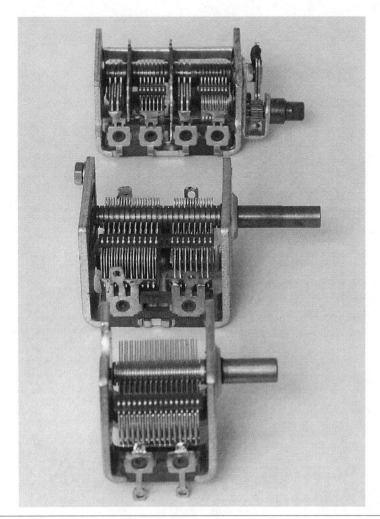

FIGURE 3-6   Air-dielectric variable capacitors.

Figure 3-6 shows various air-dielectric variable capacitors, which in general are available on the Web or via eBay.

The first capacitor shown at the bottom of the figure is a single-section 365-pF variable capacitor. No trimmer capacitor is attached. This type of variable capacitor can be used for TRF, reflex, and regenerative radios but not for superheterodyne radios, which require two or more sections.

In the center of the figure is a two-section (two gang) variable capacitor. This particular capacitor has an antenna section on the left side and an oscillator section on the right side. Notice that the number of plates on the left side (antenna section) is larger than the number of plates on the right side (oscillator section). Not shown, but on the other side of this variable capacitor are adjusting screws for each trimmer capacitor that are connected internally to the antenna and oscillator sections.

Finally the variable capacitor at the top of Figure 3-6 is a four-section capacitor. Two sections are used for the AM band, and the other two are used for the FM band. Since this book is limited to AM radios, only two will be used. Counting from left to right, section 2 is for the oscillator, and section 4 is used for the antenna of a superheterodyne receiver. Also not shown are the trimmer capacitor screw adjustments on the other side of this variable capacitor.

Transistors

In terms of transistors, the most common lead (terminal) configuration is E, B, C (emitter, base, collector), as shown in Figure 3-7.

The figure shows the most common lead configuration for the most common transistors, such as 2N4124, 2N4126, 2N3904 (as shown in the figure), 2N3906, 2N5087, and 2N5089. If other transistors are substituted, the lead configuration diagram can be

FIGURE 3-7   Common silicon transistor.

downloaded from the Web (e.g., go to Google, and type in "2Nxxxx, BCxxx, etc. data sheet"). As shown in Figure 3-7, the base and collector leads are next to each other, causing increased internal capacitance between the base and collector leads.

For high-frequency transistors, which will be used in some designs in this book, the lead terminal configuration is changed to B, E, C (base, emitter, collector), as shown in Figure 3-8.

For high-frequency performance, the interelectrode capacitance between the base and collector must be minimized. Hence, in this configuration, the base and collector leads are placed as far apart from each other as possible. An example is the MPSH10, which is used in some circuits in this book. Note that the order of its leads is B, E, and C, which is different from the more common arrangement of E, B, and C.

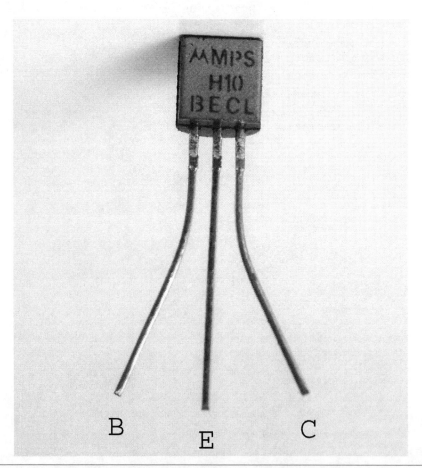

FIGURE 3-8 High-frequency silicon transistor.

Earphones

After the AM signal is demodulated and amplified, the listener can use two types of earphones or headphones, either magnetic or crystal (piezoelectric), as shown in Figure 3-9.

Most magnetic earphones or headphones have impedances from 8 Ω to 32 Ω (ohms). However, for lower-power radios, it is recommended to use impedances of 500 Ω or more. For a very low-powered radio, the crystal earphone or headphone is the best choice. It has an impedance in the thousands of ohms and is suitable for crystal radios as well.

Magnetic   Crystal

FIGURE 3-9   Earphones.

Speakers

There will be some radios in this book that are not as low powered and can drive loud speakers. A speaker with mounting holes is desired for assembling on a chassis or board via spacers and bolts. Figure 3-10 shows 3-inch and 4-inch speakers.

Passive Components

Passive components such as resistors used in the designs are in general 5 percent ¼-W resistors but 1 percent types may be used (Figure 3-11).

As seen in the figure, generally a 5 percent ¼-W resistor on the bottom will do just fine for most of the designs in this book. More precise 1 percent resistors such as the types shown at the top and center of the figure can be used for circuits requiring more precision, such as phase-shifting circuits or timing or oscillator circuits.

FIGURE 3-10 Loudspeakers.

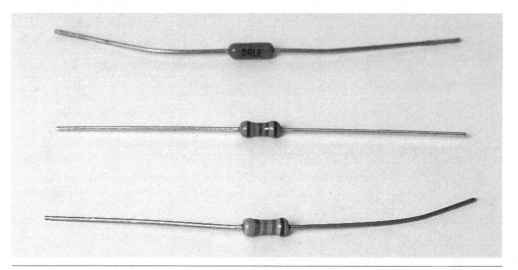

FIGURE 3-11 Resistors.

Fixed inductors also will be used in some of the designs, and some of them actually look like resistors (Figure 3-12). The color code is the same as for resistors, in microhenrys (μH). For example, yellow, violet, and red equal 4,700 μH or 4.7 mH. To make sure of an inductor's value, though, measure with an inductance meter. Generally, the resistance in ohms of the inductor will not match the inductance in millihenrys (mH) or microhenrys (μH). Therefore, if an inductor that looks like a

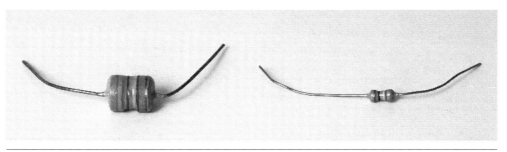

FIGURE 3-12   Inductors.

resistor is measured with an ohm meter, the resistance measurement will not match the color code and the user can then deduce that the component is most likely an inductor instead of a resistor.

Other types of inductors may come in unfamiliar shapes, such as the examples shown in Figure 3-13.

For the inductor in this figure, on the left side the value says "221," which really means 22 plus 1 zero following the 22, for 220 µH. The inductor on the right has three sets of numbers, but the only one that seems to make sense is "330." But does this mean that the inductance value is 33 plus 0 zero after 33, for 33 µH, or does it mean that the value is literally 330 µH? The measured value is 330 µH.

Therefore, in the case of inductors and capacitors, it is always advisable to measure the component first to make sure (e.g., with an inductance meter for coils and a capacitance meter for capacitors).

For small-value capacitors (5 pF to 4,700 pF), generally, ceramic disk or silver mica types are used. For stability and accuracy, though, silver mica capacitors are preferred (Figure 3-14).

Film (e.g., polyester or Mylar) capacitors are also used (Figure 3-15).

FIGURE 3-13   Other inductors/coils.

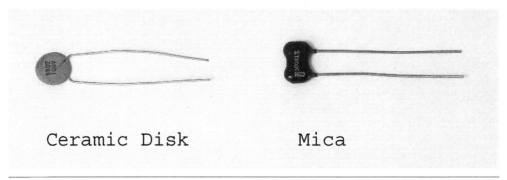

Ceramic Disk   Mica

FIGURE 3-14 Small-value capacitors.

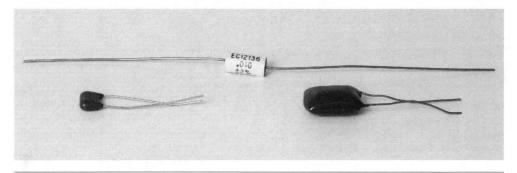

FIGURE 3-15 Film-dielectric capacitors.

The markings on many film capacitors are done in two ways. One is in microfarads. Thus the capacitor at the top of the figure is marked ".010," meaning 0.01 µF. The capacitors at the bottom of the figure are usually marked with a three-digit code in picofarads (pF), with the third digit denoting the number of zeros following the first two digits. For example, "102" means 10 plus two zeros after 10 = 1,000 pF or 0.001 µF.

Ceramic, mica, and film capacitors can be connected with the leads either way. But common electrolytic capacitors are polarized and thus are *not* the type of capacitors that can have leads switched without a problem. The schematic diagram will point out how the electrolytic capacitors should be connected. See the electrolytic capacitors in Figure 3-16, where the markings indicate negative or (–).

Audio transformers will be used in reflex, regenerative, and superheterodyne radios (Figure 3-17). These audio transformers will be used for extracting audio signals from an amplifier and providing signals suitable to drive earphones or speakers. They will have primary and secondary windings (e.g., at least four leads). Audio transformers normally are available in two types, input or driver transformers and output transformers. In newer audio amplifier designs, audio transformers are not used and are replaced by integrated circuits or discrete transistors.

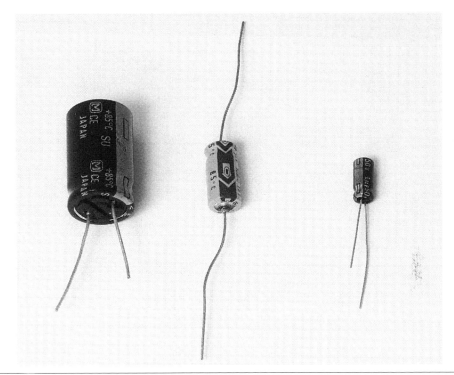

FIGURE 3-16   Polarized electrolytic capacitors.

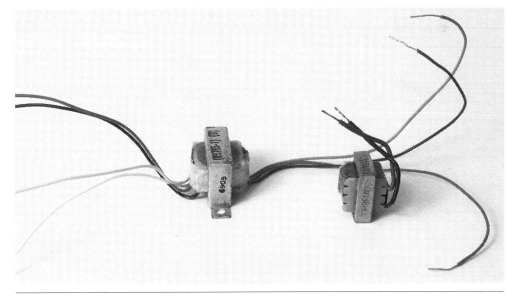

FIGURE 3-17   Audio transformers.

FIGURE 3-18   Variable resistors or potentiometers.

Single-turn variable resistors or potentiometers are shown in Figure 3-18. These generally are used for volume control, especially the ones shown on the right.

Multiple-turn variable resistors or potentiometers will be used for adjusting frequency in an oscillator or levels of a signal. They offer a more precise adjustment than single-turn types (Figure 3-19). The markings can be of two types. One is the exact value in ohms, such as 10K = 10,000 Ω or 10 kΩ. And the other can be a three-digit code where the last digit means the number of zeros added after the first two numbers, for example, 204 = 20 plus 4 zeros = 200000 = 200,000 Ω = 200 kΩ.

Finally, here are a few words about diodes (Figure 3-20). The marking on small signal diodes usually is a band or stripe that denotes the cathode. These diodes are inserted into a perforated (perf) board by bending the leads to at least ¼ inch of lead length from each side of the body of the diode to avoid stress that otherwise would crack the glass casing. See the example in Figure 3-21.

FIGURE 3-19   Multiturn variable resistors.

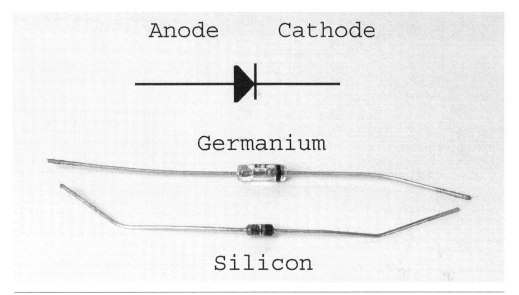

FIGURE 3-20 Diodes.

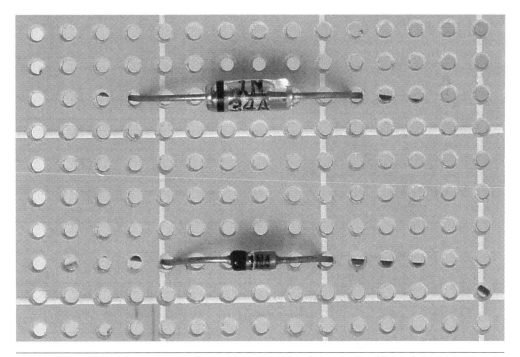

FIGURE 3-21 Diodes mounted on a board with adequate lead length prior to bending the leads.

Vector and Perforated Boards

Vector and perforated boards can be used for building radios (Figure 3-22). On the left is a perforated (perf) board. Although the radios can be built using a perf board, a vector board with a ground plane is preferred. And on the right is a vector board with the ground-plane side up. Generally, the components are placed on the ground-plane side, and wiring is done on the opposite side (the side without a ground plane).

It should be noted that vector and perf boards generally have 100-mil spacing, which is suitable for mounting dual in-line integrated circuits, headers (berg connector heads), and 7-mm coil. A 10-mm coil can be mounted as well by rotating the coil by 45 degrees.

Copper-clad boards, while not as "permanent" in building radios, are a good choice for building the first prototype radios (Figure 3-23). Debugging on copper-clad boards is much easier than on vector or perf boards.

FIGURE 3-22 Perforated (perf) and vector boards.

FIGURE 3-23 Copper-clad board.

Hardware

The hardware needed for the projects in this book includes 6-32, 4-40, and 2-56 nuts and bolts, along with washers (Figure 3-24). Not shown are 2.5-mm metric screws, which are used with poly-varicons (poly-variable capacitors).

The tie points shown in Figure 3-25 are useful for connecting wires.

FIGURE 3-24 Nuts and bolts.

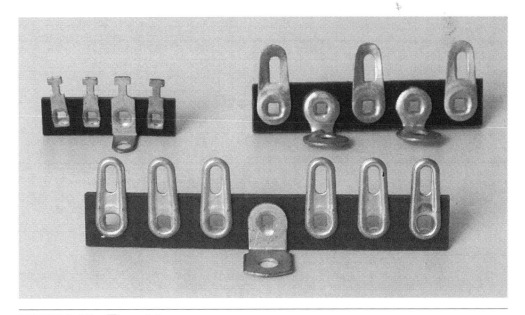

FIGURE 3-25 Tie points.

To mount the speaker, antenna coil, or vector or perf board, spacers are needed. The ones shown in Figure 3-26 use 4-40 threads, but other spacer sizes may be used.

Generally, an oscillator coil, which has a tapped primary winding and a secondary winding, is used for the local oscillator circuit for superheterodyne receivers. However, some have an inductance value suitable for an RF amplifier circuit. See the oscillator coil in Figure 3-27, which is generally used with a 60-pF variable capacitor for an oscillator circuit but is also suitable for use with a loop antenna.

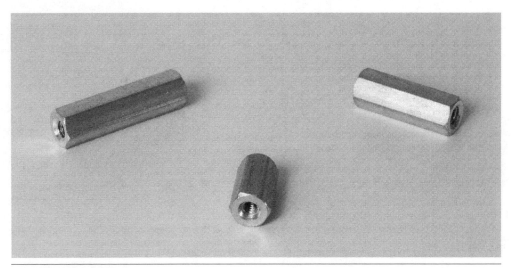

FIGURE 3-26   Spacers.

FIGURE 3-27   Oscillator coil, IF transformer, or adjustable coil/inductor.

It was found that some intermediate-frequency (IF) transformers can be converted into oscillator coils by chipping off their internal capacitors. Hacking IF transformers into oscillator or RF coils can be very useful because the inductances of IF transformers and their turns ratios make them ideal in some cases (e.g., for a low-power superheterodyne radio) (Figure 3-28).

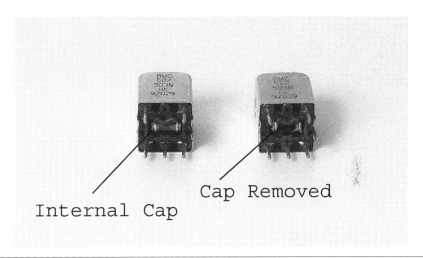

FIGURE 3-28   IF transformer with internal capacitor and IF transformer with internal capacitor removed.

Parts Suppliers

See Appendix 1.

Chapter 4

Building Simple Test Oscillators and Modulators

Before any of the radios are presented, it would be preferable to build some simple test equipment. This chapter introduces the amplitude modulated (AM) signal and various ways to generate radio-frequency (RF) and intermediate-frequency (IF) signals for testing radio projects.

The Continuous-Wave Signal

A *continuous-wave signal* is defined by a fixed amplitude and a fixed frequency. In an audio continuous-wave signal, this is better known as a *tone*.

The simplest continuous waveform is a sine-wave signal such as seen in Figure 4-1. The sine-wave signal may represent an RF signal.

However, a continuous wave (CW) signal, on its own, does not send much information, such as music or voice. A CW signal can be turned on and off for sending coded messages or for controlling a device. For example, Morse code is sent for providing alphanumeric information. Or the condition of a CW signal that is on can be used to keep a relay switch in the "on" state.

To convey voice or music, however, the CW signal must be modulated in some form. That is, the amplitude, phase, and/or frequency of the CW signal must be related to another signal. In this book, amplitude modulation (AM) is discussed because the receivers we will be building are AM radios. Figure 4-2 shows a lower-frequency signal that is a modulating signal representing an audio signal.

The Amplitude-Modulated Signal

An amplitude-modulated (AM) signal is generated by changing the amplitude of the RF CW signal (e.g., see Figure 4-1) with a modulating signal (e.g., see Figure 4-2). At first

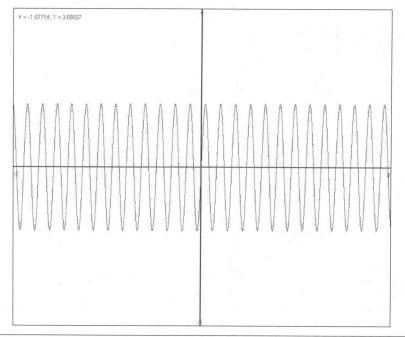

FIGURE 4-1   A continuous-wave (sine-wave) signal.

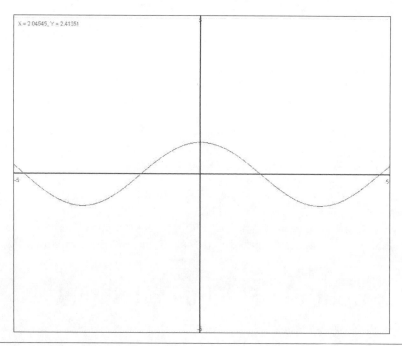

FIGURE 4-2   A lower-frequency signal such as a modulating signal.

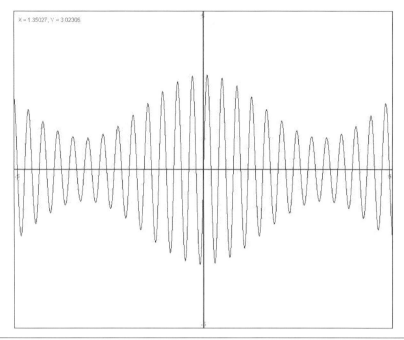

FIGURE 4-3   An amplitude-modulated (AM) signal.

glance, Figure 4-3 looks as if the CW RF signal from Figure 4-1 has been multiplied by the audio signal, as seen in Figure 4-2. Well, not quite, but almost.

If the audio signal is faded to silence, we would get zero (flat line) in Figure 4-2, which when multiplied by the RF carrier waveform of Figure 4-1 would result in a flat line or zero as well (zero times anything is zero). But we know that when the audio signal is faded to zero, we should just get the unmodulated waveform or CW signal, as seen in Figure 4-1.

So the answer really is that the RF CW waveform of Figure 4-1 is multiplied by (1 + the audio waveform of Figure 4-2) = the amplitude-modulated (AM) waveform of Figure 4-3.

The 1 added to the audio waveform is essential in preserving the shape of the modulating audio waveform on the envelope of the modulated RF CW signal in Figure 4-3. In this way, if the audio waveform fades to zero, we just get Figure 4-1, a CW RF signal, which is what we would expect.

First Project: A CW RF Test Oscillator

Before the specific schematic is shown, let's take a look at the big picture. A test generator consisting of a CW oscillator will be used for testing the tuning range, 535 kHz to 1,605 kHz, of the radios. A modulator will be added with an audio-frequency

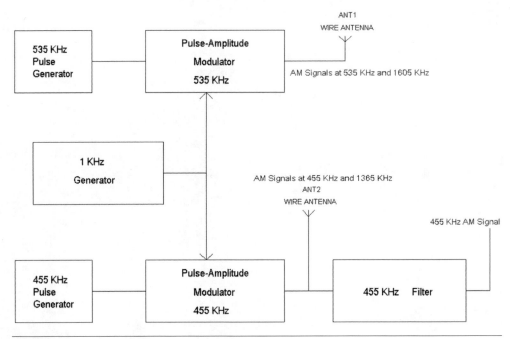

FIGURE 4-4   A block diagram of a test generator.

oscillator to produce amplitude modulation of the CW signal. Similarly, a second test oscillator is shown for tuning the 455-kHz IF amplifiers of superheterodyne radios. Figure 4-4 presents a block diagram of the test oscillator system.

In this block diagram, there are two CW oscillators at 535 and 455 kHz. Both CW oscillators produce pulses, not sine waves. Pulses also generate harmonic frequencies of 535 kHz and 455 kHz. Thus, using pulse generators avoids the need to build extra oscillators at other frequencies. For example, the third harmonic of 535 kHz is 1,605 kHz, and the third harmonic of 455 kHz is 1,365 kHz.

Each pulse generator is fed to a pulse modulator, which changes the amplitude of the 535-kHz and 455-kHz pulses via a 1-kHz audio generator. The output of the modulators then provides amplitude-modulated (AM) signals at 535 kHz and 455 kHz and their respective harmonics.

A simple wire may be placed near the radio to confirm the tuning range of the radio at the low end, 535 kHz, and at the high end, 1,605 kHz. The output of the 455-kHz modulator is connected to a 455-kHz band-pass filter to provide only an amplitude-modulated (AM) signal at 455 kHz for IF amplifier alignment in superheterodyne radios.

The entire test oscillator is shown in Figure 4-5. However, each functional section will be covered separately.

Now let's take a look at a pulse generator (CW) (Figure 4-6).

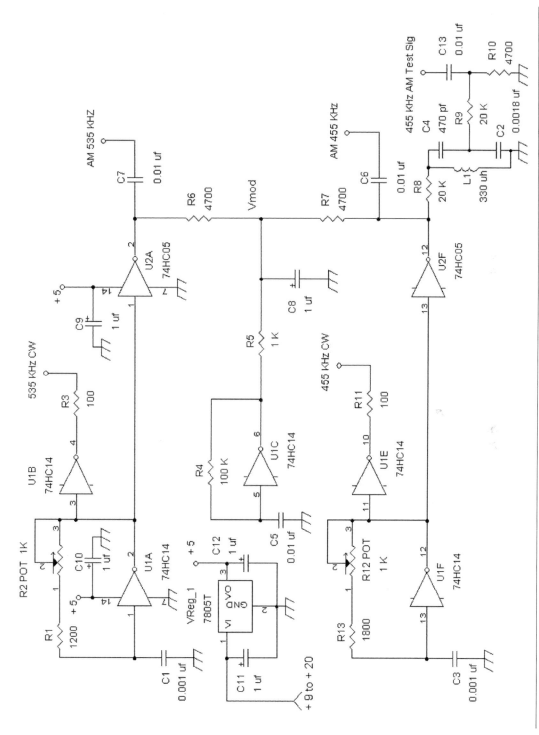

FIGURE 4-5 Schematic of the test generator.

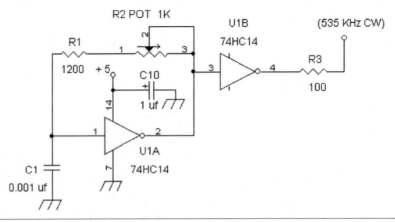

FIGURE 4-6   Pulse-generator circuit for a CW signal.

Integrated circuit U1A (74HC14) is a special logic inverter with an input circuit that has a hysteresis characteristic via a Schmitt trigger, which makes it ideal for making a relaxation oscillator for generating CW waveforms. Basically, the tripping-point voltage at the input changes depending on what the output state is.

When the output is high at pin 2, the input at pin 1 rises slowly owing to the low-pass filtering effect of R2, R1, and C1. *Note: R2 preferably is a multiturn trim pot.* Eventually, the voltage at pin 1 rises sufficiently to cause the inverter to output a low signal. Once the output is low, though, the input trigger point at the input is also lower. Thus capacitor C1 has to discharge to that lower trigger point, Vlow_trigger. When it does, the inverter goes high but also causes the input to trigger at a higher voltage, Vhigh_trigger. As a result of the dynamic nature of how the input trigger voltage changes, an oscillating signal occurs between the voltages of Vhigh_trigger and Vlow_trigger.

Both Vhigh_trigger and Vlow_trigger will vary as a function of the supply voltage. Thus a supply voltage that is not regulated will cause a shift in oscillating frequency. Therefore, it is advised to *use a regulated 5-V supply* when stability of the preset frequencies is desired.

The 74HC14 inverter gate can provide oscillating waveforms to at least 10 MHz and at nearly but not perfectly symmetric square waves. The frequency of oscillation is about $1/[(R1 + R2)C1]$ in Figure 4-5. R2 is adjusted to 535 kHz via measuring with a frequency counter (e.g., a digital voltmeter with frequency counter).

If a frequency counter is not available, use a radio with digital readout tuned to 1,070 kHz (second harmonic of 535 kHz), connect a wire to the R3 terminal, and place the wire near the radio. Listen for the radio's hiss level to go down when tuned to 1,070 kHz while adjusting R2.

For higher oscillation frequencies, one can use a 74AHC14 gate instead, but the 74HC14 type is more common. However, when using a 74AHC14 gate, values of C1, R1, and R2 may change.

Other inverters can be used, such as the 74C14, which is slower in speed than the 74HC14 but should work. Also, the 74HCT14 will work, but the frequency-of-oscillation formula is not the same as for the 74HC14, so some experimentation by the reader is required.

For really high-frequency oscillations, a 74AC14 or 74ACT14 will work up to frequencies well beyond 30 MHz (probably up to about 70 MHz or 100 MHz), but the reader will have to experiment to determine the resistor and capacitor combinations.

As a starting point, if one wants to experiment with inverter gates other than the 74HC14, try using C1 = 0.001 µF, and replace R1 with a 470-Ω resistor and R2 with a 5-kΩ multiturn pot. Also, the waveform symmetry of a 74HCT14 (or 74ACT14) inverter oscillator circuit is not as close to a 50 percent duty cycle as the 74HC14 (or 74AC14) part.

Modulator Circuit for the CW Generator

The type of amplitude modulator that is used is known as a *transistor drain modulation* (Figure 4-7). U2A (74HC05) is an inverter gate with an open-drain output. This means that the output is a short circuit to ground when the input (pin 1) is logic high and an open circuit when the input to U2A is logic low.

U1E (74HC14), another hysteresis oscillator circuit, supplies a 1-kHz signal to R5 and C8. The voltage across capacitor C8 is a triangle wave with a direct-current (DC) offset voltage to form a modulation voltage, Vmod. A CW signal (in this case from a 535-kHz hysteresis oscillator) is connected to the input of open-drain inverter gate 74HC05. By varying the voltage on pull-up load resistor R6, the output of U2A generates pulse amplitude modulation on the CW signal.

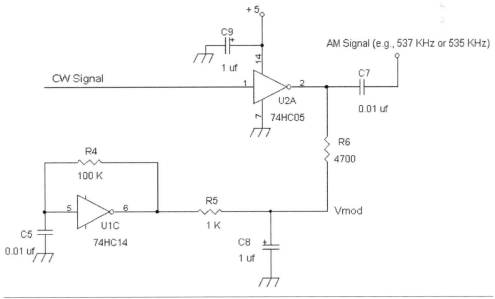

FIGURE 4-7   Pulse-amplitude-modulating circuit.

Now let's take a closer look at a pulse-amplitude-modulation circuit and its signals.

In Figure 4-8, a pulse amplitude modulator switches a CW signal on and off via a load or pull-up resistor. The CW signal causes the output to switch in two signal voltages, zero when the switch is grounded and V when the switch is in the open position.

When V, the pull resistor's voltage source, is at a fixed voltage, the output would see something like Figure 4-9.

When V is increased, we would see something at the output like Figure 4-10.

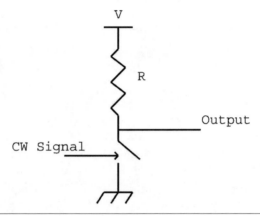

FIGURE 4-8 Modulating circuit—voltage V is varied to provide pulse amplitude modulation.

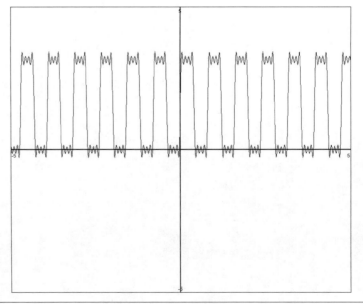

FIGURE 4-9 CW pulse signal for a fixed voltage V.

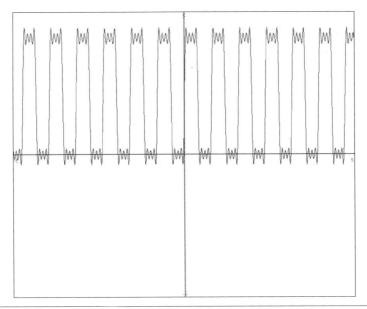

FIGURE 4-10   CW pulse signal is increased when V is increased.

Now, if we replace V with a voltage source that has both DC and alternating-current (AC) voltages, we can produce a pulse amplitude modulation, as seen in Figure 4-11.

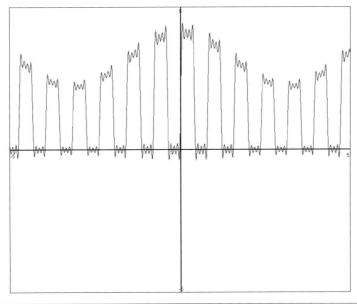

FIGURE 4-11   Pulse-amplitude-modulated signal.

Thus the output waveform shows amplitude modulation of pulses, which, although they do not exactly resemble the sinusoidal AM signal in Figure 4-3, nevertheless produce amplitude-modulated (AM) signals at the CW signal's frequency and harmonics (e.g., mainly the odd-order harmonics shown in Figure 4-11. Note that the waveform in Figure 4-11 also resembles a demodulated AM signal, such as a half-wave-rectified AM signal. This also would mean that this waveform also would contain the "audio" or modulating signal as well, which is true.

Thus the pulse-amplitude-modulated signal from Figure 4-11 is actually made up of many signals—the audio signal, the amplitude-modulated signal of the CW signal, and the harmonics of the amplitude-modulated CW signal.

Thus, if the CW carrier is at 535 kHz and the modulating or audio signal is at 1 kHz, the pulse-amplitude-modulated waveform contains a 1-kHz signal, an AM signal at 535 kHz modulated at 1 kHz, and at least an AM signal at 1,605 kHz also modulated at 1 kHz.

So now let's look at Figure 4-5 again.

Parts List

- **C1, C3:** 0.001 μF
- **C2:** 0.0018 μF
- **C4:** 470 pF
- **C5, C6, C7:** 0.01 μF
- **C8, C9, C10, C11, C12:** 1 μF, 35 volts
- **R1:** 1,200 Ω
- **R2, R12:** 1-kΩ multiturn pot
- **R3, R11:** 100 Ω
- **R4:** 100 kΩ
- **R5:** 1 kΩ
- **R6, R7, R10:** 4,700 Ω
- **R8, R9:** 20 kΩ
- **R13:** 1,800 Ω
- **U1:** 74HC14
- **U2:** 74HC05
- **Vreg 1:** LM7805 5-volt positive regulator
- **L1:** 330 μH

With the exception of the 455-kHz band-pass filter consisting of R8, L1, C4, C2, R9, and R10, we have two nearly identical circuits for the CW generator and AM modulator. So we will start with the 535-kHz CW generator. A Schmitt trigger oscillator is formed from U1A, C1, R1, and R2. The frequency is adjusted via R2, which is measured with a frequency counter at pin 2 of U1A or its buffered output at R3. As stated earlier, the frequency can be calibrated to 535 kHz via a frequency counter by grounding one lead of the frequency counter and connecting the other lead to R3.

The 535-kHz CW pulse signal is sent to the input of an open collector inverter U2A pin 1, which has a pull-up or load resistor R6 at the pin 2 output. The actual biasing voltage for R6 is provided through a filtered version of the output of the audio oscillator circuit (Schmitt trigger oscillator) U1E, R4, and C2. R5 and C8 provide a filtered version of a pulse signal, which is a triangle waveform, to act as a (modulating) voltage source for R6. Since the triangle waveform varies, the output at pin 2 of U2A provides a pulse-amplitude-modulated signal at 535 kHz and 1,605 kHz with about a 1-kHz modulating frequency. Finally, a capacitor C7 is connected to pin 2 of U2A to provide an output signal via a short-wire (e.g., <12 inches) antenna to a radio for testing.

Similarly, the 455-kHz CW oscillator circuit consisting of U1F, C3, R13, and R12 produces a pulse waveform. Variable resistor R12 is adjusted to 455 kHz with a frequency counter connected to pin 12 of U1F or R11. The 455-kHz CW pulse waveform is connected to inverter U2F's input pin 13, and the output of U2F's pin 12 is connected to a load or pull-up resistor R7. A modulating triangle waveform at 1 kHz provides a varying-voltage source to R7, and thus the output of U2B pin 12 generates a pulse-amplitude-modulated waveform at 455 kHz, 910 kHz, and 1,365 kHz with a modulating frequency of about 1 kHz. The output is also connected via capacitor C6 to a short-wire (e.g., <12 inches) antenna.

A parallel resonance band-pass filter formed by R8, L1, C2, C4, R9, and R10 is also connected to pin 12 to provide a sinusoidal 455-kHz amplitude-modulated (AM) signal at R10. The signal from R10 will be used for aligning the IF amplifiers for the superheterodyne radios later.

If a frequency counter is not available, an alternate way of adjusting R12 would be to place the short-wire antenna from C6 near a radio (with a digital readout) that is tuned to 910 kHz. Then adjust R12 until a 1-kHz tone is heard loudest on the radio.

Similarly, R2 is adjusted by placing the wire antenna connected to C7 near a radio. Tune the (digital readout) radio to 1,070 kHz, which is twice 535 kHz. Adjust R2 until a 1-kHz tone is heard loudest on the radio.

Alternate Circuits

With a little more complexity, the 455-kHz and 535-kHz circuits can be done without having to adjust for the correct frequencies. Crystals and ceramic resonators are used instead, and these components are quite stable and accurate in frequency generation.

Figure 4-12 shows that the Schmitt trigger oscillators have been replaced with a crystal oscillator and digital dividers and a ceramic resonator oscillator. The 1-kHz modulating or audio oscillator stays in this design but can be replaced with a digital frequency divider to produce modulating tone near 1 kHz.

Parts List

- **C1, C2:** 47 pF
- **C3, C4, C5, C8, C10:** 1 μF, 35 volts
- **C6, C11:** 0.01 μF
- **C7, C9:** 330 pF

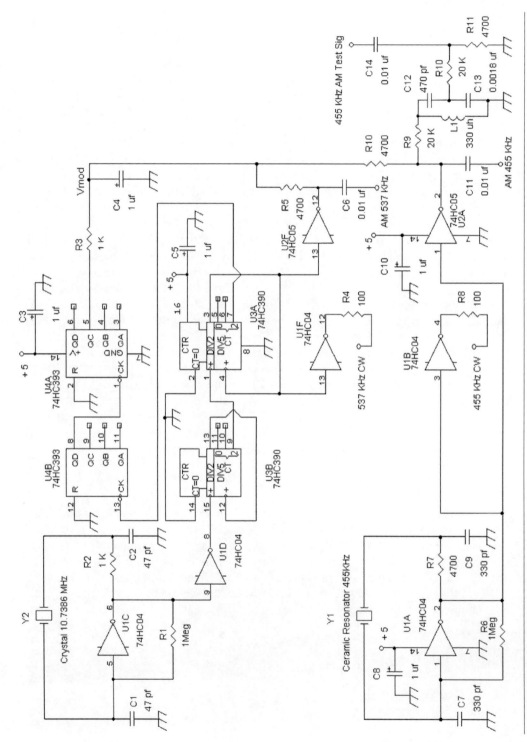

FIGURE 4-12 Test generator using a crystal and resonator.

- **C12:** 470 pF
- **C13:** 0.0018 µF
- **R1, R6:** 1 MΩ
- **R2, R3:** 1 kΩ
- **R4, R8:** 100 Ω
- **R5, R7, R10, R11:** 4,700 Ω
- **R9, R10:** 20 kΩ
- **U1:** 74HC04
- **U2:** 74HC05
- **U3:** 74HC390
- **U4:** 74HC393
- **Y1 resonator:** 455 kHz
- **Y2 crystal:** 10.738 MHz
- **L1:** 330 µH

The generator circuit in Figure 4-12 produces waveforms at 455 kHz and 537 kHz along with their harmonics by using precision resonators in place of the RC Schmitt trigger oscillators from the previous design. Because of the precise frequency generated in this design, no adjustment is necessary with a frequency counter or with a radio. Also, the frequencies 455 kHz and 537 kHz are stable even if the supply voltage varies by plus or minus 20 percent. After this circuit is built, one can confirm that there are AM signals at 455 kHz , 910 kHz , 1,365 kHz , 537 kHz, and 1,611 kHz.

The 455-kHz signal is generated by U1A, an inverter biased up as a high-gain amplifier via resistor R6. In order for oscillation to occur, there must be a 180-degree phase shift from the R7-C9 low-pass filter, the 455-kHz ceramic resonator, and C7. The R7-C9 low-pass filter has a cutoff frequency at about 100 kHz, which ensures by 455 kHz that there are at least 77 degrees of phase shift including the propagation delay of the inverter gate. The ceramic resonator acts like an inductor to resonate with its internal series capacitance within the ceramic resonator and with external capacitor C7. Thus the ceramic resonator with C7 forms a phase-shift network capable of 90 degrees of phase shifting at resonance (455 kHz) and more than 90 degrees just slightly off resonance because of its very high Q.

The output of the 455-kHz oscillator (U1A pin 2) is connected to the input of the pulse-amplitude-modulator circuit U2A pin 1, which has an open-drain output at pin 2 of U2A (see Figure 4-7 and its description for an explanation of a pulse amplitude modulator circuit).

To produce a precise frequency at the low end of the AM band, a 10.7386-MHz crystal was chosen to eventually produce a 537-kHz signal. An 11.00-MHz crystal can be used in place of the 10.7386-MHz unit as well, but the resulting frequency, when divided down, will be 550 kHz instead, a little higher than 537 kHz.

The crystal oscillator works similarly as the ceramic resonator circuit. An inverter, U1C, is biased as a high-gain amplifier via resistor R1. R2 and C2 are for a low-pass circuit with a phase shift of about 70 degrees at 10.738 MHz. The inverter itself has a propagation delay of about 8 ns, which is about 30 degrees of (lagging) phase shift at 10.738 MHz. The remaining (80 or so degrees) of phase shift is produced by the crystal's internal equivalent inductance and series capacitance inside and with external capacitor C1. The output of the inverter is connected via U1D to a decade frequency divider U2B (74HC390). The 74HC390 decade divider integrated circuit has two divide-by-10 sections. And each divide-by-10 section has a divide-by-2 and a divide-by-5 circuit. The 10.7386-MHz signal is divided by 10 to provide 1.0738 MHz and then divided by 2 to produce a 537-kHz square-wave pulse signal into pulse amplitude modulator U2F (via input pin 13 of the 74HC05), which is an open-drain output inverter. The output of U2F at pin 12 generates an AM signal via C6.

Also, the modulating signal is produced by frequency dividing down from the 10.7386-MHz crystal to generate an 839-Hz waveform at pin 5 of the 74HC393 counter. This is done by dividing the 537-kHz signal from pin 3 of the 74HC390 by 5 via its second divide-by-5 circuit at pin 7. The output signal (537 kHz/5 or 107.4 kHz) at pin 7 of the 74HC390 counter is further divided by 128 via binary counter/divider circuit 74HC393, U4B cascading to U4A to produce an 837-Hz square wave at pin 5 of the 74HC393 counter. Thus the modulating waveform is a triangle wave at 839 Hz across C4, which feeds into the load resistors R10 and R5 to provide pulse amplitude modulation at 455 kHz and 537 kHz, respectively.

Chapter 5

Low-Power Tuned
Radio-Frequency Radios

The very first radios were tuned radio-frequency (TRF) radios back in the early twentieth century. TRF radios were made commercially from the beginning of amplitude-modulation (AM) radio to the twenty-first century. As of 2012, there are still commercially made TRF radios for the AM section of an AM/FM radio, such as the Kaide Model KK-205 radio that sells for about $5 to $6 on eBay. Also the Kaide Model KK-9 AM/FM/SW (shortwave) radio that sells for about $7 on eBay has a TRF AM radio section while having a superheterodyne circuit for the FM and SW bands. This chapter will explore TRF radios that drain very little power, and a number of designs will be presented.

Design Considerations for TRF Radios

In the world of low-power design, we are talking about radios or receivers that drain less than 1 mA on one or two cells. However, we will set a goal for something in the range of less than 300 µA. Why such a low current consumption? Well, the milliampere-hour capacity of an AA alkaline cell is about 2,500 mAh, whereas an alkaline C cell has about 7,000 mAh, and a D cell has about 14,000 mAh. Thus, for a radio that drains about 300 µA continuously, one or two D cells will last about 5 years, whereas a C cell will go for about 2.5 years, and an AA cell will run for about 11 months.

As a single-cell battery drains, the voltage drops from about 1.5 volts to about 1.2 volts or 1.1 volts before becoming unusable. Thus the radios must work at the lower voltages to extend useful battery life. For the most part, low-power radios in this book shall drive crystal earphones. Conventional magnetic or dynamic earphones or headphones normally are low resistance and require milliamperes of current drive, which thus will shorten battery life considerably.

Now, what types of devices shall we use for low-powered radios? Both bipolar transistors (BJTs) and field-effect transistors (FETs) are available. For the greatest voltage gain or power gain per any given operating current, bipolar transistors always will beat field-effect transistors. This gain is usually characterized by *transconductance* or *mutual transconductance*.

The transconductance of a device can be described as the ratio of the alternating-current (AC) output current divided by the AC input-signal voltage. From Ohm's law, we know the $V = IR$, where V is the voltage in volts, I is the current in amps, and R is the resistance in ohms. Thus, via Ohm's law, $R = V/I$ or something that has volts divided by amps. Transconductance is measured by current/voltage or amps/volts. So transconductance is measured in mhos (*ohm* spelled in reverse) to signify that its unit is the reciprocal of an ohm. It should be noted that today the mho is replaced by siemens, usually denoted by S.

The higher the transconductance, the higher is the gain. So let's take a look at the transconductance of a bipolar transistor versus a field-effect transistor. For an operating collector current of 2 mA direct current (DC), the transistor has a transconductance of 0.076 mho or 0.076 S (S = siemens, where 1 S = 1 mho = 1 amp/1 volt), whereas a 2N3819 field-effect transistor at 2 mA gives out only 0.002 S. For a bipolar transistor to yield 0.002 S of transconductance, we just need to operate it at 52 μA, which is much lower than the 2N3819 FET operating at 2,000 μA or 2 mA.

An inherent advantage of a field-effect transistor is that its input terminal (gate) has close to infinite resistance; that is, a FET does not load down a signal, whereas a bipolar transistor has a finite input resistance across its base and emitter. But we will see that this finite resistance does not pose much of a problem at very low operating currents (e.g., <50 μA collector current).

To summarize a couple of the goals for low-power designs, we want the following:

1. Operating voltages from 1.2 volts to 2.4 volts or better
2. Bipolar transistors for their high gain at low operating currents (i.e., transconductance)

The TRF radio, as seen in Figure 1-1, consists of an antenna and tunable RF filter. In the very early radios from the 1920s, the TRF design had the antenna as an external long wire antenna or an external loop antenna.

One of the first TRF designs shown in this book will use ferrite bar or rod antenna coils. These antenna coils can be compact, less than 2 inches long. Or for higher sensitivity, they can be longer, on the order of 3 inches or more (see Figure 3-1).

We also will show TRF designs with an external loop antenna, which is the type used currently in stereo high-fidelity (hi-fi) receivers (see Figure 3-2). There is an advantage to using an external loop antenna, in that virtually any example can be used as long as there are sufficient windings. This loop antenna does not need to be wound to a specific inductance value to work with the tuning capacitor. Instead, the external loop antenna is connected to a "stepped down" smaller winding of an RF transformer, and the larger winding of the RF transformer is connected to the tuning capacitor.

The RF transformer may be adjusted to allow matching with tuning capacitors with various values. For example, the RF transformer may be nominally set at 330 μH to match with a 270-pF tuning capacitor. But the inductance of the RF transformer (e.g., part 42IF100) may be adjusted to an increased inductance of 480 µH, which then matches a 180-pF (or 200-pF) variable capacitor, or decreased to an inductance of 270 µH to work with a 330-pF tuning capacitor.

So, using an RF transformer with an external loop antenna allows for flexibility in designing the TRF radio with different values of variable capacitors. In addition, the external loop antenna allows for moving it away from the amplifying circuit to avoid feedback oscillation. In the past when TRF radios were built with multiple RF stages, care had to be taken to avoid oscillation. If the receiving antenna is located in the same area as the multiple RF amplifiers and filters, an oscillation can occur. In the J. W. Miller Model 570 TRF radio from the late 1930s, there are four sections of variable capacitors to form a four-stage tunable RF circuit. However, the antenna is a long wire that is external, far away, and shielded from the four-stage RF circuit. Thus, using an external antenna loop allows for easier design of multiple RF stages in TRF radios.

In this book, a design using an internal loop antenna coil (e.g., ferrite antenna coil) whose inductance matches with the tuning capacitor will be shown. Ferrite antenna coils are used commonly because of their compactness and sensitivity. A quick comparison between the ferrite antenna coils in Figure 3-1 and the loop antennas with RF transformers in Figure 3-2 shows that the ferrite antenna coils produce higher signal levels by at least twofold over the loop antenna–RF transformer combination. Although the ferrite antenna coil is generally fixed in inductance value, one can slide the coil to the middle for maximum inductance (e.g., 100 percent = maximum inductance value) to the end for minimum inductance (e.g., 80 to 85 percent of maximum inductance value) when matching with a particular variable capacitor.

Improving Sensitivity and/or Selectivity via Antenna Coils or Circuits

To receive more stations, usually a larger-area loop antenna can be made or a longer ferrite bar or rod can be used for the antenna coil. Also, the Q or quality factor of the coil determines the gain and selectivity (with selectivity being the ability to reject an interfering adjacent channel). The Q is a function of the resistance of the wire used in winding the loop antenna or antenna coil. Generally, the lower the resistance, the higher is the Q. Antenna coils are made, for example, with wire of American Wire Gauge (AWG) numbers that vary from 30 to 22. Litz wires, which consist of multiple-strand insulated wires, work better (e.g., lower resistance at high frequencies) than uninsulated stranded or solid conductor wires. For the amplitude-modulated (AM) band from 535 kHz to 1610 kHz, though, almost any type of insulated wire of low DC resistance will do.

To increase sensitivity and selectively, it is important to provide minimal loading to a parallel inductor capacitor tank circuit. For an amplifier connected to the parallel inductor capacitor circuit, this means as high an input resistance as possible. For example, an amplifier whose input resistance is at least 100 kΩ would be acceptable in not loading down the Q of the inductor capacitor tank circuit. Another alternative is to provide a tap in the coil, which allows connecting to circuits or amplifiers with less than 100 kΩ of input resistance.

As the first TRF radio design is shown, you will see the principle of tapped-down coils to preserve the high Q of the coil but also provide other beneficial characteristics, such as avoiding oscillation.

First Design of TRF Radio

Parts List

- **C1, C4, C5, C8:** 1 µF, 35 volts
- **C2, C6:** 0.1 µF
- **C3, C7:** 0.01 µF
- **R1:** 56 kΩ
- **R2:** 6,200 Ω
- **R3, R6:** 2,200 Ω
- **R4, R7:** 100 kΩ
- **R5:** 20 kΩ
- **L1 antenna coil:** 470 µH primary
- **L2:** 1 mH
- **L3:** 8.2 mH or 10 mH
- **VC1 two-gang variable capacitor:** 140 pF, 60 pF
- **D1, D2:** 1N914
- **D3:** 1N34
- **Q1, Q2:** MPSH10

At 1.5 volts, this radio drains less than 180 µA, which meets the current consumption goal mentioned previously. The radio's audio signal output is connected to a crystal earphone. Although this radio can run on 1.5 volts, 3 volts will work as well. Thus, with a single C-cell alkaline battery, there will be five years of continuous service. Not bad for an emergency radio. Also, the 1.5 volts can be obtained from three solar cells. Because of the low current consumption, one can wire a fruit/vegetable battery using two lemons or potatoes for this radio. Just be sure not to eat (consume) any of the fruits or vegetables that are used for making the battery.

In terms of construction, try to keep the base lead of transistor Q1 (MPSH10) to the L1 antenna coil's secondary lead short. The secondary winding of L1 typically has about one-eighth the turns of the primary winding. Also, to avoid "recirculation" from the amplified RF signal back to antenna coil L1, both inductors L2 and L3 should be mounted preferably at 90 degrees to the antenna coil.

This radio was deemed a "first" design, and one may wonder why the quotation marks around *first*. Well, Figure 5-1 really shows a fourth attempt to make a successful

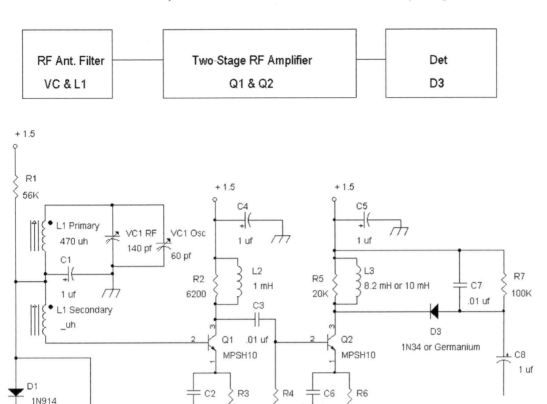

FIGURE 5-1   Block diagram and schematic of a TRF radio.

radio working off low voltage (e.g., 1.5 volts). The first three designs needed a minimum 2.4 volts and required more than two transistors.

So what made this design in Figure 5-1 successful? There are two features. First and foremost, it used inductive loading at the collectors of transistors Q1 and Q2 to raise the operating DC voltage at the collectors of these transistors. By raising the DC voltage at the collectors of Q1 and Q2, the internal-collector-to-base capacitances of the transistors are reduced.

Second was the choice of transistor to be very low-capacitance types, such as the MPSH10. When working at very low voltages such as 1.5 volts or lower, the collector-to-base capacitance is important to note because this capacitance will affect the overall gain of the amplifying stages. At higher supply voltages (e.g., 6 volts and above), general-purpose transistors (e.g., 2N4124 or 2N3904) can be used because the collector-to-base capacitance is lowered by the higher operating voltage, but the higher supply voltages would mean more batteries.

When using just resistive loads on a collector, the resistive loads form a voltage drop (collector DC current times resistance value), which, in turn, causes the collector voltage to drop as well. This collector voltage drop then causes an increase in collector-to-base capacitance, which, in turn, then causes a loss in amplification at the RF frequencies.

So in summary, the two magic ingredients are

1. Inductors connected between the supply and collector of the transistors, and
2. Using low-capacitance transistors such as the MPSH10.

Circuit Description

Antenna coil L1 has a primary winding inductance of about 470 µH, which requires about 200 pF of capacitance to receive the low end of the AM band at about 530 kHz or 540 kHz. The variable capacitor used in this example is just a common two-gang/section variable capacitor normally used for superheterodyne radios. By paralleling the RF and oscillator sections of the variable capacitor, the 200-pF capacitance is achieved.

Of course, other antenna coil and variable capacitor combinations can be used. See the following table.

Variable Capacitor	Antenna Coil
140 pF	680 µH
180 pF to 200 pF	470 µH
270 pF	330 µH
330 pF to 365 pF	250 µH
540 pF (270 pF × 2)	165 µH

For higher Q performance, it is generally better to use a combination with higher capacitance for the variable capacitor. In other types of radios, however, such as a superheterodyne radio, using a higher-capacitance variable capacitor does not offer too much of an advantage because the selectivity will be determined mainly by the IF coils or IF filters (e.g., 455-kHz ceramic filters).

The secondary of the antenna coil generally is less than 10 µH, so a smaller signal is delivered to the base of Q1. But this is okay because the high Q of the antenna coil is preserved by the step-down transformer action from primary to secondary windings. The high Q performance of the antenna coil is essential for sensitivity and, more important, for selectivity of the radio.

Signals from the stepped-down secondary winding of L1 are connected to the input of amplifier Q1, which is biased by a DC voltage reference circuit consisting of D1 and D2 to form about 1 volt. The 1 volt DC biases the base of Q1, and the 0.6-volt drop from the base to emitter of Q1 produces about 0.4 volt (to 0.3 volt) at the emitter of Q1. Thus the 1-volt reference sets up the collector current of Q1 (and Q2) to be about $0.4/2,200 = 72$ µA.

The voltage gain of a common emitter amplifier shown for Q1 or Q2 is the ratio of the AC voltage measured at the collector and the base. For example, the *magnitude* of the voltage gain = Vout@collector/Vin@base, where Vout and Vin are AC signals.

With the inductive load L1 of 1 mH and R2 of 6,200 Ω, the gain of the amplifier is about 10 at the middle of the AM broadcast band at about 1 MHz.

The output of Q1 is the collector, which is connected to the second-stage amplifier Q2 via C3. Amplifier Q2 is similar to amplifier Q1 but with a different collector load with a larger valued inductor (L3) and load resistor (R5). Thus the gain of the Q2 amplifier at its collector is about 30 or 40.

It should be noted that there are resistors connected in parallel to the load inductors. This is to ensure that the amplifiers do not oscillate by limiting the gain at high frequencies. An inductor's impedance increases with frequency. For example, removing the 20-kΩ resistor across L2 will result in an oscillating circuit instead of an amplifying circuit.

With amplifiers Q1 and Q2, the total gain is about 300 to 400. To demodulate the amplified AM signal, a germanium diode is used as an envelope detector with a peak hold capacitor C7. Note that the peak hold capacitor provides a larger-amplitude audio signal than without C7. Resistor R6 allows for a discharge path to C7 so that the audio signal is demodulated without gross distortion when the AM signal's envelope is decreasing. Without a discharge resistor on the peak hold capacitor, the demodulated signal will be "stuck" at a voltage related to the peak of the AM signal. DC blocking capacitor C8 then is connected to a crystal earphone for listening. Note: One lead of the crystal earphone is connected to ground.

Variation of the Design (Alternate Design of the TRF Radio)

A variant of this design was tried with a loop antenna with antenna transformer. Because there is less RF signal from this combination, loop antenna/transformer as compared to a ferrite antenna coil, an extra stage of amplification was added (Figure 5-2).

Parts List

- **C1, C4, C5, C8:** 1 µF, 35 volts
- **C2, C6, C9:** 0.1 µF
- **C3, C7, C10, C11:** 0.01 µF
- **R1:** 56 kΩ
- **R2, R5:** 6,200 Ω
- **R3, R6, R8:** 2,200 Ω
- **R4, R7, R9:** 100 kΩ
- **R10:** 20 kΩ
- **T1:** 42IF100 oscillator coil
- **L1, L2:** 1 mH
- **L3:** 10 mH

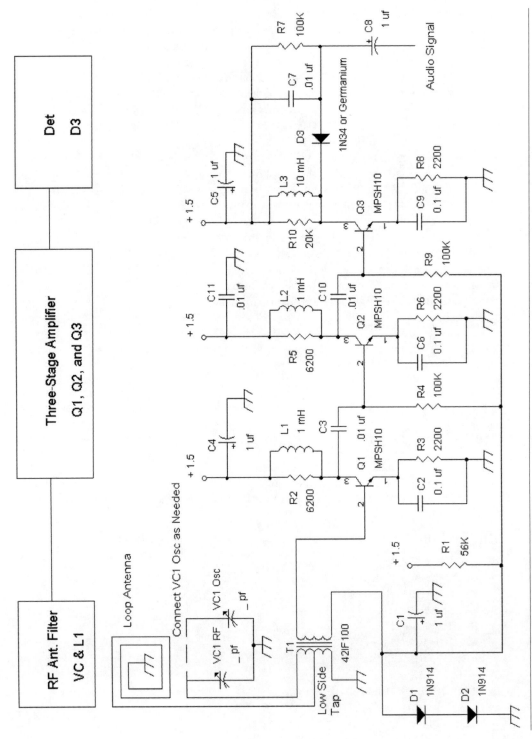

FIGURE 5-2 Block diagram and schematic of a TRF radio using a loop antenna and RF transformer.

60

- **VC1 two-gang variable capacitor:** 140 pF, 60 pF
- **D1, D2:** 1N914
- **D3:** 1N34
- **Q1, Q2, Q3:** MPSH10
- **Loop antenna** (see Figure 3-2)

The motivation for using a loop antenna and T1 is that the antenna transformer allows for matching to a wide range of capacitances for the variable capacitor. For example, T1 (part 42IF100) in Figure 5-2 has an inductance range of about 260 µH to about 570 µH. Thus the following variable capacitors may be used:

Variable Capacitor	RF Transformer
155 pF	570 µH
180 pF to 200 pF	470 µH
270 pF	330 µH
330 pF to 365 pF	260 µH

RF transformer part 42IF100 is actually an oscillator coil for superheterodyne radios. But because it possesses, by coincidence, the proper turns ratio or tap, we can use it in combination with a loop antenna to mimic a ferrite bar antenna.

The loop antenna is grounded on one of its leads, whereas the other lead is connected to the lowest tap of the 42IF100. This can be confirmed with an ohmmeter. Measuring from the center pin of the 42IF100 coil to an outer pin on the same side of this coil, the lowest tap has a smaller resistance. Coupling the antenna loop to the lowest tap allows the coil to maintain a high Q for acceptable to good selectivity. If the antenna loop is connected to the higher-resistance tap, the Q will drop, and selectivity will become unacceptable.

Because an extra stage of amplification was added, the current drain typically is less than 250 µA for the radio in Figure 5-2. However, the reader is encouraged to change the values of the emitter resistors and collector load inductors or resistors to lower the current drain while maintaining performance. For example, emitter resistors R3 and R6 may be increased from 2.2 kΩ to 3.3 kΩ or 3.9 kΩ while load inductors L1 and L2 are increased to 2.2 mH to 4.7 mH and with load resistors R2 and R5 increased from 6,200 Ω to 20 kΩ. Try one stage at a time with these suggested modifications.

Author's Earlier TRF Designs

As stated previously, the designs from Figures 5-1 and 5-2 actually came after designing three different TRF radios with a 3-volt supply. At that time, all three designs used only resistor loads at the collectors of the transistor voltage-gain amplifiers. It wasn't until the reflex radios (with transformer and/or inductive loads) were designed for Chapter 6 that I realized that the answer to lower-voltage radios was right in front of me—use inductors for the load instead of just resistors.

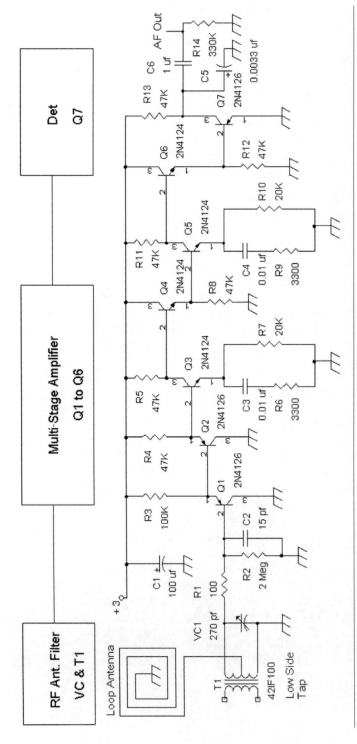

FIGURE 5-3 Block diagram and schematic of a previous TRF radio design.

So sometimes a "final" design does not always have a straight path, but the journey is still fun and exciting in terms of learning new things.

See the three designs in Figures 5-3, 5-4, and 5-5; they are for reference only. But the reader can decide if he or she would like to build any of them.

Figure 5-3 shows one of the first tries on a low-powered TRF radio. Resistive loads were tried initially, but because the voltage drop occurred, a higher voltage was required (e.g., 3 volts).

The TRF radio's block diagram is rather simple, as seen in Figure 5-3. An external antenna loop is connected to an antenna transformer T1. The antenna transformer receives RF energy from the antenna loop with a low-impedance drive at the low side tap winding of T1. The secondary winding, or the antenna transformer, is stepped up and provides the necessary inductance to resonate with the variable capacitor VC1 to form a parallel inductance-capacitance tank circuit (LC circuit).

An amplifier consisting of Q1 to Q6 must have a very high input impedance/resistance so as not to load down the parallel LC circuit (T1 and VC1) that generally has a high impedance of over 100 kΩ. For example, the input impedance/resistance of the amplifier at the base of Q1 should be on the order of 500 kΩ or more.

The voltage gain of the amplifier then delivers sufficient AC voltage levels for envelope detector Q7. Q7 acts very much like a diode but requires less driving current from the amplifier's output transistor to perform AM demodulation or detection to a crystal earphone.

One should note that even though this TRF radio uses an external loop antenna, an antenna coil such as a ferrite bar antenna coil may be substituted for the antenna transformer and external loop antenna.

Now let's take a look at another design of a low-power TRF radio, as seen in Figure 5-3. In the figure, the antenna transformer is set nominally for about 330 µH at its secondary winding, with the external loop antenna connected to a tap of the 330-µH secondary winding. Here we use a commonly available oscillator coil (42IF100) for the antenna transformer. Other commonly available oscillator coils may be used in place of the 42IF100, such as the 42IF110 or 42IF300, all available through Mouser Electronics (www.mouser.com). The 330-µH inductance is connected to a 270-pF variable capacitor. In this case, the variable capacitor has twin sections of 270 pF. Thus only one section of this variable capacitor is used. The ground terminal of the variable capacitor is always connected to its shaft. In this way, touching the shaft while tuning has no effect on adding stray capacitance.

The 330-µH inductance and the 270-pF variable capacitor then allow tuning from about 535 kHz to about 1,600 kHz. One may notice that the resonant frequency of an LC tank circuit is proportional to the square root of the inductance or capacitance. Thus, if 270 pF resonates at about 535 kHz, then at 1,600 kHz, which is about three times 535 kHz, we would need a capacitance of about $^1/_3$ squared of 270 pF, or 1/9 $\times$ 270 pF = 30 pF.

RF signals developed at the tank circuit at variable VC1 typically will be around 20 mV or more with strong stations and much less with weaker signals. Therefore, total gain of about 100 is needed to bring the level up sufficiently for envelope detection.

Transistors Q1 and Q2 form a double emitter follower circuit that has a gain of about 1 (unity) with a very high input resistance at Q1's base so as not to load tank circuit T1 and VC1 while providing a sufficiently low output resistance drive to voltage amplifier Q3. The output signal of Q3 via its collector is fed to another emitter follower Q4, which, in turn, drives a second voltage-gain amplifier Q5. And Q5's collector output signal is connected to gain-of-1 amplifier emitter follower Q6. The output of emitter follower Q6 then is connected to peak envelope detector Q7, which also looks like an emitter follower but with a peak hold capacitor C5 at its emitter.

In the first design, it was found that emitter follower stages Q1 and Q2 can oscillate in an undesirable manner when the variable capacitor is tuned to the top of the AM band (e.g., around 1,400 kHz or higher). The high impedance characteristic of the parallel tank circuit actually sets up a condition for oscillation as the variable capacitor is adjusted for minimum capacitance. Therefore, an alternative design was tried out in the next design, which uses a tapped-down coil that is fed to Q1.

A second design uses the more common 140-pF variable capacitor (Figure 5-4).

An inductance of about 680 µH is needed, though, to resonate with the 140-pF variable capacitor at the low end of the AM band (e.g., 535 kHz), and the oscillator coils mentioned earlier max out at about 500 µH. Thus a 455-kHz IF transformer (42IF104) is used instead because its inductance at the secondary winding easily can be adjusted to 680 µH. Here the primary has a suitable turns ratio of about 1:13 for a primary-to-secondary-winding ratio. Thus the external antenna is connected to the primary, and the tap of the secondary is connected to emitter follower Q1. By using the tap at the secondary winding, the driving impedance is dropped by fourfold or more compared with the impedance at the full winding (e.g., at VC2). This lowered tank impedance reduces the undesirable oscillation from Q1. But the signal is also reduced. Therefore, to increase the gain of the Q3 amplifier, Q3's emitter is bypassed to ground via C3. In Figure 5-3 C3 is in series with a gain-reducing resistor (3,300 Ω), R6.

It should be noted that T1, the 42IF104 transformer, may be replaced with the capacitor taken out from a 42IF101 or 42IF102 IF transformer. See Figure 3-28 on capacitor removal.

At 3 volts, the current drain is about 200 µA for the radio designs in Figures 5-3 and 5-4. The designs in Figures 5-3 and 5-4 had some oscillation problems, with the circuit in Figure 5-4 improved over the circuit in Figure 5-3. So, other types of circuits were tried.

The "famous" ZN414 or MK484 integrated circuit was tried as well, but oscillations also occurred similarly when a 680-µH antenna coil and a 140-pF variable capacitor were used as the tank circuit.

In terms of the emitter follower amplifiers of Figures 5-3 and 5-4, oscillation occurs when the emitter of Q1 and/or Q2 is loaded into a capacitor, which is always the case because there is at least capacitance on the board from the emitter to collector of Q1 and Q2, as well as at the base (to ground) of Q3. The capacitance loading at the emitter of Q1 or Q2 causes a phase shift in the minus direction (e.g., phase lag or negative phase shift), but the internal capacitance inherent across the base and emitter of Q1 and or Q2 forms a high-pass filter with the inductor of T1 or T2. Recall that when the variable capacitor VC1 or VC2 is tuned to the top of the AM band, there is very little capacitance across the inductor of T1 or T2. The base emitter

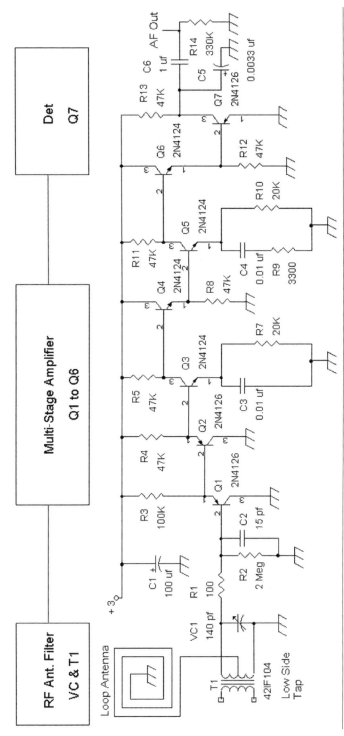

FIGURE 5-4 Block diagram and schematic of a previous TRF radio design with a loop antenna.

capacitance of Q1 and or Q2 then forms a two-pole high-pass (and high-Q) filter with amplitude gain and positive phase shift. When the positive phase shift cancels the negative phase shift with amplitude gain, oscillation occurs. In a later chapter it will be shown that this principle of introducing a phase lag via a capacitance load at the emitter and a series LC filter via an inductor connected to the input base lead of an emitter follower plus the base emitter capacitance will form an oscillator. Note that the condition for oscillation in a system requires a 0- or 360-degree phase shift with an amplitude gain of greater than 1 from input to output.

So in the quest to further reduce oscillations, a third design was done with success. Now let us turn to Figure 5-5.

The L1 inductor and VC1 capacitor tank circuit is connected to the differential pair amplifier circuit with Q1 and Q2. An emitter follower Q3 provides a gain of about 1 after the differential pair amplifier with load impedance drive into power detector Q4.

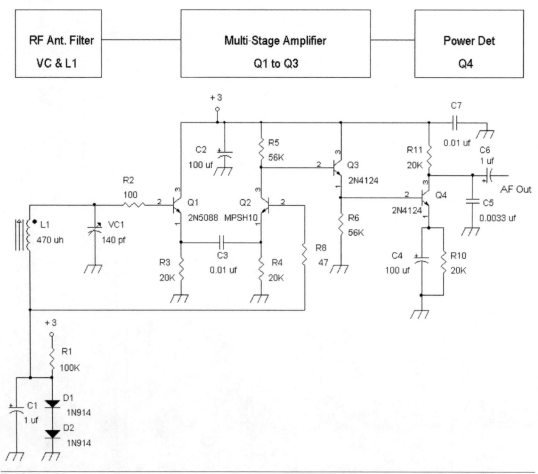

FIGURE 5-5 TRF radio with differential pair transistors to reduce unwanted oscillations.

The reason there is no oscillation is that Q1, which looks like an emitter follower, has a gain of only one-half (e.g., 0.5 at its emitter). And the emitter of Q1 is connected via capacitor C3 to the emitter of Q2. C3 is chosen to be large in capacitance and is considered to be an AC short circuit at the RF frequencies. But "looking" into the impedance of Q2's emitter is resistive, not capacitive. Therefore, there is no appreciable phase lag or negative phase shift at the emitter of Q1. Thus Q1's emitter has a gain of one-half and no phase lag, which then enables a condition for Q1 to not oscillate.

In this radio, a 250-µH ferrite antenna coil is used with a 365-pF variable capacitor. It is worth noting that the input resistance of Q1 is not as high as the emitter follower circuits of Figures 5-2 and 5-3. However, it is high enough to maintain good selectivity from the tank circuit L1 and VC of Figure 5-5.

This radio works fine at 3 volts, with a current drain of less than 200 µA.

A thought did arise—how about using complimentary metal oxide semiconductor (CMOS) gates such as hex inverter gates for the RF and audio-frequency amplifiers for a TRF radio? It turns out that when the inverter gates are biased to about one-half V_{DD} for a 74HC04 chip, *both* output transistors of the inverter gates are turned on and drain excessive DC current—on the order of tens of milliamps. CMOS gates normally have *either but not both* of the output transistors turned on, which drains much less current. So the idea of using CMOS gates for low-power radios had to be scrapped.

Chapter 6

Transistor Reflex Radios

This chapter introduces a method of using a form of signal recirculation to reduce the number of active components for amplification in radios. Unlike the tunable radio-frequency (TRF) radio circuits shown in Chapter 5, where each amplification stage amplified only one type of signal (e.g., an RF signal), reflex radio circuits allow a single transistor stage to amplify both radio-frequency (RF) and audio-frequency (AF) signals.

Motivation Behind Amplifying Both Radio-Frequency and Audio-Frequency Signals

When the first transistor radios were designed and sold commercially in the mid-1950s (e.g., the four-transistor radio Regency TR-1 in 1955), transistors were very expensive. In the mid-twentieth century, audio small-signal transistors were about $1 to $2 each. For example, in the 1956 Allied Radio Catalog, the very famous Raytheon CK722 (PNP audio germanium) transistor sold for $2.20. Back in 1956, a loaf of bread cost less than 25 cents. If you wanted a "high-frequency" transistor in 1956, Raytheon sold its CK760 (aka 2N112) for $6.35. Back in those days, anything in the 2-MHz to 4-MHz (or more) range was considered a high-frequency transistor.

Of course, by the 1960s, transistor prices dropped, but they were still somewhat expensive. A quick look at the 1966 Allied Radio Catalog shows that a PNP audio transistor such as the RCA 2N408 was only 38 cents, and its "high-frequency" transistor 2N412 (16.5 MHz) went for only 43 cents. Remember, though, that in 1966 a gallon of gasoline was less than 40 cents.

So the motivation to design a radio that amplified both radio frequencies and audio frequencies was economics. Back in those days of building transistor radios, the transistors were considered costly. Thus, minimizing the number of transistors in a design allowed the radio to be sold at a cheaper price.

And throughout the 1950s and 1960s there were basically two types of transistor radios one could buy. The most common was the superheterodyne radio, usually a six-transistor radio, and the other (less common) was a reflex design using only two transistors.

It should be noted that there were some superheterodyne radios that included a reflex circuit as well. But these radios used one of the intermediate-frequency (IF) transistors to amplify both the IF signals (e.g., a 455-kHz amplitude-modulated [AM] signal) and the low-level audio signal. One of these radios, the Sylvania four-transistor Model 4P19W used the IF stage as an emitter follower amplifier (gain of about 1) for buffering the detected audio signal and providing a low-impedance drive to the audio output stage.

However, in the typical transistor reflex radios, the RF/AF amplifier provides voltage gain (e.g., >1). These two-transistor radios sometimes were called "boy's radio" because they were more like toys. The performance of the two-transistor radio was very poor in sensitivity, selectivity, and audio output compared with the superheterodyne types, but it was adequate for listening to local stations.

Typically in the 1960s, two-transistor radios cost about $3 to $4, whereas superheterodyne radios cost at least $6 or $7. The average price for a superheterodyne radio was about $9 to $10 for imported versions (e.g., from Japan) and at least $14 to $15 for those made in the United States.

Now let's look at a typical two-transistor reflex radio made in the 1960s, the Windsor radio, as shown in Figure 6-1.

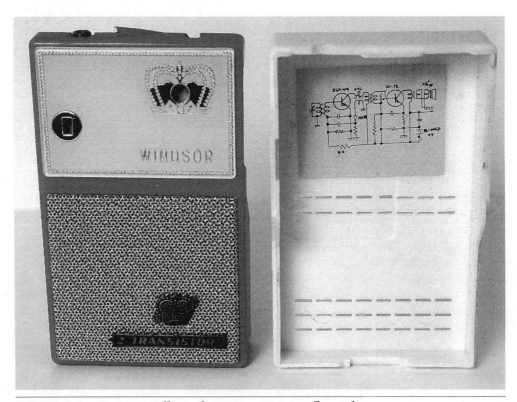

FIGURE 6-1   Commercially made two-transistor reflex radio.

Figure 6-1 shows the front side and its back cover with a schematic for the Windsor radio. This radio drained about 8 mA to 10 mA at 9 volts. When tuned to a strong local station, it delivered sufficient volume to fill a room.

If one looks carefully at the schematic, there is a radio-frequency choke (RFC) transformer and two audio transformers in this radio. The audio transformers are still available today, but the RFC transformer is not.

The primary of the RFC transformer is connected to the collector of the first transistor, whereas the secondary is connected to a detection diode. A measurement of this RFC transformer is as follows:

- Primary inductance is 936 µH.
- Secondary inductance is 3,300 µH.

The turns ratio of the RFC transformer is about 1:2 because the secondary inductance is almost four times the inductance of the primary winding.

$$\text{Turns ratio} = \sqrt{\text{primary inductance/secondary inductance}}$$

which means that

$$\text{Turns ratio} = \sqrt{936/3{,}300} = 1/1.877$$

With the back cover taken off, Figure 6-2 shows the circuits of the Windsor radio, and the RF transformer (aka RFC transformer) is pointed out.

RF Transformer

FIGURE 6-2 An RF transformer in the commercially made two-transistor radio.

The figure shows the Windsor radio's RF transformer with primary and secondary windings indicated. Unfortunately, though, such an RF transformer is not readily available today.

One-Transistor TRF Reflex Radio

Some of the goals for a single-transistor reflex radio are to work at 3 volts or less, drain about 250 µA using a crystal earphone, and provide workable performance at 1.5 volts. And yet another goal is to design this radio without the type of RF transformer seen in Figure 6-2.

As Figure 6-1 shows, inductive and/or transformer loads are used in both transistor stages. Inductive and transformer loads to the collectors of the transistors allow for the collectors actually to swing a higher voltage than the battery supply. Those who are familiar with flyback direct-current–direct-current (DC-DC) switching converters would know this feature.

In addition, using inductive or transformer loads in class A power amplifiers allows for a theoretical efficiency of up to 50 percent. In contrast, a resistive-load class A power amplifier yields at most about 25 percent efficiency.

In the follow-up to the one-transistor reflex radio, the multiple-transistor reflex radio, a second amplifier will be added for driving low-impedance headphones or speakers and will use an output transformer.

So now let's take a look at the one-transistor reflex radio in Figure 6-3.

Parts List

- **C1, C7:** 0.01 µF
- **C2, C6:** 1 µF, 35 volts
- **C3:** 0.0022 µF
- **C4:** 100 µF, 16 volts
- **R1:** 47 kΩ
- **R2, R4:** 1 kΩ
- **R3, R5:** 100 kΩ
- **R6:** 200 kΩ
- **T1 audio transfomer:** 10 kΩ primary, 10 kΩ or 2 kΩ secondary
- **L1 antenna coil:** 470 µH primary, 23 µH secondary
- **L2:** 3.9 mH or 4.7 mH
- **VC1 two-section variable capacitor:** 140 pF, 60 pF
- **D1, D2:** 1N914
- **D3:** 1N34
- **Q1:** MPSH10

In this particular design, a 680-µH ferrite antenna coil was used. It has a tap at 470 µH that matches a variable capacitor of about 200 pF. Thus both sections of a variable capacitor with 140 pF and 60 pF work fine with the primary winding of L1.

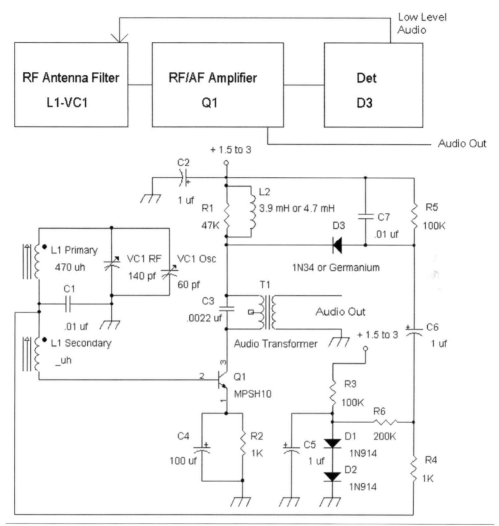

FIGURE 6-3   Block diagram and schematic of a one-transistor reflex radio.

The secondary winding of L1 is stepped down by about 5- to 10-fold (e.g., turns ratio from primary to secondary winding is 5 to 10:1), which allows connecting into the base of RF/AF amplifier Q1. C1 provides an alternating-current (AC) short circuit at high frequencies (e.g., RF frequencies) but a low-pass filtering effect at audio frequencies. The base of Q1 is biased to about 0.8 volt by the voltage reference circuit D1 and D2 and C5 by way of R6, a 200-kΩ resistor. There is a slight voltage drop, about 200 mV, across R6. Probably R6 can be reduced to 100 kΩ if needed.

The forward voltage drop across the base and emitter of Q1 is about 0.6 volt, so thus the operating collector current of Q1 is about 200 µA [= (0.8 volt − 0.6 volt)/1 kΩ].

Amplified RF signals are passed through the primary of T1 and C3, an RF bypass capacitor to the RF load inductor L2, and resistor R1. As stated previously, L2 should be mounted 90 degrees from the antenna coil L1 to avoid recirculation or feedback of the amplified RF signal (back) to the antenna coil.

R1 at 47 kΩ is actually optional. R1 is paralleled across the inductor to avoid possible oscillation. In the prototype radio, R1 was not used, and there were no oscillation problems. It may not be needed, for example, if R1 is removed and oscillations do not occur.

The amplified RF signal then is demodulated via D3, C7, and R5, which provides a low-level audio signal (along with a residual amplified RF signal). This low-level audio signal is fed back to the base of Q1 via C6 and R4 and the secondary winding of L1. Normally, one would not think that R4 is necessary and that the signal from C6 can be connected directly to C1. In many cases, connecting C6 to C1 would be fine in terms of not having oscillations. But to further avoid recirculating any amplified RF signals back to Q1, R4 at 1,000 Ω forms RF filtering with C1. To reiterate, there are actually some residual RF signals at the anode of D3, which when coupled back to the base of Q1 will cause some oscillations. So R4 is needed to reduce or eliminate oscillations.

The amplified audio signals then are extracted via T1's primary winding (10 kΩ), and the amplified audio signal is provided via its secondary winding (e.g., 2 kΩ). The secondary winding may be connected to a high-impedance earphone (e.g., crystal earphone or 2000-Ω headset).

The approximate current drain at 1.5 volts is less than 200 μA, and at 3 volts, the drain is about 250 μA. To extend the life of the battery, an On-Off switch may be used in series with the battery. Voltage reference circuits D1 and D2 provide a somewhat stable operating DC collector current for Q1 as the supply voltage is varied.

A couple of extra notes:

1. Q1 was tried originally with a high current gain (beta or Hfe) transistor, a 2N5089. This transistor performed poorly with very bad selectivity in this reflex radio circuit.
2. Q1 then was changed to a general-purpose transistor, the 2N4124, which gave satisfactory results but with a little less gain than the low-capacitance MPSH10 transistor.

Multiple-Transistor Reflex Radio Circuit

For the two-transistor reflex radio, we will just add an audio power amplifier to drive low-impedance transducers such as a loudspeaker or standard headphone. Figure 6-4 shows a two-transistor reflex radio that uses a 3-volt to 4.5-volt power supply.

Parts List

- **C1, C7, C9:** 0.01 μF
- **C2, C4, C10:** 100 μF, 16 volts
- **C3:** 0.0022 μF
- **C5, C6:** 1 μF, 35 volts

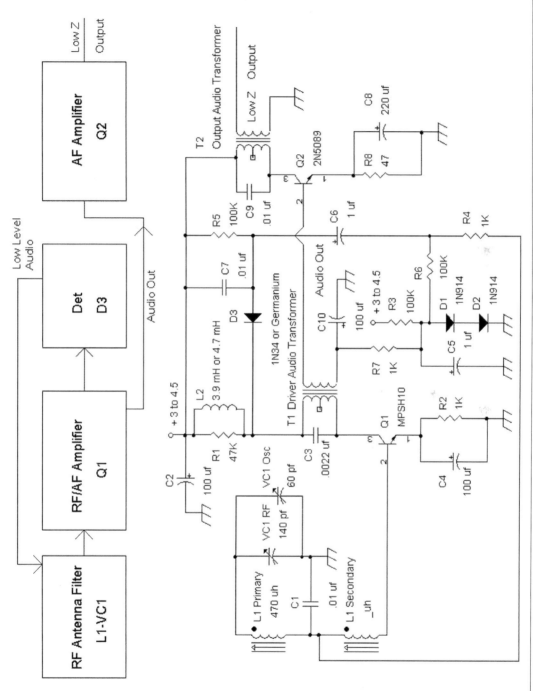

FIGURE 6-4   Block diagram of a two-transistor reflex radio.

- **C8:** 220 µF, 16 volts
- **R1:** 47 kΩ
- **R2, R4, R7:** 1 kΩ
- **R3, R5, R6:** 100 kΩ
- **R8:** 47 Ω
- **L1 antenna coil:** 470 µH primary, 23 µH secondary
- **L2:** 3.9 mH or 4.7 mH
- **VC1 two-gang variable capacitor:** 140 pF, 60 pF
- **T1 audio transformer:** 10 kΩ primary, 2 kΩ secondary
- **T2 audio transformer:** 120 Ω to 500 Ω primary, 8 Ω secondary
- **D1, D2:** 1N914
- **D3:** 1N34
- **Q1:** MPSH10
- **Q2:** 2N5088, 2N5089, or 2N3904

For the two-transistor reflex radio, an audio power amplifier is added via Q2, R8, C8, T2, and C7. The secondary winding of T1 is biased to about 0.9 volt DC such that the emitter of Q2 is about 0.3 volt. This results in a DC collector current of about 6 mA for Q2. *At this point we are not as concerned about building a low-power radio.*

Depending on the operating voltage, the optimal primary impedance of T2 is determined. The secondary impedance of T2 is 8 Ω for driving a loudspeaker or 8-Ω to 32-Ω headset. For a given DC current and supply voltage, the optimal primary impedance Z of T2 is roughly equal to Vsupply/Q2 collector DC current. So, if we use two cells for 3 volts and the DC collector current is 6 mA, T2's primary impedance is about 3/0.006 = 500 Ω. If the DC current is raised to 10 mA by changing R8 to about 30 Ω, then T2's primary impedance is about 3/0.010 = 300 Ω. Because of the "high" current drain, a power switch in series with the battery will prolong battery life. See the following table.

Supply Voltage, V	Q2's DC Current, mA	R8, Ω	T2 Primary Z, Ω
3	6	47	500
3	10	30	300 (250)
4.5	6	47	750 (800)
4.5	10	30	450 (500)

The values in parentheses show the closest alternative value that can be used instead. Remember that the secondary impedance of T2 is 8 Ω.

There is one "trick" to learn about transformers with center taps. The impedance from the center tap to any other lead is one-quarter the total impedance of that winding. Thus, if the T2 transformer's primary impedance from end to end of the winding is 1,200 Ω, the impedance from the center tap (CT) to either end of the winding is 1,200 Ω/4 = 300 Ω. Likewise for a 1,000-Ω primary impedance, the center tap winding to one end of the transformer will be 1,000 Ω/4 or 250 Ω. Figure 6-5 shows T2's primary winding connection.

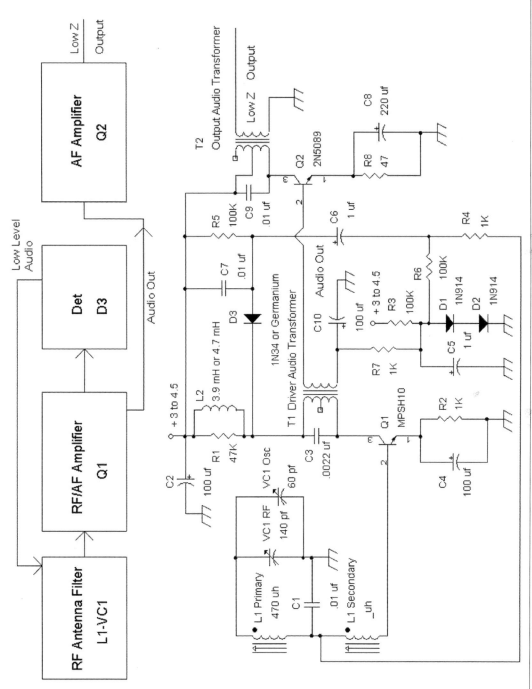

FIGURE 6-5 Two-transistor reflex radio using a center tap of the audio output transformer.

Chapter 7

A Low-Power Regenerative Radio

When I suggested writing a chapter on regenerative radios, I thought this was going to be a piece of cake. It had to be easy, right? Just add some positive feedback to the system, and magically you get higher gain and selectivity. And after all, I had already designed a vacuum-tube version without a problem back in 2007 that had been featured in *EDN Magazine* on the Web (www.edn.com/blog/Designing_Ideas/41377-A_super_het_radio_runs_5_years_on_a_C_cell_plus_a_pentode_radio.php).

It turns out that my first regenerative radio was designed with a little luck. Yes, luck sometimes plays a part in successful designs.

However, in trying to duplicate that success from a pentode vacuum tube to transistors—well, that's a different story. The pentode tube design had just the correct amount of transconductance, the antenna coil chosen received the amplitude-modulated (AM) stations with good signal strength, and the number of turns added to the antenna coil for regeneration resulted in a simple high-performance radio that did not go into premature oscillation. One may say that out of beginner's luck I hit the perfect triad in my first design of a vacuum-tube regenerative radio.

For the transistorized version, the transconductance is much higher than in the pentode, the antenna coil is less sensitive than the one in the pentode, and the number of turns used for regeneration in the antenna coil also was different. So the first few versions of the transistorized regenerative radio actually failed—yes, they failed.

In the following sections of this chapter, some of best features of regenerative radios will be stated, some of the problems encountered will be mentioned, and finally, solutions to these problems will be presented. For now, let's see what a regenerative radio ideally is supposed to do.

Improving Sensitivity by Regeneration

In Chapter 6 on reflex radios we saw that a single transistor can function as a radio frequency (RF) and audio frequency (AF) amplifier at the same time. But the RF amplification factor of this transistor is fixed by the biasing current of its collector (see Figure 6-3). The direct-current (DC) bias voltage across resistor R2 determines the transconductance of the transistor, and therefore, its gain is fixed.

For regenerative radios, we try to feed back part of the RF energy of the RF amplifier to be reamplified without running into oscillation. This is very tricky because one wants the highest gain possible but also a lack of oscillation within the RF amplifier. *Also the leads of the secondary may have to be reversed. Refer to the note at the end of this chapter.*

Figure 7-1 presents a "sketch" of a possible transistorized regenerative radio.

In the figure, the primary of antenna coil L1 is connected to an RF amplifier circuit consisting of Q1 and Q2. Q1 is an emitter follower amplifier designed to

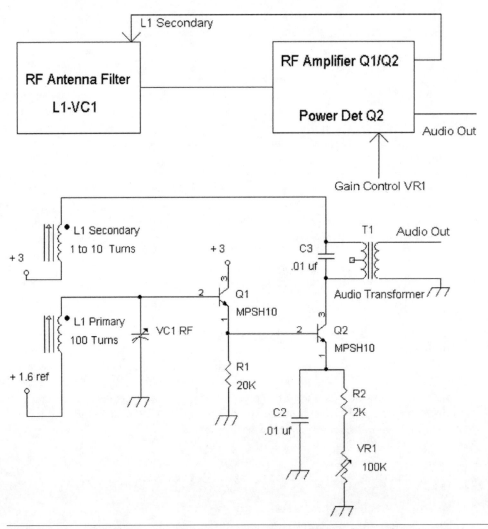

FIGURE 7-1   Block diagram and schematic of a "possible" design for a regenerative radio.

give near-unity gain while also providing a high-impedance input resistance for L1's primary, which will not load down or degrade the Q of the antenna coil. The secondary of L1, which has a smaller number of turns compared with the primary of L1, is used differently in tunable radio-frequency (TRF) radios and reflex radios. Here the secondary winding of L1 is connected to the amplifier output of Q2 to magnetically induce some of the amplified RF signal back to the antenna coil. *The secondary winding of the coil is not reversible, so positive feedback occurs only when the windings are connected in the correct manner. If connected in any other manner, negative feedback occurs, and the gain cannot be increased as desired.*

The amount of RF signal fed back to the antenna coil is related to how variable resistor VR1 is set. The lower the resistance set to VR1, the higher is the DC collector current of Q2, which increases the gain of the Q2 RF amplifier. Conversely, with VR1 set to a maximum resistance value, a minimum DC collector current is set for Q2, thus setting the gain of the RF amplifier to a minimum. Thus VR1 controls the amount of positive feedback to the antenna coil. The more positive feedback sent back to the antenna coil, L1, the higher is the overall gain of the system. At some point, the RF amplifier Q1 and Q2 will oscillate when VR1 is set to a gain that is sufficient to induce oscillation.

In Figure 7-1, Q2 also works as an AM detector by being a power detector. With a sufficient RF signal level into the base of Q2, power detection or AM demodulation occurs. The audio transformer T1 then extracts the audio signals (e.g., demodulated AM signals).

In operating a regenerative radio, one turns the regeneration control below the threshold of oscillation and then tunes for the stations desired. Once a station is tuned in, turn up the regeneration control until gain is increased, but back off the regeneration as soon as an oscillation is heard (e.g., a whistling or squealing sound).

Improving Selectivity by Q Multiplication via Regeneration

The Q determines the selectivity of an inductor capacitor circuit. Selectivity can be determined by the bandwidth of a tank circuit. The narrower the bandwidth of a resonant circuit, the higher is the selectivity. One measurement of bandwidth is determined by tuned frequency divided by Q.

Thus, if a station is tuned to 1,000 kHz and the Q of the antenna coil is 50, the bandwidth is 1,000 kHz/50 or 20 kHz.

However, AM stations are spaced 9 kHz or 10 kHz apart depending on what part of the world you are in. Thus a bandwidth of 20 kHz theoretically can receive two stations that are adjacent to each other. If the antenna coil has a Q of 100, the bandwidth of the antenna coil variable capacitor tank circuit is 10 kHz, which would be a minimum requirement to separate channels from each other.

In some cases, a Q of 100 is achievable for a particular antenna coil. But getting a Q factor of 200 out of an antenna coil is rare. A limitation on the Q of a coil, antenna coil, or inductor is the internal coil resistance. For example, a typical coil may have a coil resistance of from less than 1 Ω to at least 10 Ω. Therefore, one way to effectively reduce or at least partially null or cancel out the internal coil resistance is to provide a negative resistance in parallel with the coil. This negative resistance is generated by an active-amplifying device such as a transistor along with applying positive feedback.

What happens is the amplifying device pumps energy back (via positive feedback) into the tank circuit to overcome the resistive loss. When a short pulse excites an inductor capacitor tank circuit, the tank circuit will "ring" at its resonant frequency but will decay and fail to ring after a period of time. The Q multiplier effect from a positive-feedback circuit pumps energy back to the coil to sustain a longer ringing effect when excited by a short pulse.

Thus the Q multiplier increases the original Q of the coil by a factor that is determined by the amount of positive feedback applied. With no positive feedback, the Q of the coil is still the original Q. The more positive feedback is applied, though, the higher is the multiplying effect on the original Q of the coil. (There will be a follow-up on the subject of Q in the later chapters of this book such as Chapters 17 and 20.)

Design Considerations for a Regenerative Radio

Two specific characteristics are needed to design a regenerative radio. They are:

1. The RF signal that is being amplified must be "strong" enough to work on its own without positive feedback to raise the amplitude level. That is, if the RF signal is too small to begin with, trying to raise its level via positive feedback may lead to oscillation.
2. The positive feedback must be controllable such that the gain can be raised easily while not causing oscillation of the RF amplifier.

If you take a look at Figure 7-2, you will see a tickler oscillator circuit, which looks like a regenerative radio.

In the figure, a resonant circuit is formed by the L1 primary and VC1, which resonates at a high impedance with 0 degrees of phase shift. The resonant circuit is amplified by Q1 and Q2, with output current from Q2 fed to the L1 secondary. Current flowing into the L1 secondary (10 turns) creates a positive-feedback condition, which causes a sustained oscillation. Any RF signal picked from the L1 primary winding is now small compared with the continuous-wave (CW) signal it is generating at the base of Q1 via oscillation. Thus the RF signal is basically "washed out" by the oscillation signal.

For a larger picture of what's going on, let's take a look at Figure 7-1 again. Note that the number of turns on L1's secondary winding is in a range of 1 to 10 turns. The reason is that when the gain control is set by changing the gain of the RF

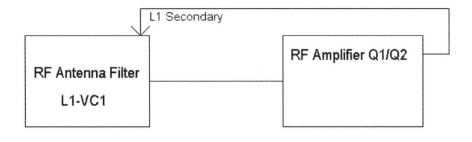

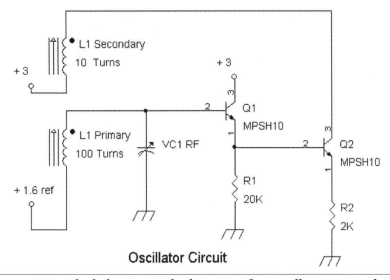

Oscillator Circuit

FIGURE 7-2   Block diagram and schematic of an oscillator circuit that resembles a regenerative radio circuit.

amplifier of Q2 via variable resistor VR1, there will be a "sweet" or optimal number of turns on L1's secondary. This optimal number of turns on L1 allows the positive-feedback regenerative system to achieve a balance between having Q2 provide further amplification of the RF signal (without oscillation) and also providing Q multiplication to increase the Q or selectivity of L1. For example, if the number of turns is too many, the positive-feedback regenerative system will break into oscillation before the received signal can be demodulated in a satisfactory manner.

Figure 7-3 shows the relationship of the RF amplifier's gain and threshold of oscillation as a function of the gain control setting. The top drawing in the figure shows the relationship of the RF gain in an amplifier such as the voltage gain of Q2 in Figure 7-1. As the gain control VR1 is adjusted for minimum resistance, the RF gain is increased via increasing the DC collector current of Q2, which also increases regeneration or the amount of positive feedback.

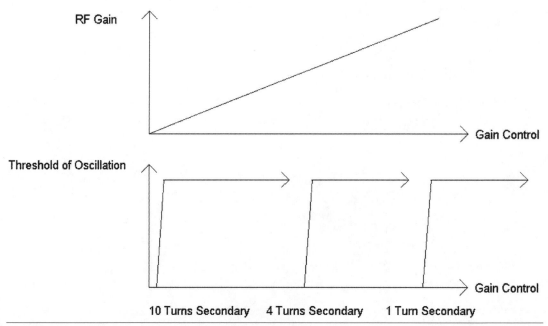

FIGURE 7-3 Relation of the threshold of oscillation and the RF gain based on the number of secondary winding turns.

However, as the gain control is adjusted for increased RF gain, the RF amplifier itself will start to oscillate. Thus, at the bottom of the figure we see that for 10 turns on the antenna coil L1 of Figure 7-1, oscillation occurs even when the RF gain is low. Under this condition, the overall gain of the RF amplifier is insufficient to raise the incoming or received RF signal to a high enough level for demodulation because oscillation breaks in too early. Once the RF amplifier breaks into oscillation, amplification of the RF signal received from the antenna coil is too low or not useful.

It should be noted that many commercially made antenna coils have a secondary winding of 10 or more turns. Thus, when these antenna coils are tried in the circuit of Figure 7-1, oscillation breaks in much too early before the RF signal can be amplified sufficiently for demodulation. If the antenna coil has four turns for its secondary on L1 of Figure 7-1, then we see from the bottom drawing of Figure 7-3 that there is some useful voltage gain before oscillation starts. And if just one turn of wire is wrapped around the antenna coil, we see that a much larger voltage gain can be provided to amplify the RF signal from the antenna coil before the regeneration (via increasing the gain of Q2) breaks into oscillation. Thus the key to designing a successful regenerative radio is to first have sufficient RF gain to allow demodulation of the AM signal and then second to have the capability of increasing the RF gain, which includes adding positive feedback to further increase the RF gain of the system while also increasing the effective Q or selectivity of the antenna coil without oscillation occurring.

So here is a first somewhat successful transistorized regenerative radio using an internal antenna.

Parts List

- **C1, C5, C6, C7:** 1 µF, 35 volts
- **C2:** 7 pF or 6.8 pF
- **C3:** 0.01 µF
- **C4:** 0.0033 µF
- **R1:** 100 kΩ
- **R2, R6:** 1 MΩ
- **R3:** 1 kΩ
- **R4, R7:** 56 kΩ
- **R5:** 10 kΩ
- **VR1:** 100 kΩ
- **T1 audio transformer:** 10 kΩ primary, 10 kΩ or 7 kΩ secondary
- **L1 antenna coil:** 470 µH with 1 turn wrapped
- **VC1 two-gang variable capacitor:** 140 pF, 60 pF
- **D1, D2, D3, D4:** 1N914
- **Q1, Q2:** MPSH10
- **Q3:** 2N4124 or 2N3904

This was really the first circuit that worked in a similar way to the pentode regenerative radio featured in *EDN Magazine*. The trick here was that only one turn was used for the secondary winding to provide the positive feedback for regeneration. And that one turn of wire [e.g., 30 American Wire Gauge (AWG)] was located in the center of L1's primary winding.

Diodes D1 to D4 form a voltage reference of about 1.6 volts to 2.0 volts for providing a DC bias voltage to Q1. Resistor R3 (1 kΩ) and capacitor C2 form a network to ensure that Q1 does not oscillate parasitically. As seen here, the RF signal is picked up by the antenna coil L1 primary and is connected to Q1, a near-unity gain amplifier with high-resistance input, so as not to load down the Q of L1's primary winding.

Q2 serves two purposes: One is to amplify the RF signal from L1's primary winding and send back the signal to L1 via the one-turn secondary winding for regeneration or positive feedback. The second purpose of Q2 is to provide demodulation of the AM signal via power detection. When AC signals of greater than 10 mV peak are coupled to a common emitter amplifier such as Q2, distortion occurs in a manner that demodulates AM signals.

Gain (or regeneration) control variable resistor VR1 is used to set the amount of regeneration such that the positive feedback increases the RF gain and increases the Q of L1. The audio signals are extracted from Q2 via audio transformer T1's primary. Capacitor C4, which is across the primary of T1, filters out the RF signal from Q2, and C4 also provides an RF signal path to the secondary winding of L1.

Audio amplifier Q3 is a common emitter amplifier that further amplifies the demodulated AM signal. This audio amplifier was needed because of the very low levels of audio signals from power detector Q2.

The radio design in Figure 7-4 was one of the first transistorized regenerative radios that finally worked for me—after a few other tries.

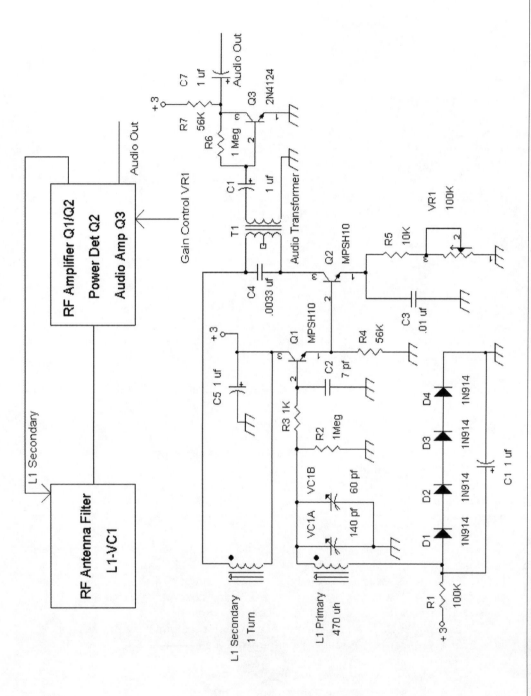

FIGURE 7-4 Block diagram and schematic of a regenerative radio using a secondary winding of one turn around the primary winding.

But what if one does not want to go through the trouble of winding wire around the antenna coil and instead wants to use the existing secondary winding? Well, there is a way to use the existing secondary winding. And two methods will be shown.

Figure 7-5 shows almost the same circuit as Figure 7-4, but it uses the existing secondary winding of 10 to 20 turns.

Parts List

- **C1, C5:** 1 μF, 35 volts
- **C2:** 7 pF or 6.8 pF
- **C3:** 100 μF, 16 volts
- **C4:** 0.0033 μF
- **R1:** 100 kΩ
- **R2:** 1 MΩ
- **R3:** 1 kΩ
- **R4:** 56 kΩ
- **R5:** 10 kΩ
- **VR1:** 1 kΩ or 2 kΩ
- **T1 audio transformer:** 10 kΩ primary, 10 kΩ or 7 kΩ secondary
- **L1 antenna coil:** 470 μH primary
- **VC1 two-gang variable capacitor:** 140 pF, 60 pF
- **D1, D2, D3, D4:** 1N914
- **Q1, Q2:** MPSH10

In this version, VR1, the regeneration control, is connected across the secondary winding of L1. The regeneration signal current from Q2 is fed to the wiper of VR1, which can divert this signal current to the +3-volts supply voltage *and/or* to the secondary winding. That is, part of the signal current from Q2 can be fed to the secondary winding of L1 in a variable manner. This variable signal current into the secondary winding allows for good control of the regeneration. For example, if the wiper of VR1 is set toward pin 3 of VR1, most of the regeneration signal current will be diverted to the secondary of L1. However, if the wiper of VR1 is set toward pin 1 of VR1, most of the regeneration signal will be diverted to the +3-volts supply, leaving little or no signal current to the secondary winding. And setting the wiper of VR1 in between will send a fraction of the regeneration signal to the secondary winding and a fraction of this regeneration signal to the +3-volts supply.

Note that the +3-volts supply is also an AC signal ground. That is, other than having +3 volts DC, the +3-volts rail has no AC signals on it because AC-wise, it is the same as ground. Also note that Q2 now is a fixed-gain amplifier and does not have a variable resistor, as seen in Figure 7-4.

It should be noted that in the prototype, VR1 is a multiturn potentiometer. The audio signal from T1 may be amplified further by using the Q3 amplifier circuit from Figure 7-4.

By placing a variable resistor across the secondary winding of L1, the primary winding is loaded down with an equivalent parallel resistor of (VR1's value) $\times N \times N$,

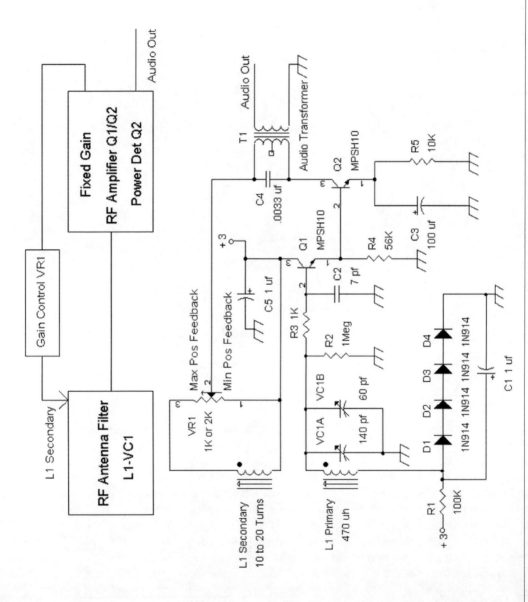

FIGURE 7-5 Block diagram and schematic of a regenerative radio using a potentiometer to adjust and control the amount of positive feedback.

where N is the turns ratio of primary to secondary windings. Thus, if $N = 20:1$, $N \times N = 400$, VR1 = 1 kΩ, and the equivalent parallel resistor is 400 × 1 kΩ, or 400 kΩ. Should $N = 10:1$, the equivalent parallel resistor is 100 kΩ.

Depending on the turns ratio of the antenna coil, the value of VR1 can affect Q. Thus the last regenerative radio design shown will not affect the Q of the antenna coil by not loading the secondary winding with resistance while allowing a range of number of turns on the secondary winding for regeneration. When all the previous designs (Figures 7-4 and 7-5) were analyzed and assessed for performance, there seemed to be something missing. In Figure 7-4, one had to experiment in winding a number of turns to hit the optimal spot in terms of RF gain before oscillation broke out. In Figure 7-5, there was the potential problem of loading down the Q of the antenna coil.

Certainly a higher-performance design could be made, but how? It occurred to me that maybe I could separate the regeneration from the RF amplifier and have a separate second-stage amplifier to increase the RF level. The block diagram for this idea is shown in Figure 7-6.

By using a dedicated transistor Q3 for the positive-feedback circuit, the regeneration can be controlled for a range of turns on the secondary of the antenna coil, whereas a separate gain stage in Q2 can further amplify the signal for even better sensitivity. This scheme worked pretty well considering that only a 2-inch-long ferrite antenna coil was used and the radio in the San Francisco Bay Area picked up radio station KNX (1,070 kHz) from Los Angeles.

Parts List

- **C1, C3, C9, C10:** 1 µF, 35 volts
- **C2:** 7 pF or 6.8 pF
- **C4, C5, C6, C11, C12:** 0.01 µF
- **C7:** 0.15 µF
- **C8:** 0.0015 µF
- **R1:** 56 kΩ
- **R2:** 1 MΩ
- **R3, R7, R14:** 1 kΩ
- **R4, R8:** 4,700 Ω
- **R5, R9, R12:** 100 kΩ
- **R6:** 2,200 Ω
- **R10, R13:** 10 kΩ
- **R11:** 20 kΩ
- **L1 antenna coil:** 470 µH
- **VC1 two-gang variable capacitor:** 140 pF, 60 pF
- **D1, D2:** 1N914
- **Q1, Q2, Q3:** MPSH10
- **Q4:** MPSH10 or 2N3904
- **VR1:** 5 kΩ

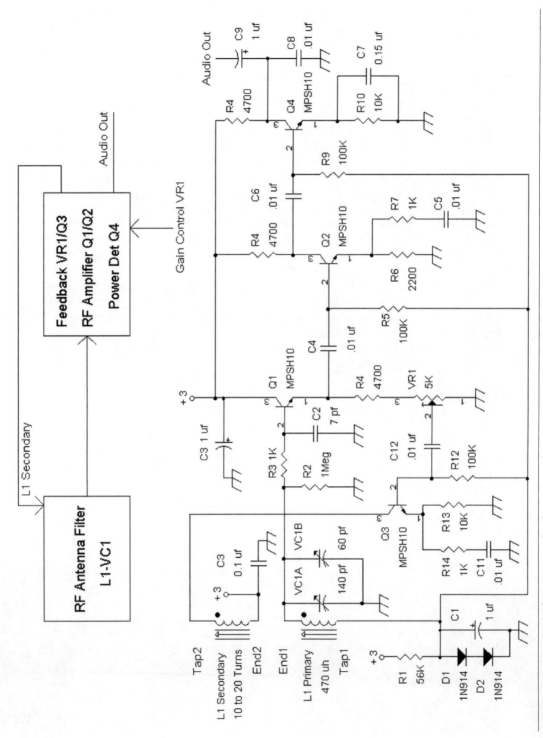

FIGURE 7-6 Block diagram and schematic of a preferred design for a regenerative radio.

RF signals are amplified with near-unity gain from Q1, which has an emitter load resistor of R4 and VR1, the regeneration control. The wiper of VR1 allows for up to 50 percent of the signal from Q1 to be sent to the input of the regeneration amplifier Q3.

Thus, in this design, a separate transistor Q3 is used for providing the regeneration signal back to L1 via the secondary winding. The signal from the collector of Q3 is a current source, which means a high resistance to ground. This high output resistance (e.g., >100 kΩ) from the collector of Q3 therefore does not load the secondary winding with resistance and thus further allows the antenna coil L1 to retain a high Q.

By turning up the signal from VR1, regeneration occurs, and increased RF voltage gain is achieved at the output of Q1's emitter. The RF signal from Q1 is further amplified by Q2. The output of Q2 then is connected to a power detector Q4, which demodulates the AM signal.

At 3 volts for the supply, this radio drained less than 300 µA. So two D cells should last about 5 years continuously. It should be noted that this radio will work down to 1.4 volts as well. Of course, three vegetable/fruit cells or about four to six solar cells can be used to power this regenerative radio.

Depending on how the antenna coil is wound, regeneration may or may not occur. Thus, the secondary winding connections should be connected one way or reversed to achieve regeneration. That is, if regeneration does not occur the first time, try reversing the secondary winding's leads, which should then lead to regeneration or positive feedback of the RF signal.

Chapter 8

Superheterodyne Radios

In Chapter 7 on regenerative radios we found that positive-feedback systems can enhance the selectivity and sensitivity for improved performance in a radio. Also, we found that for a regenerative radio, best performance was achieved when we added as much positive feedback or regeneration as possible, but just below the threshold of oscillation. When oscillation occurs in a regenerative radio, we lose performance.

The regenerative radio performs much better than "regular" tuned radio-frequency (TRF) radios, but it still has limitations because the selectivity is determined by the amount of regeneration added. For strong signals, the selectivity is less than when a weaker signal is tuned. The reason is that the weaker signal will require more regeneration, and thus more Q multiplication occurs for the weaker signal versus the stronger one. Thus the selectivity is variable in a regenerative radio.

So is it possible to design a radio in which the selectivity can be determined by some other means and such that the selectivity is "constant"? Yes, it is with a superheterodyne radio (Figure 8-1).

In the figure, the superheterodyne radio has a two-gang variable capacitor. Signals from the radio stations via the tuned filter (VC_RF) and a local oscillator (VC_Osc) are combined into a frequency-translation circuit (aka mixer circuit) that produces an amplitude-modulated (AM) signal at an intermediate frequency (IF). Essentially, the mixer circuit "maps" the incoming RF signal's frequency to a new frequency such as 455 kHz, an IF frequency. The IF circuit includes a band-pass filter centered around the IF. This filter passes signals at or near the IF while rejecting or attenuating all signals outside the IF band. Demodulation of the AM signal at 455 kHz from the output of the IF amplifier is done with conventional envelope or power-detector circuits.

In a superheterodyne receiver, there are four new elements, a two-ganged (or more) variable capacitor, a local oscillator, a mixer, and an IF filter/amplifier. Demodulation is still done with envelope or power detectors. Now let's go over these four elements briefly:

1. A ganged variable capacitor has a common shaft to turn the rotor plates of two or more sections. The two sections may be identical, or more commonly, the

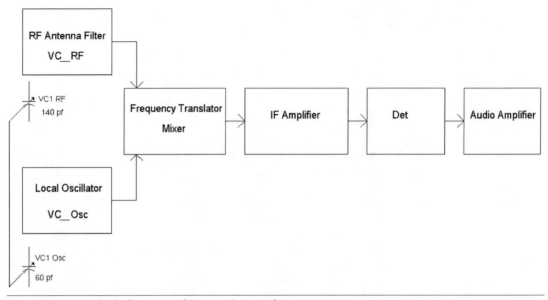

FIGURE 8-1   Block diagram of a superheterodyne receiver.

two sections are not identical. One section serves as the RF tunable filter (e.g., 540 kHz to 1,600 kHz), whereas the other section resonates with an oscillator coil and varies from about 1 MHz to 2 MHz for a 455-kHz IF.

2. A local oscillator generates a range of frequencies and varies depending on which station is tuned to the AM band. With the ganged variable capacitor, this local oscillator generates a frequency that typically is 455 kHz above the station that is being tuned. For example, when the variable capacitor is at maximum capacitance, the RF filter will be tuned to 540 kHz, whereas the local oscillator will generate a 540 kHz + 455 kHz = 995 kHz signal. Or when the ganged variable capacitor is at a minimum capacitance and is tuned to 1,600 kHz for the RF tunable filter, the local oscillator generates a 1600 kHz + 455 kHz = 2,055 kHz.

3. A mixer circuit really means that there is some type of multiplication of two signals going on. When multiplication occurs in two signals of two different frequencies F1 and F2, the output (of the mixer) will provide signals that have frequencies of (F1 – F2) and (F1 + F2), along with other signals of different frequencies from F1 or F2 as well. This is different from an *additive* mixer, which simply sums two signals together. In an additive mixer for two signals with frequencies F1 and F2, the output will only give signals of frequencies F1 and F2.

4. The IF filter generally passes a signal whose frequency is the difference frequency of the local oscillator and the incoming RF signal. For example, from element 2, if the radio is tuned to 540 kHz, the local oscillator will be at 995 kHz, and thus the difference between the two frequencies is 995 kHz – 540 kHz = 455

kHz. Similarly, for a radio tuned to 1,600 kHz, the local oscillator will be at 2,055 kHz, so the difference frequency signal will be 2,055 kHz – 1,600 kHz = 455 kHz. An IF filter therefore passes a signal whose frequency is around the IF (e.g., 455 kHz) and attenuates or removes signals whose frequencies are outside the vicinity of the IF. For example, the IF filter also will attenuate the signals from the oscillator and the RF signal input. *It should be noted that most superheterodyne radios employ simple mixers that will output sum and difference frequency signals but also will output signals from the local oscillator (1MHz to 2 MHz) and the RF signal input (540 kHz to 1,600 kHz). The (455-kHz) IF filter then also will remove/attenuate signals related to the local oscillator or incoming RF signal.*

Commercially Made Transistorized Superheterodyne Radios

One of the first transistor AM radios, the TR-1, was made by I.D.E.A., Inc., Regency Division, in 1955. The circuit topology of the TR-1 followed closely that of the vacuum-tube radios that used converter circuits instead of a separate local oscillator circuit and a mixer circuit (Figure 8-2A).

In Figure 8-2B, an antenna coil is connected to a grid of the 1R5 pentagrid (five-grid) converter tube, whereas an oscillator coil is coupled to three grids of the 1R5 tube. Thus the local oscillator's signal and the incoming RF radio station signal are

FIGURE 8-2A   Portable four-tube radio.

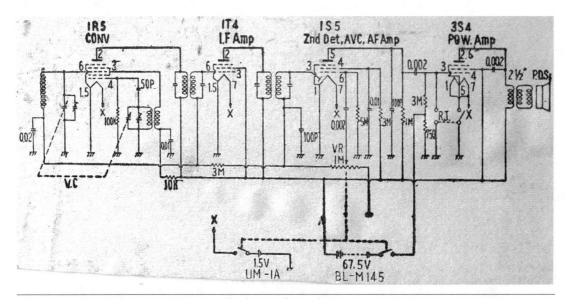

FIGURE 8-2B Schematic diagram of a four-tube radio.

connected to just one vacuum tube. Connected to the plate of the 1R5 tube is the first IF transformer to extract AM signals that have been "mapped" to 455 kHz.

Now let's turn to a block diagram of a typical transistor radio such as the Regency TR-1 (Figure 8-3).

The built-in antenna coil and variable capacitor VC_RF tune into a desired radio station to produce an RF signal to the input of the converter circuit. The converter

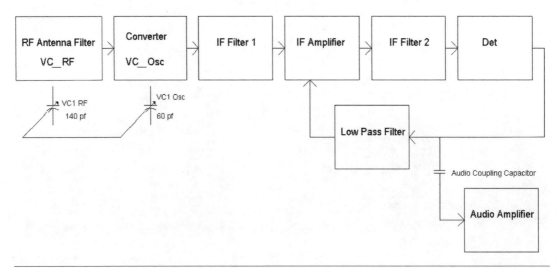

FIGURE 8-3 Block diagram of a four-transistor radio.

circuit consists of a transistor oscillator circuit in which the oscillation signal is combined with the RF signal. When the two signals, the oscillation signal and the RF signal, are added together, the nonlinear characteristic of the transistor also generates a multiplying action between the two signals. As a result, the output of the converter circuit then produces signals that have frequencies of sums and differences from the RF signal and the oscillation signal.

The output of the converter is connected to IF Filter 1, which only extracts signals that are of the difference frequency. Further amplification of this difference frequency (e.g., 455 kHz) is done with an IF amplifier, and the output of the IF amplifier is connected to a second IF filter.

From the second IF filter comes a 455-kHz AM signal, which is demodulated by an envelope detector (or power detector). The envelope detector performs half-wave rectification on the negative half (lower half) of the AM envelope for two functions. One function is to provide a demodulated AM signal or an audio signal. The second function is to provide a negative direct-current (DC) voltage for automatic volume control (AVC). The negative DC voltage is level shifted up by adding a positive bias voltage such that when a strong signal is received, the net total voltage is a positive voltage that decreases in proportion to signal strength. And when a weak signal is received, the net total voltage is still positive and increases in proportion to the weakness of the received signal.

The DC bias voltage and the half-wave-rectified negative half of the AM envelope are passed through a low-pass filter to remove IF and audio signals, passing only a DC voltage to change the gain of the IF amplifier. For example, for a strong signal received, the constant DC voltage is +1.0 V, and when a weak signal is received, the constant DC voltage is +1.2 V. The AVC system allows a more even audio volume between weak and strong stations received.

To raise the level of the demodulated AM signal, an audio amplifier (AF Amp) is used to supply sufficient signal to drive a loudspeaker or low-impedance earphone. Figure 8-4 provides a more detailed look at the converter circuit.

The converter circuit in this figure essentially is an oscillator circuit. Normally, the oscillator circuit has the negative (−) input terminal connected to a voltage source, which is an alternating-current (AC) ground. For now, let's take a look at just the oscillator section. The output of the amplifier is connected to a secondary winding of the oscillator coil/transformer (e.g., 42IF100). The primary winding of the oscillator coil/transformer has the required inductance with a variable capacitor to span a range of about 1 MHz to 2 MHz for the oscillator signal. The inductance in the primary winding is tapped down to a smaller voltage, as denoted by the K, a scaling factor of less than 1. The tapped-down signal then is connected to the positive input terminal of amplifier to form a deliberate positive-feedback system so as to provide oscillation.

With the circuit oscillating, the transistor is driven with a relatively large oscillating signal voltage at the positive (+) terminal, usually greater than a 200-mV peak-to-peak sine wave. This large signal also causes the transistor amplifier to distort and produce harmonic distortion at the output of the amplifier. When a low-level RF signal from the antenna coil is combined with the amplifier via its negative (−) input terminal, the combination of an RF signal and an oscillating signal

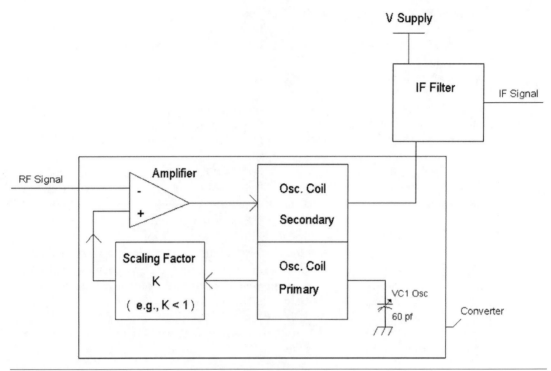

FIGURE 8-4   Block diagram of a converter circuit.

produces intermodulation distortion signals at the output of the amplifier. A couple of the intermodulation distortion signals happen to have frequencies that are the sum and difference frequencies of the RF signal and the oscillating signal. These intermodulation distortion signals are attenuated (e.g., ignored) by the oscillator's tank circuit and do not affect the oscillator. But one of these intermodulation distortion signals is extracted via an IF filter tuned to the difference frequency or the IF frequency. And the output of the IF filter then is connected to the input of an IF amplifier (not shown).

A Four-Transistor Radio Schematic

In 1955, the Regency TR-1 was the first commercially manufactured superheterodyne radio. It used four transistors for the following functions: the converter, the first IF amplifier, the second IF amplifier, and the audio amplifier. This radio ran off a 22.5-volt battery, which was common back then and up to the 1970s but is rare today.

For our first superheterodyne radio, we shall use a 1.5-volt to 3-volt source instead a 22.5-volt battery as well as four transistors, but as the converter, the IF amplifier, the audio driver, and the audio power amplifier.

Parts List

- **C1, C4, C10:** 33 µF, 16 volts
- **C2, C6, C9, C13:** 0.01 µF
- **C3, C7, C8:** 1 µF, 35 volts
- **C5:** 0.15 µF
- **C11, C12:** 220 µF, 16 volts
- **R1:** 3,300 Ω
- **R2:** 2,200 Ω
- **R3:** 1 kΩ
- **R4:** 10 kΩ
- **R5, R6:** 20 kΩ
- **R7:** 470 Ω
- **R8:** 20 Ω
- **VR1:** 100 kΩ
- **D1, D2, D3, D4:** 1N914
- **Q1, Q2, Q3, Q4:** 2N3904
- **L1 antenna coil:** primary 600 µH to 680 µH, secondary 10 to 20 turns
- **T1 oscillator coil:** 42IF100
- **T2 IF transformer:** 42IF101
- **T3 IF transformer:** 42IF103
- **T4 audio driver transformer:** primary 10 kΩ CT, secondary 600 Ω CT or primary 2.5 kΩ, secondary 600 Ω CT
- **T5 audio output transformer:** primary 120 Ω, secondary 8 Ω
- **VC1:** 140 pF RF and 60 pF oscillator

Figure 8-5A shows a block diagram of the four-transistor radio, and Figure 8-5B shows the schematic of the four-transistor radio.

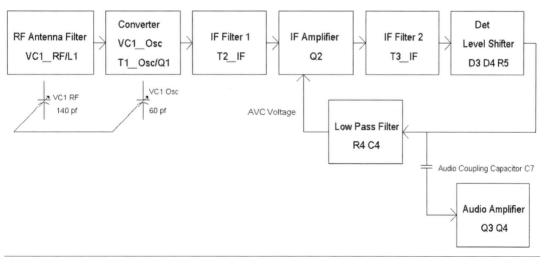

FIGURE 8-5A   Block diagram of a four-transistor radio.

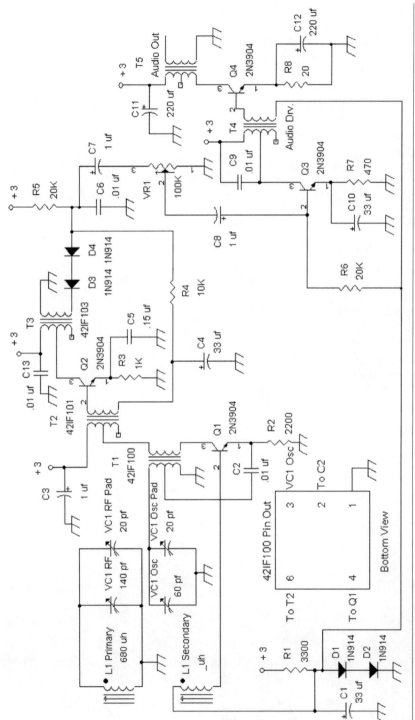

FIGURE 8-5B Schematic diagram of a four-transistor radio.

Radio signals are received via antenna coil L1. L1 resonates with variable capacitor VC1_RF. The tuned RF signal is stepped down via the L1 secondary winding, which is connected to the base of Q1, an oscillator circuit. Oscillator coil T1 has a secondary winding that serves as a tickler winding to couple the output signal from the collector of Q1 back to the input emitter of Q1 to form a positive-feedback system that ensures oscillation. The primary winding of T1 is connected to the second gang or section of variable capacitor VC1_Osc to allow varying the oscillation frequency in the range to 1 MHz to 2 MHz. It should be noted that the primary winding of T1 has a tapped winding that is connected to the emitter of Q1 via capacitor C2. The tapped winding allows connection to the low-input resistance at the emitter of Q1, which is typically less than 500 Ω to avoid degrading the Q of the parallel resonant tank circuit, consisting of the inductance in the primary winding of T1 and capacitor VC1_Osc. The base of Q1 is biased by diodes D1 and D2 to provide a DC bias voltage at the base of Q1 of about 1 volt, which also sets the DC collector current for Q1. R2 is chosen to set a sufficient emitter current (e.g., ~ 200 µA) to provide enough voltage gain in transistor Q1 to reliably produce oscillation.

Because the secondary winding of L1 provides the RF signal to the base of oscillator transistor Q1, multiplication of the oscillator signal with the RF signal occurs. And the output signal at the collector of Q1 includes an amplified frequency-translated signal version of the RF signal to 455 kHz.

Note that T2, the first IF transformer's primary, via its tapped winding is in series with the secondary winding of T1, the oscillator coil. By this series connection, the 455-kHz (IF) signal is extracted from the collector of Q1.

The secondary winding of T2 is also stepped down to allow connection to the base of the IF amplifier's transistor Q2, which has a moderate input resistance (e.g., in the few thousands of ohms) so as not to degrade the Q at the primary of T2.

At the output of the IF amplifier's transistor Q2, the collector is connected to the second IF transformer. The AM signal from second IF transformer's secondary is demodulated via diodes D3 and D4. One may ask why two diodes in series? The second diode performs the function of level shifting up the voltage by about 0.5 volt DC, which is needed in an automatic volume-control system with transistor Q2. Thus diodes D3 and D4 perform envelope detection, and through biasing resistor R5, the two diodes also level shift the detected lower half of the AM envelope to a voltage of about 1 volt DC when the RF signal is zero or weakly received. This 1-volt signal is further filtered by R4 and C4 and applied to the secondary winding of T2 to bias the IF transistor Q2. When a signal is received, the demodulated signal at C6 is sitting at 1 volt DC with a negative-going AC audio signal. Low-pass filter R4 and C4 filter out the AC audio signal and provide a DC voltage that is 1 volt minus the average carrier level of the received signal. The stronger the received RF signal, the more the average carrier level is "subtracted" from the 1-volt DC signal at C4. By lowering the biasing voltage for Q2 depending on how strong the received signal is, Q2 performs an automatic gain or volume control for the demodulated audio signal. Note that if there were only one diode instead of D3 and D4, maximum DC voltage would be 0.5 volt, which is insufficient to turn on the base of Q2 (e.g., Q2 requires at least 0.6 volt on the base).

It should be noted that the second IF transformer here or, in general, the last IF transformer before diode detection usually has a lower turn ratio from primary to secondary windings compared with any of the other IF transformers. The reason is to provide more IF signal to the diode detector, which also generates sufficient AVC voltage. For example, if T3 is changed from a 42IF103 part to a 42IF101 or a 42IF102 transformer (both with about a 20:1 turns ratio) instead, there will be lower amplitude-demodulated signals (compared with the 42IF103 transformer with a 6:1 turns ratio) because of the higher turns ratio in these two parts, which then results in insufficient AVC voltage for good automatic volume control.

The demodulated signals from D4 are AC-coupled to the base of the audio driver transistor Q3 via volume control VR1. The output of Q3 is connected to an interstage step-down transformer T4 (2.5-kΩ primary to 150-Ω secondary) to generate increased current drive into output audio transistor Q4. Transistor Q4 is biased at about 20 mA, which results in an input resistance of about 130 Ω to 150 Ω at the input of Q4. The amplified audio signal via Q4's collector is connected to step-down transformer T5 to drive a speaker or low-impedance earphone.

Although this radio runs off two cells for 2.4 volts to 3 volts, it will operate down to 1.2 volts as well. Because of the current drain of this radio, a power switch in series with the battery is suggested.

An Eight-Transistor Radio

For designing an eight-transistor radio, a more "traditional" approach was taken. That is, in most common descriptions of a superheterodyne radio, there is a tuned RF circuit, a separate local oscillator, a mixer, IF filters and amplifier(s), a detector, and the audio amplifier (Figure 8-6A).

In Figure 8-6A we see that there is a ganged variable capacitor for the tuned RF stage and local oscillator. The antenna coil receives radio stations via magnetic energy and resonates or tunes with the RF section of the variable capacitor. RF signals from the antenna coil are connected to an input of the mixer circuit. The local oscillator is also connected to the mixer circuit such that there is a combination of both RF and oscillator signals at the input(s) of the mixer. Because the oscillator's signal into the mixer is very large, the mixer generates intermodulation distortion products at its output. *It should be noted that the intermodulation distortion products or signal from the mixer also can be thought of as a result of the oscillator signal multiplying with the RF signal.* And one of the intermodulation distortion products is a frequency-translated version of the RF signal to an IF such as 455 kHz. Thus the first IF filter extracts the 455-kHz signal from the output of the mixer.

Signals from the first IF filter are amplified by the first IF amplifier. The output of the first IF amplifier Q3 is connected to a second IF filter, and the output of the second IF filter is further amplified by a second IF amplifier Q4. Finally the output of the second amplifier is connected to a third IF filter. From the third IF filter, demodulation occurs via a detector circuit. But also from the detector circuit is a DC level-shifting circuit that is connected to a low-pass filter with a resistor capacitor

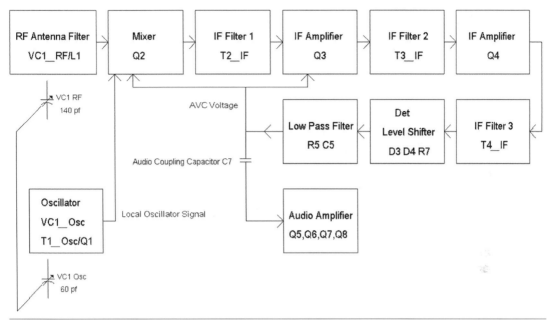

FIGURE 8-6A   Block diagram of an eight-transistor radio.

(RC) time constant of 0.1 second to 0.5 second. The output of the low-pass filter then provides an AVC voltage, which controls the conversion gain of the mixer circuit and the voltage gain of the first IF amplifier. By having gain control over two stages, the mixer and first IF amplifier, there is more AVC range compared with applying an AVC voltage to one stage such as the previously described four-transistor radio.

An audio driver stage amplifies the demodulated AM signal with sufficient current drive to a push-pull audio amplifier for listening via a loudspeaker or low-impedance earphone.

Parts List

- **C1, C6, C8:** 0.15 μF
- **C2, C5, C12, C15:** 33 μF, 16 volts
- **C3, C4, C7, C9, C18, C19, C20:** 0.01 μF
- **C10, C11, C17:** 1 μF, 35 volts
- **C13, C14, C16:** 220 μF, 16 volts
- **R1, R13:** 2,200 Ω
- **R2, R4, R6:** 1,000 Ω
- **R3, R10:** 4,700 Ω
- **R5, R12:** 10 kΩ
- **R7, R8:** 20 kΩ
- **R9:** 100 kΩ

- **R11:** 470 Ω
- **R14:** 3.0 Ω
- **D1, D2, D3, D4, D5:** 1N914
- **D6:** 1N270 or 1N34
- **Q1 to Q8:** 2N3904
- **L1 antenna coil:** 600 µH to 680 µH with tap or secondary winding
- **VC1 variable capacitor:** 140 pF and 60 pF
- **T1 oscillator coil:** 42IF100
- **T2 IF transformer:** 42IF101
- **T3 IF transformer:** 42IF102
- **T4 IF transformer:** 42IF103
- **T5 audio transformer:** primary 1.5 kΩ, secondary 600 Ω CT
- **T6 audio transformer:** primary 48 Ω CT, secondary 8 Ω
- **VR1:** 100 kΩ

Figure 8-6B presents a schematic diagramof the tuner section of the eight-transistor radio, while Figure 8-6C presents a schematic diagram of the audio section of the eight-transistor radio.

Oscillator circuit Q1 generates a 1-MHz to 2-MHz signal of more than 300 mV (peak-to-peak sine wave) at its emitter via oscillator coil T1 and variable capacitor VC1_Osc. Positive feedback is established by coupling the output signal from the collector of Q1 back to its emitter terminal via a tapped winding from T1's primary winding. The oscillation signal at the emitter then is connected to the emitter (input) of mixer circuit Q2. Low-level RF signals are connected to the base input terminal of the mixer Q2 via a tap in antenna coil L1. As stated earlier, VC1_RF tunes with the antenna coil (680 µH) to resonate or give maximum signal level when tuned to a radio station.

At first glance, it looks as though the signal at the collector of mixer transistor Q2 will be an amplified version of the two signals, the low-level RF signal and the oscillator signal, connected to its inputs. But because the oscillator signal's amplitude is very large, Q2 is driven into deliberate distortion, which causes its output at the collector to generate signals that are the sum and difference frequencies of the low-level RF signal (540 kHz to 1,600 kHz) and the oscillator signal (1 kHz to 2 MHz), as well as providing amplified versions of the RF and oscillator signals. However, because the difference frequency signals (e.g., IF signals) are at 455 kHz and all other signals are at frequencies above 455 kHz, the first IF filter, T2, is able to extract the IF signal and reject all the other signals from the mixer's output.

The extracted IF signal from the secondary winding of T2 is amplified by the first IF amplifier Q3 via a connection to the base of Q3. The amplified signal from the collector of Q3 is then connected to a second IF filter T3 for further band-pass filtering around 455 kHz. A second-stage IF amplifier Q4 further amplifies signals from the secondary winding of T3. A third IF filter T4 is connected to the output of Q4. Demodulation of the 455-kHz signal at the secondary winding of T4 is done by diodes D3 and D4, which also provide a DC level-shifted voltage of the demodulated signal. The demodulated and DC level-shifted signal is further low-pass filtered to remove

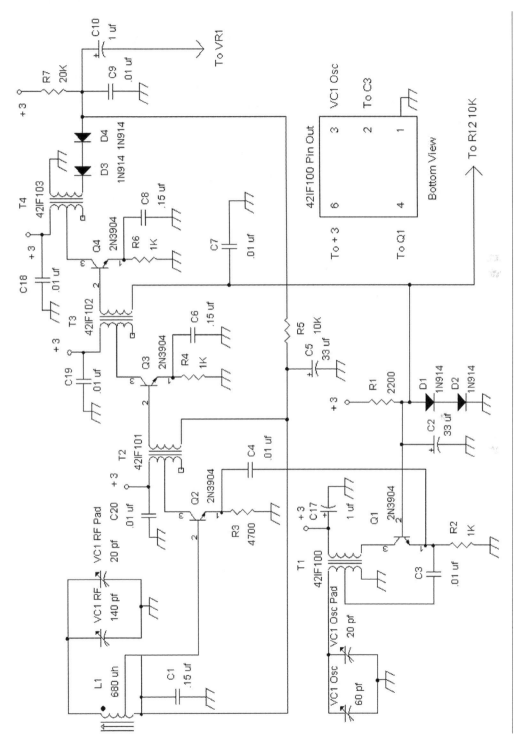

FIGURE 8-6B Schematic diagram of the eight transistor radio's tuner section.

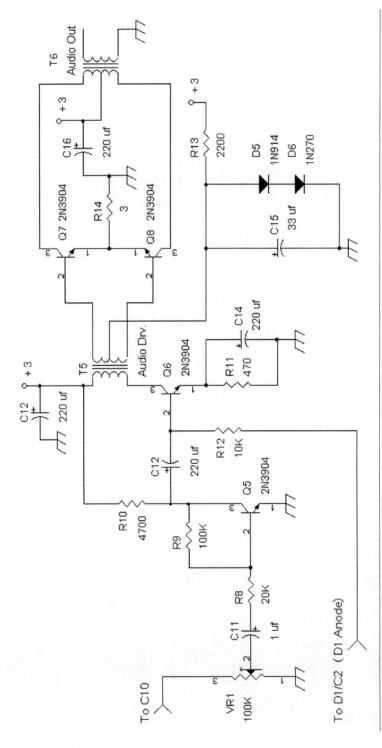

FIGURE 8-6C Audio amplifiers for the eight transistor radio.

audio information via R5 and C5. Thus the voltage at C5 represents a DC shifted-voltage version of the average RF carrier level of the signal from the radio station. And thus this voltage at C5 provides an AVC voltage that is connected to the bases of Q2 and Q3 via the secondary winding of L1 and primary winding of T2. Since both bases of Q2 and Q3 require at least 0.6 volt of biasing voltage, the voltage from D3, D4, and R7 provides about 1.0 volt of DC to start with that allows biasing of Q2 and Q3.

Automatic volume control then is accomplished by varying the signal output from mixer 2 and first IF amplifier Q3 in an inverse manner. The higher the received signal, the lower is the AVC voltage to turn down the gain. Conversely, the lower the received signal, the higher is the AVC voltage to turn up the gain.

The demodulated signal from D4 is further audio amplified by Q5 and Q6, and its volume is controlled by VR1. Amplifier Q5 is an inverting-gain amplifier that is connected to a driver amplifier Q6. The output of amplifier Q6 is connected to an interstage transformer T5 to boost the audio signal current into transistors Q7 and Q8 in a push-pull manner. The phase of the audio signal into the base of Q7 is 180 degree in reference to the phase of the audio signal at Q8. Biasing of Q7 and Q8 is accomplished by the voltage generated by diodes D5 and D6 and R13. Signal currents in a push-pull manner from Q7 and Q8 are connected to output transformer T6 so that its secondary winding can drive a loud speaker or low-impedance earphone.

This radio will work fine off 2.4 volts to 3 volts but also will run off 1.2 volts. Also because of the current drain, a power switch in series with the battery is suggested.

Alternative Oscillator and Antenna Coil Circuit

In Chapter 3 it was noted that one of the sections of a twin variable capacitor can be used for the oscillator circuit. A series capacitor of 110 to 120 percent of the variable capacitor's maximum capacitance in an oscillator circuit allows for accurate tracking with the tuned RF stage. For example, with a twin variable capacitor of 270 pF, a 330-pF series capacitor is needed for the oscillator circuit. And for a twin 335-pF variable capacitor, a 390-pF series capacitor is required. Figure 8-7 shows an alternative oscillator and antenna coil circuit for the eight-transistor radio.

In this circuit, instead of using a nonsimilar two-gang variable capacitor for the tuned RF and oscillator circuit, a twin 270-pF variable capacitor is used. Twin variable capacitors are used not only for superheterodyne radios covering the broadcast AM band or medium-wave band, but they are also used for long- and short-wave radio bands as well.

For an AM radio with a twin 270-pF capacitor with a series 330-pF capacitor C16, we need about 140 µH of inductance. Because the 42IF100 oscillator coil T1 has nominally about 330 µH, a 220-µH or 270-µH fixed inductor is paralleled with the main primary winding of T1 to provide a net inductance of about 140 µH. Also, a fixed 15-pF capacitor C22 is paralleled with the primary winding of T1 or with L2 for better tracking with the tuned RF stage. In addition, the emitter resistor R2 (for the oscillator

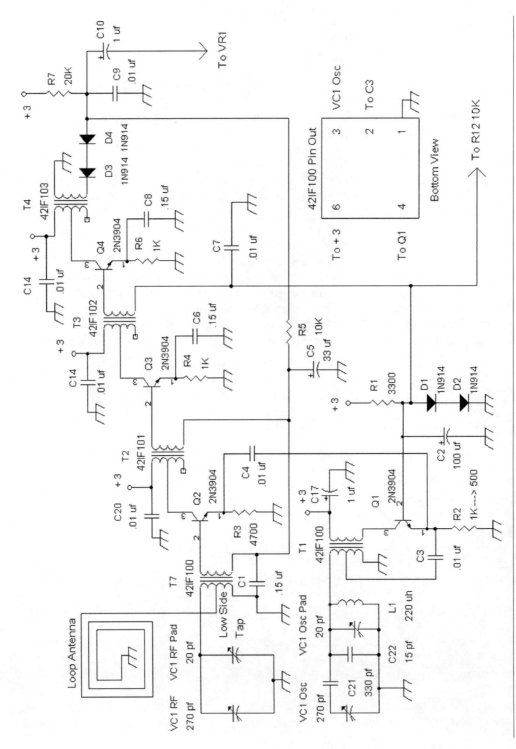

FIGURE 8-7 Alternative tuner section for the eight-transistor radio.

108

circuit Q1) is changed from 1 kΩ to 510 Ω to ensure more than 300 mV peak to peak of the oscillation signal (at Q1's emitter). Note that C3 is connected to the low-side tap of T1. The output of the oscillator is connected to mixer transistor Q2's emitter terminal through C4.

As mentioned in Chapters 3 and 5, using a loop antenna allows for more flexibility because loop antennas can be made with about 10 turns around a 6-inch rectangle, square, or ellipse or bought readily at MCM Electronics or on the Web. Thus a loop antenna is connected to a second 42IF100 coil T7; it is connected to the low-side tap that is used as an RF transformer. The other section of the twin 270-pF variable capacitor is connected to the main primary winding of T7, whereas the secondary winding of T7 provides RF signals to the mixer transistor Q2 via its base terminal. And for automatic volume control, one side of the secondary winding of T7 is connected to the AVC voltage source at C5. Figure 8-7 shows the alternative oscillator and antenna coil circuit, including part of the mixer circuit Q2 and first IF amplifier circuit Q3. This circuit can be "dropped in" or substituted for the oscillator and tuned RF section circuits of the eight-transistor radio circuit shown in Figure 8-6B.

Figure 8-8 shows another type of oscillator, a differential-pair transistor oscillator.

As stated earlier, the loop antenna is connected to the low-side tap of T7 (42IF100).

It was found that an IF transformer (the 42IF104) has about 140 µH from the low-side tap of its primary winding. Because of interelectrode capacitances of the oscillator transistor, it is undesirable to connect this low-side 140-µH tap directly to a transistor. Instead, therefore, the stepped-down secondary winding is used. At resonance, resistance "looking" into the 140-µH tap is very high, on the order of 100 kΩ or more, but looking into the secondary, this 100 kΩ gets divided by $N \times N$, where N is the turns ratio from the low-side tap to the secondary winding. For the 42IF104, N is in the range of about 14, which means that at resonance the equivalent resistance across the secondary winding is about 500 Ω (100 kΩ divided by 14^2).

The differential-pair transistor oscillator is formed by Q1A and Q1B. The collector of Q1A is connected to the secondary winding of the 42IF104 IF transformer, which is like a 500-Ω load. To ensure oscillation by positive feedback, the collector of Q1A is connected to the base of emitter follower transistor Q1B, where its emitter is connected back to the emitter of Q1A. With sufficient DC collector currents set for both transistors, there will be enough gain to sustain a reliable oscillation. And the oscillator's signal is taken from the collector of Q1A and added in series with the RF signal via the secondary winding of T7.

It should be noted that the mixer circuit is modified slightly by grounding the emitter of Q2 with a 0.15-µF capacitor and summing both oscillator and RF signals into the base of Q2. Again, the oscillator signal is large enough to cause a modulation or multiplying effect on the RF signal from the secondary winding of T7. Thus the output signal from Q2's collector into the IF filter, T2, is a 455-kHz AM signal.

The AVC voltage from C5 is sent to the base of Q2 via resistor R20 and to the base of the IF amplifier transistor Q3 via the secondary winding of the second IF transformer T2 (42IF102). As we will see in Chapter 9, the differential-pair oscillator will play an important role in low-powered oscillator circuits.

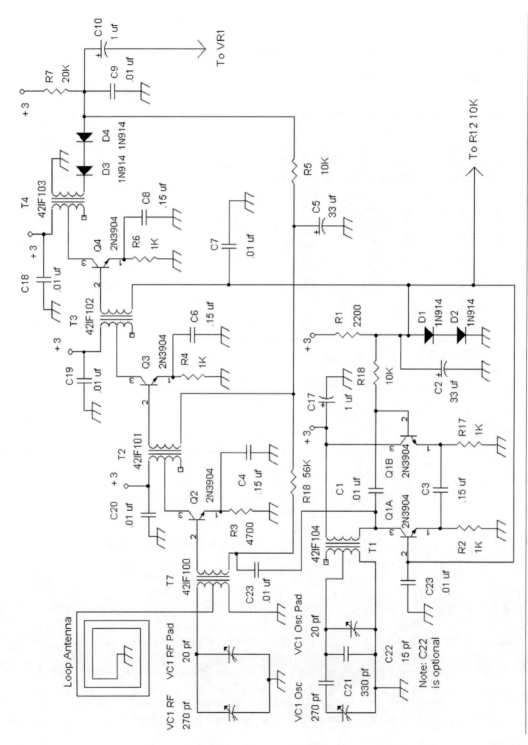

110

FIGURE 8-8 Differential-pair oscillator (Q1A, Q1B) for alternative tuner section of the eight-transistor radio.

An Item to Note

Generally, the 42IF100 oscillator coil can be substituted with the 42IF300 or the 41IF110 for use as an RF transformer (with a loop antenna) or as an oscillator coil. However, some experimentation may be needed to achieve equivalent results to the 42IF100. For example, in an oscillator circuit, the converter or oscillator transistor's emitter resistor *may* have to be reduced in value (e.g., in half) or increased (e.g., up to twofold) when substituting the 42IF300 or 42IF110 for the 42IF100 coil.

Chapter 9

Low-Power Superheterodyne Radios

Chapter 8 presented superheterodyne radios of moderate current drain (e.g., >10 mA) that powered loudspeakers. This chapter will explore superheterodyne receivers with extremely low current draw that allow years of continuous playing on a single battery. What we will find out is that the superheterodyne circuit topology shown in Chapter 8 will not apply for low-power design, and instead, a modified circuit topology is required.

Design Goals for Low Power

The design goals are as follows:

1. Current drain of less than 150 µA with a crystal earphone
2. Operational supply voltage of 1.2 volts
3. Sensitivity and selectivity performance at least equal to that of the four-transistor design in Chapter 8

As with the low-power circuits in earlier chapters, the goals in terms of current drain and supply voltage remain the same. However, because the circuit topology includes an oscillator, mixer, and two intermediate-frequency (IF) amplifiers, one would think that power consumption would go up. Instead, we design these circuits, the mixer, and IF amplifier to run each at about 20 µA or less. Moreover, a low-power superheterodyne radio with less than 150 µA of drain will last about 5 years continuously on a single alkaline C cell.

Low-Power Oscillator, Mixer, and Intermediate-Frequency Circuits

In Chapter 8 we found out that with commercially made oscillator coils such as the 42IF100 or 42IF300, the oscillator or converter transistor's collector current requires about 200 μA to ensure reliable oscillation over the tuning range of 1 MHz to 2 MHz. The step-down ratio of the oscillator coil thus requires a minimum transconductance or gain from the transistor for oscillation to occur. Transconductance of a bipolar transistor is roughly equal to the direct-current (DC) collector current divided by 0.026 volt. For example, if the DC collector current is 1 mA, then the transconductance of the transistor is 0.001mA/0.026 volt = 38 mA/volts. As you can see, transconductance is proportional to DC collector current.

For a low-power oscillator, the transconductance of the transistor will be much lower than the transconductances of the oscillator/converter transistors shown in Chapter 8. Therefore, the oscillator/converter circuits will not oscillate reliably at DC collector currents of less than 100 μA.

The turns ratio of the 42IF100, 42IF300, or 42IF110 oscillator coil is at least 20:1, which is required for the one-transistor oscillator/converter circuit of Chapter 8. However, if an extra transistor is added to form a two-transistor oscillator circuit, then the turns ratio can be lower, something like 4:1, 3:1, 2:1, or even 1:1. The extra transistor "buffers" the oscillator signal to allow less loading on the oscillator's tank circuit such that the oscillator signal is not stepped down or attentuated as much. By avoiding stepping down or attenuating the oscillator signal too much, the two transistors with lower transconductance still will provide sufficient gain to sustain reliable oscillation.

Another way to analyze the oscillator/converter circuits of Chapter 8 is that the load resistance for the collector of the transistor at resonance is low because of the high step-down ratio. This low-valued load resistance (in the few kiloohms) requires higher transconductance.

In contrast, if the load resistance at resonance is much higher, a lower transconductance is required because gain is related to the load resistance multiplied by the transconductance. Thus, using a differential-pair oscillator circuit allows for a higher load resistance at resonance to "make up" for the lower transconductance of the transistors.

Turning to the mixer, the same type of mixer will be used as in Chapter 8, but at a much lower current. And the IF amplifier circuits will be similar to the ones used in Chapter 8 but again at a lower current. Because the IF amplifier circuits will be running at much lower transconductance or gain, coupling from one stage of the IF signal to another will not use the secondary winding of the IF transformer. The signal voltage from the secondary is stepped down from the primary winding so as to allow loading into a lower input resistance of the IF amplifier. However, when the IF amplifier is run at a lower operating current such as 20 μA, the input impedance is sufficiently high (e.g., approximately 50 kΩ to 100 kΩ) to allow coupling from the transistor's collector output terminal of the previous stage to the input of the next amplifying stage. By skipping the secondary winding and using the signal voltage at the primary of the IF transformer, more IF signal voltage is provided.

Low-Power Detector and Audio Circuits

A germanium diode will be used for demodulation or detection of the amplitude-modulated (AM) signal from the last IF stage. However, since the secondary winding of the IF transformer provides a lower signal voltage, the diode will rectify the IF signal at the primary winding instead.

To drive the crystal earphone, a low-power audio amplifier will be used. This audio amplifier will have an input resistance of at least 100 kΩ to maintain the Q or selectivity characteristic of the last IF transformer.

"First" Design of a Low-Power Superheterodyne Radio

Actually, the real first design was built a couple of years ago with a current drain of about 140 μA for five years of continuous service on a single C cell. This radio was mentioned in *EDN Magazine* on the web in October of 2011 (www.edn.com/blog/ Designing_Ideas/41377-A_super_het_radio_runs_5_years_on_a_C_cell_plus_a_ pentode_radio.php). But the "first" design for this book is a refined version based on the radio just mentioned that features even lower power. Figure 9-1 provides the block diagram.

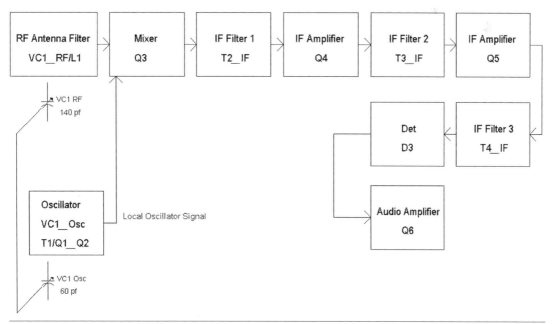

FIGURE 9-1   A low-power superheterodyne radio block diagram.

There are only a couple of differences between this radio and the superheterodyne radio block diagrams in Chapter 8. One is that the oscillator circuit requires two transistors instead of one. However, the two-transistor oscillator circuit drains only about 36 µA versus the hundreds of microamperes consumed by the one-transistor oscillator circuits in Chapter 8.

The second difference is that the automatic volume control (AVC) system was eliminated. Instead, there will be an IF gain control that will act as a volume control.

Parts List

- **C1, C2, C3, C7, C8, C9, C10, C11, C12, C16, C17:** 0.01 µF
- **C4, C5, C15:** 33 µF, 16 volts
- **C6:** 0.15 µF
- **C13:** 0.0022 µF
- **C14:** 1 µF, 35 volts or 33 µF, 16 volts
- **VC1 variable capacitor:** 140 pF and 60 pF
- **R1, R2, R5, R6, R7:** 12 kΩ
- **R3, R4, R12:** 56 kΩ
- **R8, R10, R13, R14:** 100 kΩ
- **R9:** 1.1 MΩ
- **R11:** 1 MΩ or 1.1 MΩ
- **VR1:** 100-kΩ pot
- **D1, D2:** 1N914
- **D3:** 1N34 or 1N270
- **Q1, Q2, Q3, Q4, Q5:** MPSH10
- **Q6:** 2N5089 or 2N5088
- **L1 antenna coil:** 600 µH to 680 µH primary, 10 to 20 turns for secondary winding
- **T1:** 42IF103 IF transformer with internal cap removed or 42IF106 IF transformer
- **T2, T3, T4:** 42IF101 IF transformer

Figure 9-2 provides a schematic diagram of our "first" low-power superheterodyne radio, which at 1.25 volts drained about 87 µA, or *four score and 7 microamps!*

For the oscillator circuit, a different approach is needed to achieve oscillations at low transistor collector currents. Also, because a conventional or off-the-shelf oscillator coil has too high a turns ratio, some other type of oscillator coil must be used for a low-power oscillator circuit.

Since the third IF transformer (the 42IF103) has a turns ratio of 6:1 instead of 20:1, it is possible to use it as an oscillator coil for a low-power circuit. The first step is to remove the internal capacitor by using a small-blade screwdriver to break out the ceramic material on the bottom of the 42IF103 IF transformer. The leads of the capacitor can stay and do not need to be removed. Alternatively, Xicon, which manufactures the IF transformer and oscillator coils, also makes a version of the 42IF103 without the capacitor, which is the 42IF106. The required inductance is about 330 µH, which is achieved across the high-side tap. Thus the variable capacitor VC1 is connected to the high-side tap connection of the primary of T1. This time the

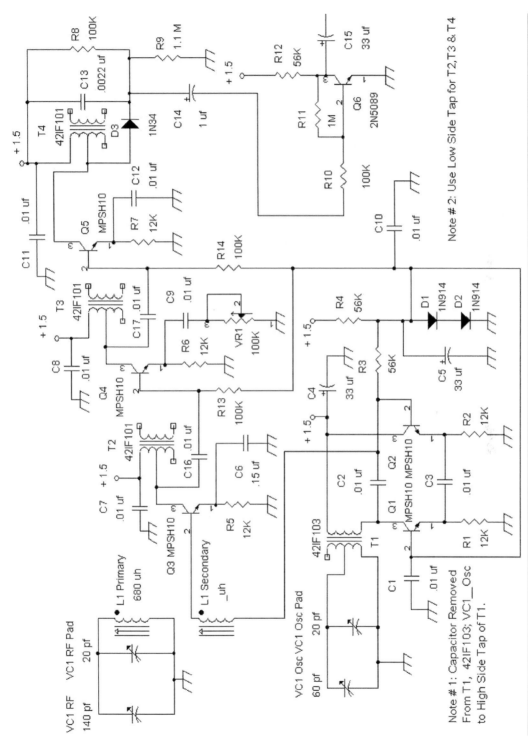

FIGURE 9-2 Schematic diagram of a six-transistor low-power superheterodyne radio.

117

low-side tap connection of the 42IF103 or 42IF106 primary winding is left alone as an open circuit.

At resonance via VC1_Osc and the primary high-side tap winding of T1, the secondary winding has a sufficiently high load resistance for Q1 to set up a sustained oscillation via positive feedback via Q2. The oscillator's output signal is about 170 mV peak to peak at C2. This oscillation signal is added to the radio-frequency (RF) signal from the secondary winding of L1. Many antenna coils come with a secondary winding. But some antenna coils come with just a primary winding. Thus a secondary winding can be made by winding about 10 to 20 turns of wire near or on the primary winding of the antenna coil.

It should be noted that the secondary winding can be made with more turns than usual because the input resistance to the mixer transistor Q3 is higher than usual owing to the lowered collector current. For example, most antenna coils come with a primary-to-secondary turns ratio of about 10 to 15:1. For this project, a turns ratio of 4 to 5:1 will provide more RF signal into the mixer transistor while still maintaining the high Q of the antenna coil.

With the RF signal added to the 170 mV of oscillator signal, mixer transistor Q3 is driven into gross distortion such that there is a multiplying effect of the RF signal and the oscillator signal. Thus the collector of Q3 includes a signal that is a frequency-translated (455-kHz) version of the RF signal.

IF transformer T2 works as an inductor capacitor tank circuit tuned to 455 kHz. The collector of Q3 is fed to the low-side tap of T2 so as to form a lower-impedance load. At the primary, the turns ratio from the whole winding to the low-side tap is about 3:1. The equivalent parallel resistance across the whole winding of the primary winding is about 500 kΩ. Thus the resistance at resonance at the low-side tap is 500 kΩ divided by $3 \times 3 = 9$, or about 55 kΩ.

By C16, the low-side tap of T2 is connected to the input of the first IF amplifier transistor Q4, which has an input resistance of greater than 100 kΩ.

Note   The input resistance to a common or grounded emitter amplifier is the current gain divided by (DC collector current/0.026 volt). Assuming a current gain of 50 and a DC collector current of 12 µA, then the input resistance is about 50/0.000046mho = 108 kΩ.

IF amplifier Q4 also has a gain control, VR1, that provides a gain reduction of 8:1. When VR1 is adjusted for maximum gain (VR1 = 0 Ω), the input resistance of Q4 is about 108 kΩ. But when VR1 is set to 100 kΩ, the gain of amplifier is lowered, but the input resistance is raised from 108 kΩ to 108 kΩ + (50 + 1) $\times$ 10.7 kΩ = 654 kΩ. The 10.7-kΩ number is the parallel resistance of 100 kΩ and 12 kΩ.

The output of Q4 via its collector is connected to the low-side tap of the primary winding of T3 and coupled to the second IF amplifier Q5 via C17. The second IF amplifier's input resistance is similar to that of the Q4 amplifier and equals about 100 kΩ or more.

A germanium diode (D3) demodulates the 455-kHz AM signal at the collector of Q5. R8 and R9 provide about 100 mV of forward voltage bias to the germanium diode detector. Audio signals from the cathode of D3 are amplified further by Q6, an inverting audio amplifier, to drive the crystal earphone via C15.

Alternative Low-Power Superheterodyne Radio Design

Figure 9-3 presents a slight modification to the preceding design. By using a twin variable capacitor, the two-transistor oscillator circuit can use an IF transformer with its internal capacitor removed or, better yet, a standard variable inductor.

Parts List

- **C1, C2, C3, C7, C8, C9, C10, C11, C12, C17, C18, C19, C20:** 0.01 µF
- **C4, C5, C15:** 33 µF, 16 volts
- **C6:** 0.15 µF
- **C13:** 0.0022 µF
- **C14:** 1 µF, 35 volts or 33 µF, 16 volts
- **C16:** 330 pF
- **R1, R2:** 22 kΩ

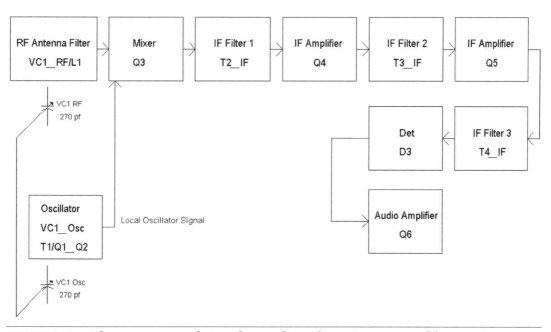

FIGURE 9-3   A low-power superheterodyne radio with a twin-gang variable capacitor.

- **R3, R4, R12:** 56 kΩ
- **R5, R6, R7:** 12 kΩ
- **R8, R10, R13, R14, R15:** 100 kΩ
- **R9:** 1.1 MΩ
- **R11:** 1 MΩ or 1.1 MΩ
- **D1, D2:** 1N914
- **D3:** 1N34 or 1N270
- **Q1, Q2, Q3, Q4, Q5:** MPSH10
- **Q6:** 2N5089 or 2N5088
- **L1 antenna coil:** 330 µH primary, 10 to 20 turns for secondary winding
- **T1:** 42IF101 IF transformer with internal capacitor removed, or a 120-µH or 150-µH variable inductor
- **T2, T3, T4:** 42IF101 IF transformer
- **VC1 variable capacitor:** twin gang 270 pF and 270 pF
- **VR1:** 100-kΩ pot

Figure 9-4 presents a schematic diagram of the alternate low-power superheterodyne radio. In this configuration, the oscillator tank circuit formed by VC1, C16, and T1 is not stepped down via a secondary winding but rather is connected via C2 to the base of Q2. This means that at resonance, Q1's collector load is a resistance of at least 10 kΩ. This higher load resistance at the collector of Q1 provides more overall gain in the circuit for oscillation. However, because the collector of Q1 is sensitive to stray capacitance that will shift the oscillation frequency, the output signal from the oscillator is taken from the emitter of Q2, a lower-resistance point. Any small stray capacitance loading into the emitter of Q2 does not cause the oscillator's frequency to shift.

The waveform at the Q2 emitter is a half-wave-rectified sine wave, which resembles a positive-going pulse. This positive-going pulse then is added to the RF signal from L1's secondary winding and fed to the mixer Q3. From here on out, the circuit works identically to the one shown in Figure 9-2.

This radio drained about 97 µA at 1.25 volts, but the reader is encouraged to change the values of R1 and R2 from 22 kΩ to 33 kΩ (or 39 kΩ) and change R3 from 56 kΩ to 75 kΩ or 100 kΩ to lower the power consumption further.

It should be noted that the reader also can replace the ferrite antenna coil L1 in Figure 9-4 with a loop antenna and an RF transformer (actually an oscillator coil used as an RF coil), such as the 42IF100, 42IF110, or 42IF300 coil. The loop antenna will be connected to the low-side tap of the RF transformer at the primary winding, the variable capacitor VC1's RF section will be connected to the primary winding, and the other end of the primary winding will be grounded. The secondary winding of the RF transformer will be connected in the same way as the secondary winding of L1 in Figure 9-4. For a reference, see Figure 8-8 for the schematic pertaining to the loop antenna and T7.

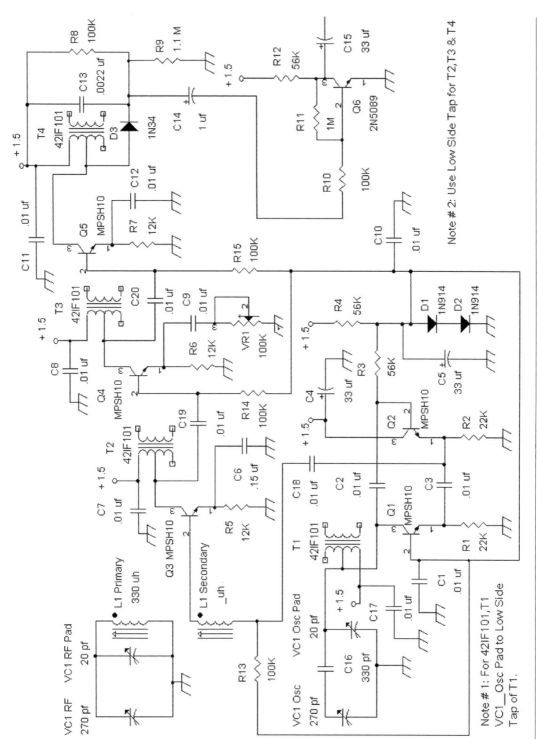

FIGURE 9-4 Schematic diagram for the alternate low-power superheterodyne radio.

121

Photos of Low-Power AM Superheterodyne Radios

Figure 9-5 is a picture of a prototype of the circuit from Figure 9-2. Figure 9-6 shows the original low-power AM superheterodyne radio that was featured in *EDN Magazine*. This radio lasted about five years on a C cell. It drained about 140 µA to 150 µA depending on the setting of the IF gain control (on the right side of the radio). This radio used a similar circuit to that shown in Figure 9-2. However, the oscillator coil was hacked or modified extensively to match the variable capacitor and to ensure that the low-power differential-pair transistor oscillator circuit performed as expected. For making this oscillator coil, an IF transformer was taken apart, and the primary winding was unwound and then rewound to achieve the correct inductance and tapping ratio. Excess wire was cut off because an oscillator coil generally has less inductance than an IF coil. Also, this procedure of unwinding and rewinding the coil was very challenging because the thickness of the wire in the IF transformer was about the same thickness as a human hair.

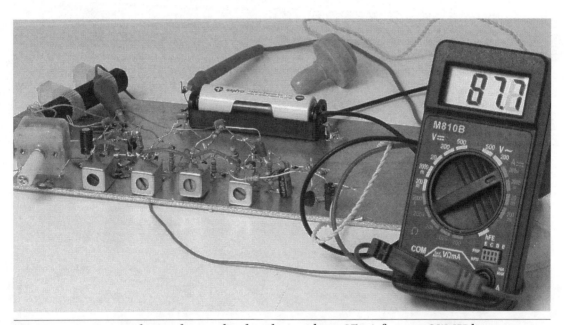

FIGURE 9-5 A superheterodyne radio that drains about 87 µA from an NiMH battery.

FIGURE 9-6   Original low-power superheterodyne radio.

See below for the schematic diagram of Figure 9-6.

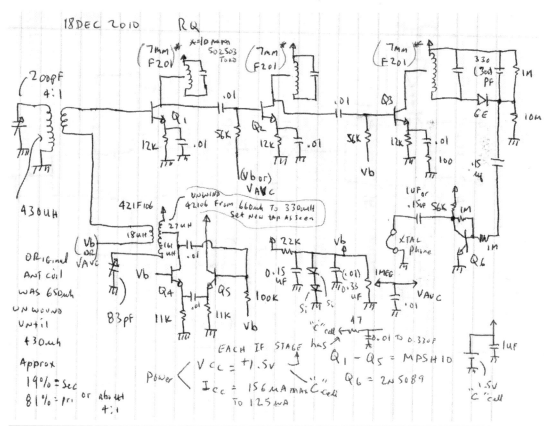

FIGURE 9-7   Schematic diagram of the original low-power superheterodyne radio.

Chapter 10

Exotic or "Off the Wall" Superheterodyne Radios

In Chapters 8 and 9, the superheterodyne radios were designed in a straightforward manner. The radio-frequency (RF), intermediate-frequency (IF), and audio signals were amplified with dedicated amplifiers. This chapter presents reflex superheterodyne radios. In one design, the mixer oscillator converter circuit doubles as an audio power amplifier. And in another design, the oscillator converter stage and IF amplifier circuit also operate as audio driver and audio output amplifiers.

A One-Transistor Superheterodyne Radio

Chapter 6 introduced a single-transistor reflex radio. This one-transistor radio was a tunable radio-frequency (TRF) circuit with the detected output recirculated back to the RF amplifier to increase the audio level. The selectivity of this reflex radio was adequate, and the audio output was suitable for driving generally high-impedance earphones of 2,000 Ω or more. The selectivity of the reflex radio is limited by the loading of the antenna coil via the input to its transistor circuit.

Selectivity can be increased by adding a second stage of RF tuning at the collector of the one-transistor circuit. However, more often than not, undesirable oscillations will occur with a tuned circuit (e.g., antenna coil and variable capacitor) at the input and another tuned circuit at the output. The reason is that the antenna coil is receiving the RF signal everywhere, including the amplified RF signal from the collector. So is there another way to increase selectivity? Yes, one can try designing a superheterodyne radio with one transistor.

However, because we are using just one transistor, there is a limitation as to how much gain is available for amplifying RF and audio-frequency (AF) signals. One objective of this radio is to drive a low-impedance earphone (e.g., 32 Ω). Thus, in order to provide sufficient audio drive to a low-impedance earphone, this radio will drain current on the order of many milliamps (i.e., not a very low-power receiver).

Design Considerations for a One-Transistor Superheterodyne Radio

The converter oscillator circuit, which usually runs at a couple or few hundred microamperes now will operate at about 10 times the current, anywhere from a few to about 10 milliamperes. The higher operating current is required because the converter oscillator circuit will pull the additional duty of amplifying audio signals. However, the higher collector current increases the gain of the converter oscillator circuit, which can lead to parasitic oscillations. In addition, the higher collector current leads to higher oscillation signals at the collector of the converter transistor, which robs the audio-signal's voltage swing. Therefore, it is desirable to limit the oscillator's output voltage.

Another design consideration for a one-transistor superheterodyne receiver is demodulation after the IF filter. Because the converter oscillator transistor is running at such a high collector current, there is a tremendously large oscillator signal at input of the IF filter that does not get filtered out. This large oscillator signal added on top of the IF signal then hampers envelope detection of the 455-kHz IF signal. In reality, for single-tuned IF transformers, the signal from the output of the first IF transformer contains both signals from the IF of 455 kHz and also the oscillator frequency, which ranges from 1 MHz to 2 MHz. Usually, the first IF amplifier does not have a problem handling both signals, and the output from the first IF amplifier is fed to a second IF transformer, which then filters out almost completely the oscillator signal while passing the 455-kHz IF signal.

Thus at least two stages of IF filtering may be required to reject the oscillator's signal from the converter oscillator circuit while allowing the 455-kHz IF signal to be envelope detected.

Finally, the operating voltage of a one-transistor superheterodyne radio is normally in the range of 9 volts to 18 volts. At the transistor's collector terminal, three signals are superimposed. They are the oscillator signal, the 455-kHz IF signal, and the audio signal. Thus, for the following design, a 9-volt supply will be used. Figure 10-1 shows a block diagram for a one-transistor superheterodyne radio.

The radio station's RF signal is received by antenna coil L1, which couples the RF signal via a secondary winding of L1 to converter circuit Q1. The output of Q1 feeds an IF transformer and an audio transformer. The output of the first IF T2 transformer then is coupled with a second IF transformer T3, which provides sufficient rejection of the oscillator's signal for envelope detection of the 455-kHz IF signal. Detector D3 then provides an audio signal that is fed back to the converter oscillator transistor's input for audio amplification. Audio transformer T5 is connected to the output of the converter oscillator transistor Q1 for extraction of the amplified audio signal. This amplified audio signal then has sufficient current to drive a low-impedance earphone.

Parts List

- **C1:** 100 µF, 16 volts
- **C2, C3, C4, C9:** 0.01 µF

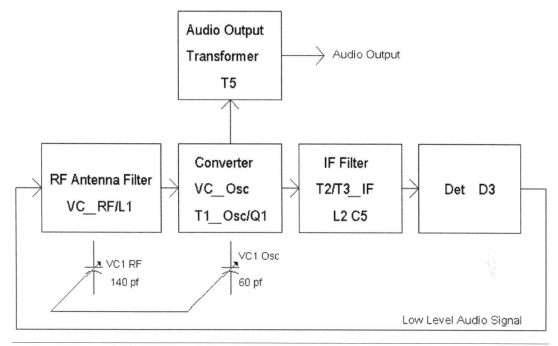

FIGURE 10-1 Block diagram of a one-transistor reflex superheterodyne radio.

- **C5:** 820 pF, 1% silver mica
- **C6:** 0.0039 µF
- **C7, C8:** 1 µF, 35 volts
- **R1:** 180 Ω
- **R2:** 390 kΩ
- **R3:** 47 kΩ
- **R4:** 20 kΩ
- **VR1:** 50 kΩ
- **T1 oscillator coil:** 42IF110
- **T2, T3 IF transformer:** 42IF103
- **T4 audio-drive transformer:** 10 kΩ primary, 10 kΩ secondary
- **T5 audio output transformer:** 1 kΩ primary, 8 Ω secondary
- **L1 antenna coil:** 600 µH to 680 µH primary, 10 turns for secondary winding
- **L2:** 150 µH
- **D1, D2:** 1N914
- **D3, D4, D5:** 1N270 or 1N34
- **Q1:** 2N5089
- **VC1:** two-gang variable capacitor 140 pF and 60 pF

Figure 10-2 presents the schematic for the one-transistor radio. As shown in the schematic, the RF signal from the tuned circuit consisting of VC1 RF and L1 is

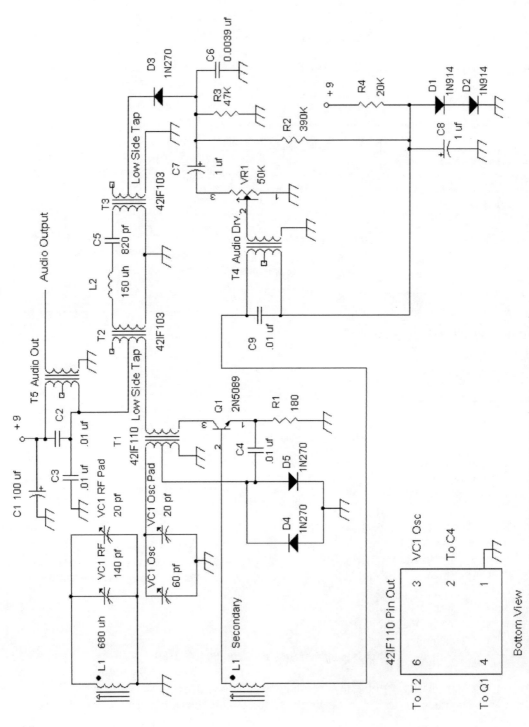

FIGURE 10-2   Schematic diagram of the one-transistor reflex radio.

128

connected to the base of Q1. The collector of Q1 is connected to a secondary winding of the oscillator coil T1 that feeds energy back to its primary winding to provide a reliable oscillation signal from 1 MHz to 2 MHz at the emitter of Q1. At the low side of the primary winding of T1, two diodes are wired back to back to provide amplitude limiting of the oscillator signal because the operating collector current of Q1 is biased to a much higher than "normal" direct-current (DC) current. With a local oscillator signal at the emitter of Q1 and an RF signal at the base of Q1, a multiplying effect of the two signals occurs to provide an IF signal at 455 kHz at the collector of Q1. The first IF transformer T2 provides extraction of the IF signal but still contains some of the local oscillator's signal. For further attenuation of the local oscillator signal while passing the IF signal, a series resonant circuit formed by L2 and C5 is connected to a second IF transformer T3 that passes signals around 455 kHz and provides further attenuation of the local oscillator signal, whose frequencies range from 1 MHz to 2 MHz. The signal from the IF transformer T3 provides further attenuation of the local oscillator signal and is connected to D3 for envelope detection of the 455-kHz amplitude-modulated (AM) signal. The detected signal now is an audio signal that is connected to the primary winding of audio transformer T4, whose secondary winding is connected in series with the secondary winding of L1.

Thus the base of Q1 has both RF signals from L1 and AF signals via T4. Amplified audio signals at the collector of Q1 are extracted via audio transformer T5. Sufficient audio signal current is provided by the secondary winding of T5 to drive a low-impedance earphone. Also, to conserve battery life, a power switch may be connected in series with the battery.

Note   Although the one-transistor superheterodyne radio provides higher selectivity than the one-transistor reflex radio, the sensitivity is about the same. Thus the one-transistor superheterodyne reflex radio generally performs not that much better than a one-transistor TRF reflex radio such as the one shown in Chapter 6. This radio is more of an exercise in circuit design and is neither low power nor high performance.

If time permits, a second one-transistor reflex superheterodyne radio will be designed. But there is only so much one can do with just one transistor to convert RF signals into an "amplified" IF signal while trying to provide audio amplification as well. See Chapter 23 for an update of the one-transistor reflex superheterodyne radio. In the next design using two transistors, the performance is improved significantly, although this design is not considered to be low power.

A Two-Transistor Superheterodyne Radio

Although neither one- nor two-transistor superheterodyne radios generally were made commercially, there were three-transistor designs. For example, in 1960, the Truetone Model DC3090 was sold. This was a three-transistor superheterodyne design with a converter circuit, an IF/AF amplifier, and an audio power amplifier. The detected

AM signal from the IF amplifier was fed back to the IF amplifier for further AF amplification, and the audio signal from the IF amplifier then was connected to the audio power amplifier for driving a loudspeaker. Thus the Truetone radio was a reflex radio using the IF amplifier for amplifying audio signals and 455-kHz IF signals from the converter circuit.

Thus the Truetone radio had the following characteristics:

1. A converter oscillator circuit for translating the RF signal from an antenna coil to an amplified IF signal using a first transistor
2. An IF amplifier with two IF transformers using a second transistor
3. Audio amplification with two stages, a first stage via the IF amplifier and a second stage via an audio output amplifier using a third transistor to drive a speaker

However, the design for a two-transistor radio can have the same characteristics if the converter oscillator circuit works as the first-stage audio amplifier. Thus the two-transistor superheterodyne radio has the following characteristics:

1. A converter oscillator circuit for translating the RF signal from an antenna coil to an amplified IF signal *and* for amplifying audio signals from the detector circuit using the first transistor
2. An IF amplifier with two IF transformers that also provides large audio signal amplification to drive a speaker using the second transistor

In both cases, the three- and two-transistor designs have a converter stage, an IF stage, and two stages of audio amplification. But the two-transistor superheterodyne radio design has sort of a two-stage reflex circuit, whereas the three-transistor radio has a one-stage reflex circuit. Figure 10-3 presents the block diagram for a two-transistor superheterodyne radio.

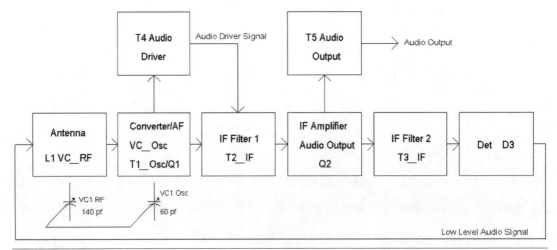

FIGURE 10-3   Block diagram of the two-transistor superheterodyne radio.

RF signals are connected to the input of converter oscillator circuit Q1 and T1 to provide a 455-kHz signal to the first IF filter T2. The output of T2 is connected to an IF amplifier (Q2) for further amplification of the 455-kHz IF signal. The output of Q2 is connected to the second IF transformer T3, whose secondary winding provides an AM signal for envelope detection via diode D3. Audio signals via D3 then are fed back to the input of Q1 to further amplify the detected AM signal. The amplified audio signal from Q1 via audio driver transformer T4 then is coupled to the input of Q2 for providing large audio signals that are extracted by audio output transformer T5 to drive a loudspeaker.

Parts List

- **C1, C3, C4, C6:** 0.01 μF
- **C2:** 0.0047 μF
- **C5:** 100 μF, 16 volts
- **C7, C10, C11:** 1 μF, 35 volts
- **C8, C9:** 0.0033 μF
- **R1, R4:** 22 kΩ
- **R2:** 1,000 Ω
- **R3:** 24 Ω
- **R5:** 510 kΩ
- **VR1:** 50-kΩ pot
- **T1 oscillator coil:** 42IF100
- **T2 IF transformer:** 42IF101, use low side tap on primary winding
- **T3 IF transformer:** 42IF103, use low side tap on primary winding
- **T4 audio driver transformer:** 10 kΩ primary, 10 kΩ secondary; use the CT in the secondary for 2.5 kΩ, or 10 kΩ primary, 2 kΩ secondary
- **T5 audio output transformer:** 500 Ω primary, 8 Ω secondary
- **D1, D2:** 1N914
- **D3:** 1N270 or 1N34
- **Q1, Q2:** 2N3904
- **VC1:** two gang-variable capacitor 140 pF and 60 pF
- **L1 antenna coil:** 600 μH to 680 μH primary, 10 to 20 turns for secondary winding
- **L2:** 4.7-mH inductor

Figure 10-4 presents the schematic diagram of the two-transistor two-stage reflex superheterodyne radio. As seen in this diagram, a 9-volt battery will be used as the power supply, and the radio will drain around 10 mA to 20 mA for driving a loudspeaker.

RF signals are received via antenna coil L1, which is tuned with variable capacitor VC1 RF. These RF signals are connected to the base of converter oscillator transistor Q1, which has a DC collector current at the typical hundreds of microamperes. The oscillator circuit consists of the collector output of Q1 coupling back to the input (emitter) of Q1 via the secondary winding of T1 and its tapped primary winding, respectively.

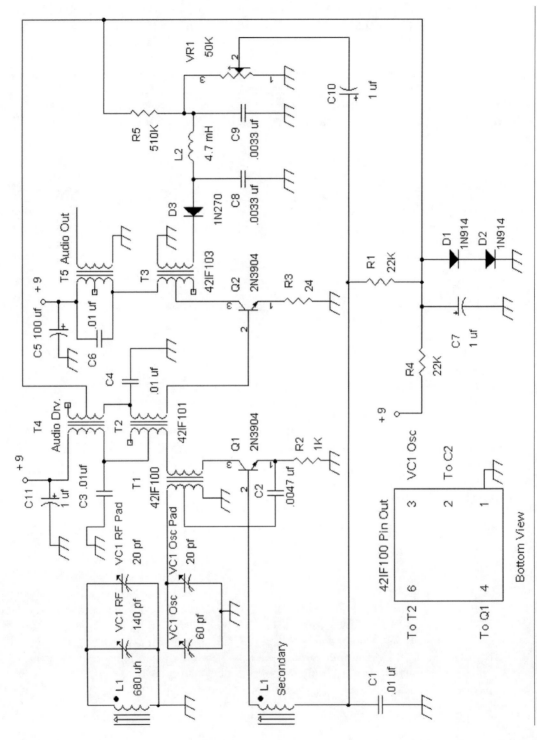

FIGURE 10-4 Schematic diagram of the two-transistor superheterodyne radio.

Because the emitter of Q1 has a large oscillator signal, Q1 is driven into sufficient distortion that the small RF signal coupled with the base of Q1 is effectively multiplied with the oscillator signal. Stated in another way, the large-amplitude oscillator signal results in a time-varying transconductance, where the time-varying function is the oscillator's signal.

From the collector output terminal of Q1 then is a 455-kHz IF signal that is connected to the primary winding of the first IF transformer T2. The signal from the secondary winding of the IF transformer then is connected to the base of Q2 for amplification of the 455-kHz signal.

Because Q2's collector is operating at a direct current of about 15 mA, generally one would think that the input resistance at the base of Q2 would be in the hundreds of ohms (e.g., 250 Ω or less), which can excessively load down the signal from T2's secondary winding. However, because the emitter of Q2 has a 24-Ω resistor to ground, the actual input resistance is about the transistor's current gain (beta or Hfe, typically 100) times 24 Ω, which is about 2,400 Ω. Thus the emitter 24-Ω resistor raises the input resistance while essentially providing about the same voltage gain as if Q2 were biased at (typically) 1 mA of collector current with the emitter bypassed to ground with a capacitor.

The amplified IF signal then is connected to the second IF transformer T3, and the secondary winding of T3 is connected to D3 for demodulation of the 455-kHz AM signal. Because the demodulated AM signal is going to be recirculated into the converter oscillator transistor Q1, it is essential to remove any 455-kHz signal prior to connection to the base of Q1 via the secondary winding of L1, the antenna coil. Therefore, a low-pass filter formed by C8, L2, and C9 removes substantially the 455-kHz IF signal component from the detector diode D3 while coupling an audio signal to the base of Q1 via the secondary winding of L1.

At audio frequencies, the converter oscillator circuit has an input resistance of about 100,000 Ω (beta times R2 = 1,000 Ω, where beta = 100). The primary impedance of T4 is in fact about 10 kΩ because the tapped secondary of T4 is being loaded by the input resistance of Q2, which is 2,400 Ω. Recall that the tapped secondary winding of T4 should be 2,500 Ω. Thus there is close to an optimal power transfer in impedance matching for T4 into Q2, and the audio-voltage gain from the base of Q1 to the base of Q2 is about 5. The audio gain from the base of Q1 to its collector is 10 kΩ/1 kΩ = 10. But because the center tap of the secondary of T4 is used, we only get half the audio-signal voltage, which is connected to the base of Q2.

With the amplified (T4) audio signal coupled to the base of Q2, the collector of Q2 supplies sufficient current to output transformer T5 for driving a loudspeaker at its secondary winding. The DC collector current of Q2 is set at about 15 mA with a 500-Ω impedance load to allow for 7.5 volts peak or 15 volts peak to peak of alternating-current (AC) signal swing. Note that the term *impedance* is used instead of resistance. Transformer T5 allows the audio-signal voltage at the collector of Q2 to swing *above* the 9-volt power supply. Thus the 15 mA of collector current allows the collector to swing about 7.5 volts above the 9 volts. For power conservation, the reader may connect a power switch in series with the battery.

Chapter 11

Inductor-less Circuits

In previous chapters there was extensive use of coils, inductors, or transformers for radiofrequency (RF) and intermediate-frequency (IF) signals. This chapter presents alternatives to coils or inductors.

Ceramic Filters

Ceramic filters emulate high-Q inductor/capacitor (LC) circuits as band-pass filters. These filters generally are three-terminal devices with input, ground, and output terminals. However, ceramic filters also are made as two-terminal devices.

The earliest use of ceramic filters for amplitude-modulated (AM) transistor radios can be found in the 1960s. For example, the Motorola Model XP3CE and Model XP4CE 9- and 11-transistor radios from 1965 to 1967 each used a two-terminal ceramic filter in the first IF amplifier. Normally, the first IF amplifier has an RF bypass capacitor from the emitter to ground. In these Motorola radios, the RF bypass capacitor is replaced with a two-terminal ceramic filter, which acts like a series resonant 455-kHz tank circuit. Thus at 455 kHz, the two-terminal ceramic filter has a minimum impedance or resistance, and at frequencies outside 455-kHz band, such as 470 kHz, the ceramic filter has a high impedance. By replacing the RF bypass capacitor with the two-terminal ceramic filter at the emitter of the common-emitter IF amplifier, the output current from the IF amplifier is peaked at the IF frequency and lowered at frequencies outside the IF band.

Two-terminal ceramic filters should *not* be confused with the very common two-terminal ceramic resonators. A ceramic resonator has a very narrow bandwidth that generally is not suitable for an IF filter used in a standard AM radio. However, because ceramic resonators have very narrow bands, they can be used as a continuous-wave (CW) IF filter for receiving Morse code. Ceramic resonators generally are used for generating frequencies such as seen in one of the generator circuits in Chapter 4. For this chapter, we will concentrate on (wider-bandwidth) ceramic filters instead because resonators exhibit very narrow band characteristics that are not suitable for broadcast AM receivers.

Today, two-terminal ceramic filters are not that readily available, but three-terminal ceramic filters are found commonly. Three-terminal ceramic filters for the AM band could be found in stereo receivers in the 1970s, such as the 1973 Sony STR-222.

As the cost of these three-terminal devices dropped, they were very commonly designed into portable or pocket radios by the middle to late 1970s or early 1980s. For example, in 1981, the Sony ICF-200 pocket radio had three-terminal ceramic 455-kHz and 10.7-MHz filters for AM and frequency-modulated (FM) bands.

One characteristic of a ceramic filter is that unlike an inductor, the direct-current (DC) resistance is very high, just like an open circuit. Another feature of a ceramic filter is that it is meant to be driven and terminated by a specified resistance. Most AM band ceramic filters, for example, have driving and terminating resistances from 1,000 Ω to about 3,000 Ω. And ceramic filters may be specified for a wide bandwidth such as 10 kHz to a narrow bandwidth of 3 kHz.

Although ceramic filters have a response like a typical LC band-pass filter near the pass-band frequencies, they do suffer from allowing out-of-band signals to pass through. And, in general, the specified frequency response of a ceramic filter includes an IF transformer preceding it to help remove spurious or out-of-band signals such as the local oscillator signal that is from 1 MHz to 2 MHz.

Manufacturers of ceramic filters include Murata, Toko, and Token. Note that AM (and FM) band ceramic filters can be obtained from www.mouser.com.

Figure 11-1 shows a selection of ceramic filters. In the figure, the ceramic filter on the left is a single-element filter, whereas the item on the right is a multiple-element filter. A multiple-element ceramic filter is essentially two or more single-element ceramic filters that are cascaded.

FIGURE 11-1 455-kHz ceramic filters.

Note   The pin out for the single-element ceramic filter in the lead on the left is the output terminal, and the center lead is the ground or common terminal, and finally, the lead on the right is the input terminal.

Figure 11-2 is a block diagram of a radio with 455-kHz IF ceramic filters. Antenna coil L1 and variable capacitor VC1 provide an RF signal to the converter circuit Q1. Oscillator coil T1 provides a local oscillator signal for converting the incoming RF signal to an IF signal. Ceramic filters extract the IF signal from the output of the converter circuit. The IF signal is filtered and amplified by an IF amplifier circuit. The output of the last IF amplifying transistor Q3 is fed to a diode detector D3 for demodulation.

Parts List

- **C1, C3, C7:** 0.01 µF
- **C2, C4, C8:** 1 µF, 35 volts
- **C5, C6:** 0.15 µF
- **R1:** 2,200 Ω
- **R2, R5:** 3,000 Ω
- **R3, R6, R10:** 22 kΩ
- **R4, R7, R8:** 2,200 Ω
- **R9:** 47 kΩ
- **L1 antenna coil:** 600 µH to 680 µH, 10 to 20 turns for secondary winding
- **VC1:** two-gang variable capacitor 140 pF and 60 pF
- **D1, D2:** 1N914
- **D3:** 1N34 or 1N270
- **T1 oscillator coil:** 42IF100
- **CF1, CF2, CF3:** Ceramic filter SFU455A, Murata or equivalent
- **Q1, Q2, Q3:** 2N3904

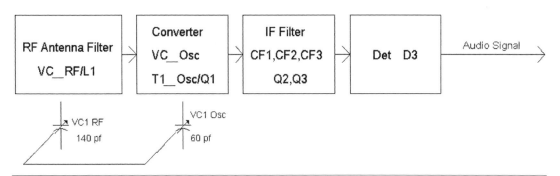

FIGURE 11-2   Block diagram of a radio using ceramic filters.

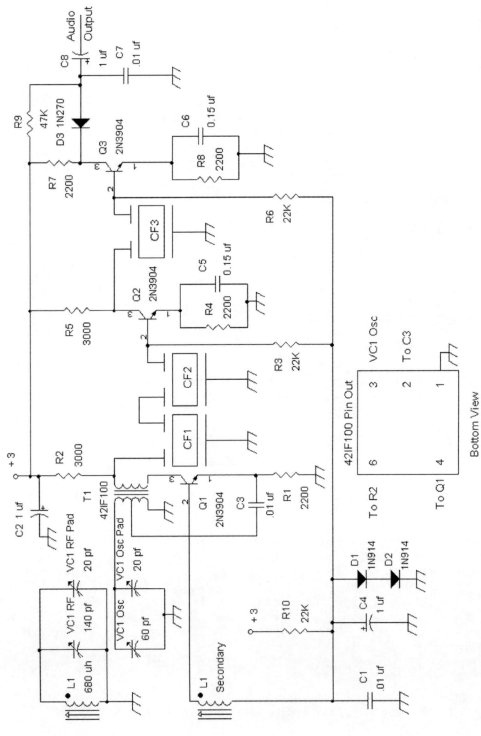

FIGURE 11-3 Schematic diagram of an AM radio with ceramic filters using a 3-volt supply.

Figure 11-3 shows a schematic for an AM radio that uses ceramic filters to replace IF transformers. Received radio station signals are tuned by L1 and VC1 RF to provide RF signals to the base of converter oscillator transistor Q1. Oscillator coil T1 provides a local oscillator signal from about 1 MHz to 2 MHz via coupling the oscillation signal from the secondary to primary windings. A multiplying effect between the RF signal and the local oscillator signal occurs via a large oscillator signal at the emitter of Q1 and a small RF signal at the base of Q1. Thus at the collector of Q1 there is an IF signal as well as a very large-amplitude oscillator signal. Because the ceramic filters have poor rejection for some out-of-band signals, including the oscillator signal, two 455-kHz ceramic filters (CF1 and CF2) are used to extract the 455-kHz IF signal. From the output of ceramic filter CF2 is a usable IF signal for further amplification via the first IF amplifier transistor Q2. The collector terminal output of Q2 is then fed to a single ceramic filter CF3 whose filter output is connected to the input of the second IF amplifier Q3. The output collector terminal at Q3 now provides an IF signal for envelope detection via diode D3. Thus audio signals are provided at C8 for a crystal earphone or the audio signals may be connected to an audio amplifier. Note that D3 is rectifying the negative portion of the AM envelope because the positive portion actually goes into clipping when a strong signal is received. However, the negative portion of the AM envelope at the collector of Q3 is not affected and does not clip. Also, to conserve energy, a power switch may be connected in series with the battery or power source.

Gyrators (aka Simulated or Active Inductors)

When generating FM was in its infancy, there were no varactor tuning diodes. The FM transmitter back in the 1930s or 1940s used vacuum tubes for amplification, for oscillation, and for modulation of the frequency of the oscillator. The choice circuit to generate FM of that era was the reactance modulator. The reactance modulator circuit consisted of a vacuum tube that acted as a variable inductor. With no audio signal into this vacuum tube, an inductance of fixed value was formed in parallel with the oscillator's main coil. If the vacuum tube changed in plate current due to an audio signal, the inductance changed as well. So for a fixed plate current in the vacuum tube, a fixed inductor was synthesized from the plate to the cathode of the tube. Today, the reactance modulator can be designed with field-effect transistors, bipolar transistors, and/or integrated circuits.

For this book, we will be working with simulated inductors or gyrators consisting of solid-state devices, and the inductance will be fixed.

Figure 11-4 shows a very simple gyrator. The gyrator in the figure is a "simulated" inductor with one lead that is grounded. So looking into the C1 and R2 input terminal of the gyrator is equivalent to looking into a regular inductor as in the lower portion of the figure with the other lead grounded.

An off-the-shelf inductor or coil always has at least two characteristics. The first is the inductance, and the other is the DC resistance. So any coil can be modeled as a perfect inductor with a series resistance. The lower the DC resistance of the

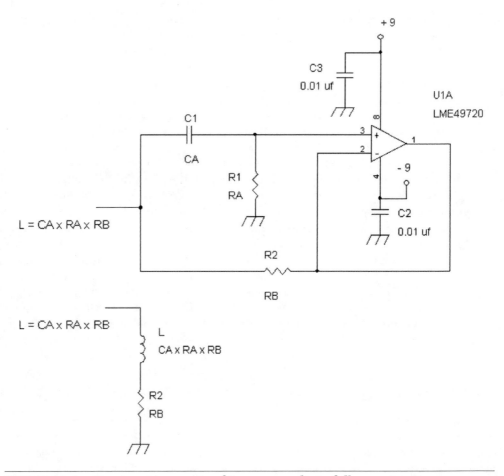

FIGURE 11-4 Gyrator using a gain-of-1 op amp voltage follower circuit.

coil, the closer this coil comes to an ideal inductor. At a low frequency or at DC, the impedance of the coil is just its internal resistance. For example, an antenna coil's primary winding may be anywhere from 1 Ω to 10 Ω in terms of internal resistance. And at high frequencies, the coil has a very high impedance if there is no appreciable capacitance in the coil.

Now let's look at Figure 11-4. The input of the gyrator circuit is C1 and R2. C1 and R1 form a high-pass filter network, which means that at low frequencies there is reduced amplitude or no low-frequency signals across R1. At high frequencies, C1 acts closer to a short circuit, which means that high-frequency signals appear at R1. The signal at R1 is amplified by a gain of 1 and fed back to the input via R2.

So let's take a look at what happens if the signal at the input is a DC voltage. C1 will block the DC voltage, and the voltage at R1 will be zero. The output of the amplifier then will be zero as well, which grounds one end of R2. This means that at

DC the input of the gyrator is just a resistor, R2. Note that a coil at DC is also just the internal resistance, which means that a coil can be modeled at DC as a resistor as well.

For a high-frequency signal at the input of the gyrator, C1 passes most or all of the input signal into R1. One can say that the high-frequency signal at the input and output of the amplifier is really close to or the same as the signal appearing at the input of the gyrator circuit. This then means that signal voltage on R2 that is connected to the amplifier is really about the same signal voltage at the input of the gyrator. But R2 is also connected to the input. If we assume that R1 is very large and does not load the input signal, then most or all of the signal current flowing to the gyrator is through R2. But, if the signal voltage on both ends of R2 is almost the same or is the same, then there is no net current flowing into R2. Recall Ohm's law, which says that it is the potential difference (voltage difference) across a resistor that determines the current flowing into the resistor. If the voltage is the same on both ends of R2, then the potential difference is zero, and thus there is no current through R2. So, at high frequencies, there is little or no current flowing into R2 or the gyrator. At high frequencies the gyrator circuit looks like an open circuit or a very high impedance circuit.

But isn't this the same as the effect on a coil at high frequencies? A high-frequency signal into a coil will measure as a very high-impedance device or as an open circuit. Thus the gyrator circuit is "equivalent" to a coil or inductor.

The gyrator circuit in Figure 11-4 has an inductance of L = C1 × R1 × R2. For example, if C1 = 1,000 pF, R1 = 1,000 Ω, and R2 = 50 Ω (R2 is the equivalent internal resistance of a coil), then this gyrator has an inductance of 50 μH. However, in practice, the Q of this gyrator circuit usually does not exceed 10. Note that the Q values of antenna coils and IF transformers are at least 50 typically. So the gyrator circuit in Figure 11-4 is not quite suitable for an IF filter, for example.

But there is another type of gyrator circuit that has two amplifiers and works as a generalized impedance converter (GIC). Gyrators using a GIC topology generally have a Q of 50 or more. (Figure 11-6B shows a gyrator circuit using op amps U7A and U7B.)

Inductor-less (aka Coil-less) Superheterodyne Radio

Here is a problem to solve: How do you design a superheterodyne radio without any coils? One way is to substitute every inductor with a gyrator. Or if a band-pass filter is needed, one can use a ceramic filter or a gyrator.

But what if you want to design an inductor-less superheterodyne radio with just a one-section variable capacitor or, better yet, tune the radio without a variable capacitor at all? So this was what I was thinking during the last four months before I was going to write this book. The initial design was to use ceramic filters for the IF stages and no tuned RF stage. This design would work, except that using a 455-kHz ceramic filter with a single conversion RF mixer circuit would result in receiving radio signals at image frequencies.

For example, without a variable band-pass filter tuned to the desired frequency and fed to the converter or mixer circuit, an image signal will convert (or mix down) to the same IF frequency at twice the IF frequency (e.g., twice the IF frequency = 2 × 455 kHz = 910 kHz) plus the frequency of the desired station. See the following table for some examples.

Desired Station Frequency	IF Frequency	Image Frequency
540 kHz	455 kHz	1,450 kHz
610 kHz	455 kHz	1,520 kHz
680 kHz	455 kHz	1,590 kHz

However, there is another technique called *double conversion* that eliminates receiving stations at the image frequencies. But a double-conversion radio requires a second mixer or converter circuit and other IF filters. The double-conversion radio is workable, but it was decided that the inductor-less radio design will stick to just one converter circuit instead.

Therefore, to get around the problem of receiving stations at image frequencies, the IF frequency was raised to about 600 kHz (or higher), and a broadband low-pass filter was used to remove signals higher than 1,600 kHz. Any signal above 1,600 kHz would be filtered out and not able to mix or convert down to the new 600-kHz IF frequency. See the following table for some examples.

Desired Station Frequency	IF Frequency	Image Frequency
540 kHz	600 kHz	1,740 kHz
610 kHz	600 kHz	1,810 kHz
680 kHz	600 kHz	1,880 kHz

As stated previously, any signal above 1,600 kHz will be attenuated or removed by a low-pass filter. Figure 11-5 is a block diagram for an inductor-less superheterodyne radio. For the antenna, a short wire of less than 2 or 3 feet or a telescoping antenna will be sufficient to pick up radio stations in the AM broadcast band. The antenna is connected to a high-pass filter to attenuate signals below the AM band so that the radio minimizes picking up hum or other low-frequency signals. From the output of the high-pass filter, the wide-band RF signal is amplified by a low-noise circuit so that any noise contribution from the succeeding low-pass filter is negligible. The output of the low-noise amplifier is connected to the input of a 1.6-MHz low-pass filter, and the output of the low-pass filter is fed to a unity gain inverting amplifier and to an input of a balanced RF mixer. The output of the unity gain inverting amplifier is fed to the other input of the balanced RF mixer. A local oscillator formed by the same type of hysteresis oscillator circuit as shown in Chapter 4 provides twice the local oscillator frequency. The output of the local oscillator is connected to a flip-flop divide-by-two circuit that provides a "perfect" square wave at the local oscillator frequency signal to the balanced mixer.

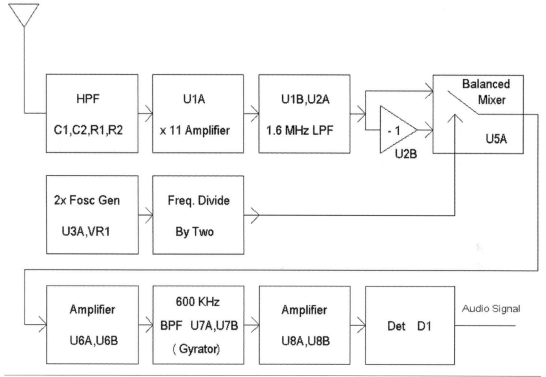

FIGURE 11-5   Block diagram of an inductor-less radio.

The mixer is balanced because neither the oscillator signal nor the RF signal from the low-pass filter appears at the output of the balanced mixer. Only a multiplied signal (RF signal times the oscillator signal) appears at the output. From the output of the balanced mixer, the sum and difference frequency signals are amplified and then fed to a single high-Q gyrator band-pass filter. The output of the 600-kHz IF band-pass filter is further amplified and then finally demodulated with a detector (e.g., diode envelope detector).

Parts List

- **C1:** 0.0056 µF
- **C2:** 0.0018 µF
- **C3, C4, C27:** 0.15 µF
- **C5:** 0.001 µF 5%
- **C6:** 0.0022 µF 5%
- **C7:** 100 pF 5%
- **C8:** 0.0027 µF 5%

- **C9, C19:** 300 pF 5%
- **C10:** 30 pF 5%
- **C11, C12, C13, C14, C15, C17, C18, C24, C25, C28, C32, C33, C34:** 0.01 µF
- **C16, C22, C31:** 100 µF, 16 volts
- **C20:** 24 pF 5% or 22 pF 5%
- **C21, C26, C35, C36, C37, C38, C39:** 1 µF, 35 volts
- **C23:** 47 pF 5%
- **C29, C30:** 510 pF 5%
- **R1, R4:** 100 Ω
- **R2, R6, R7, R8, R9, R10:** 301 Ω 1%
- **R3, R13, R14, R15, R16, R34:** 10 kΩ
- **R5, R11, R12, R20, R22, R30, R31:** 1,000 Ω 1%
- **R17:** 910 Ω or 820 Ω
- **R18:** 2,000 Ω
- **R19:** 47 kΩ
- **R21, R23, R29, R32:** 3,900 Ω
- **R24, R33:** 100 kΩ
- **R25, R26, R27, R28:** 511 Ω 1%
- **U1, U2, U7:** LME49720 or LM4562
- **U3:** 74AHC14 or 74HC14
- **U4:** 74HC74 or 74HCT74
- **U5:** 74HC4053
- **U6, U8:** NE5532 or LM833
- **U9:** LM78L05 or LM7805
- **VR1:** 1-kΩ pot
- **VR2:** 10-kΩ pot
- **D1:** 1N270 or 1N34 or 1N914

Figures 11-6A and 11-6B provide schematic diagrams for the inductor-less radio. A short wire or telescopic antenna is connected to a high-pass filter circuit consisting of R1, R2, C1, and C2, which has a cutoff frequency at about 280 kHz. This high-pass filter removes or attenuates low-frequency signals that would be demodulated as hum or extraneous noise. The output of the high-pass filter at C2 then is coupled to a five-pole 1,600-kHz active low-pass filter consisting of capacitors from C5 to C10, U1B, and U2A. The output of the low-pass filter is inverted in phase via U2B so that the inputs of the mixer U5A have a balanced or push-pull signal. Thus the inputs of the balanced mixer U5A are connected to a noninverting output of the low-pass filter and an inverting output of the low-pass filter. By commutating between the noninverting and inverting outputs of the low-pass filter, multiplication occurs with the local oscillator signal from the flip-flop frequency divider circuit U4A.

The local oscillator signal is generated by a hysteresis oscillator circuit much like the one seen in Chapter 4, which was used as an RF generator. Since the hysteresis oscillator in Chapter 4 is sufficiently stable in frequency for generating an RF signal to test a radio, it is also stable enough to be used as a local oscillator in this inductor-less receiver.

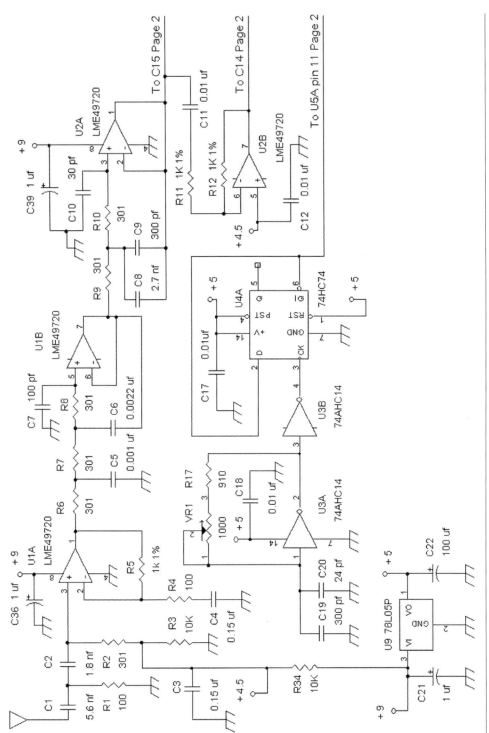

FIGURE 11-6A Inductor-less radio schematic diagram part 1—amplifiers and mixer.

145

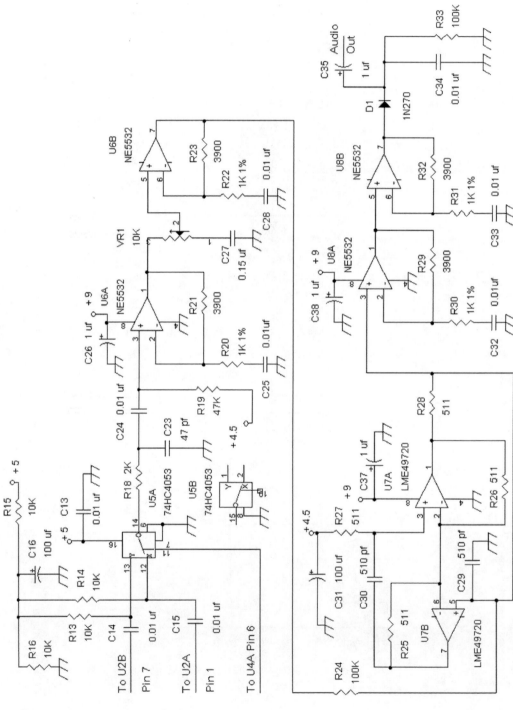

FIGURE 11-6B Inductor-less radio schematic diagram part 2—gyrator, amplifiers, and detector.

146

With an IF frequency set at 600 kHz, the desired local oscillator signal's frequency to the balanced mixer is 540 kHz + 600 kHz to 1,600 kHz + 600 kHz, or 1,140 kHz to 2,200 kHz. The hysteresis oscillator U3A must generate twice the frequency because a divide-by-two circuit, U4A, halves the frequency of the oscillator. Therefore, the output signal frequency of U3A must be adjustable via VR1 to range from 2,280 kHz to 4,400 kHz.

You may ask why not skip the divide-by-two circuit and connect U3A directly to the mixer U5A. It turns out that by dividing the oscillator's frequency by two via U4A, a perfect square wave is used as the commutating signal for mixer U5A. The square wave is symmetric, which then allows the mixing or multiplying effect to be balanced such that the RF signal from the low-pass filter does not appear at the output of U5A. The hysteresis oscillator does not generate a perfectly symmetric waveform. That is, the logic high duration is not equal to the logic low duration. Sending such a nonsymmetric waveform to the mixer will cause leakage of the RF signal to the output of the mixer, which is not preferred in a balanced-mixer configuration.

In Chapter 10 concerning the one-transistor superheterodyne radio, it was mentioned that having a huge-amplitude local oscillator signal added to the IF signal hampers proper envelope detection of the IF signal. And thus multiple IF filters are needed to sufficiently remove or attenuate the oscillator signal.

In this inductor-less radio, there is only one IF filter, so great care must be taken at the design stage to provide a mixer/converter circuit that outputs essentially no local oscillator signal nor the RF signal. This is the reason for using a balanced mixer.

The output of the balanced mixer is fed to a second low-pass filter at around 1,600 kHz to remove the very high-frequency components from the balanced mixer owing to the square-wave switching. The output of this second low-pass filter is fed to amplifiers U6A and U6B. VR2 serves as a variable gain control to adjust the IF gain. The output signal from U6B then is band-pass-filtered via R24, resonating capacitor C29, and gyrator (active inductor) circuit U7A and U7B that has an inductance of $[(C30 \times R28 \times R25 \times R27)/R26] = L$. R24 sets the Q of the parallel tank circuit to greater than 50 at a resonant frequency of 600 kHz. The 600-kHz IF signal from the output of the band-pass circuit is further amplified via U8A and U8B before being demodulated by D1. Audio signals via C35 then are connected to a crystal earphone or to an audio amplifier. Also, a power switch may be connected in series with the battery or power source to conserve energy.

Note that the inductor-less superheterodyne radio is simple to use. There are no coils to adjust, and there are no multiple-gang variable capacitors to deal with. And this radio uses very common parts that are more available than the antenna coils, IF transformers, ceramic filters, and/or variable capacitors that were used in previous superheterodyne radio designs.

Chapter 12

Introduction to Software-Defined Radios (SDRs)

Chapter 11 presented a different type of superheterodyne radio—a coil-less receiver. Two new concepts were introduced—image signals and switching-type radio-frequency (RF) signal mixers. It was also in Chapter 11 that image signals were shown to be eliminated or attenuated by providing a band-pass filter prior to RF mixer so that the desired signal is output from the mixer and not the image signal. In the coil-less radio, a higher intermediate frequency (IF) of 600 kHz was used with a 1.6-MHz low-pass filter to provide image signal rejection. The coil-less radio is rather simple to implement with integrated circuits.

It should be noted that in Chapters 8 through 11, the selectivity of the superheterodyne radio was influenced by the band-pass characteristics of the IF filter. If more selectivity is required, then usually more IF transformers, coils, and/or ceramic filters must be added to the design. And if the IF needs to be changed to some frequency other than 455 kHz, then the IF transformers or ceramic filters need to be replaced.

So one of the main motivations for a using a computer or digital processing in superheterodyne radios is to replace the IF section, whereby the IF filters can be generated in the digital domain. Once an analog signal is digitized by an analog-to-digital converter, the digital domain can take over to mimic many of the past analog functions, such as filtering, RF mixing, delaying signals, or amplifying the IF signals. In the digital domain, the frequency and band-pass characteristic (e.g., bandwidth) can be designed with flexibility. For example, with digital filtering, the band-pass filter characteristics can be tailored to any specific bandwidth without changing the hardware.

So, if one needs a 455-kHz filter in the digital domain to have narrower or broader bandwidth, just setting some registers in the digital filter program changes the filter characteristics. So there is no need to change a physical part. The filter's characteristic is determined by software.

And as technology progresses, more and more circuits can be mimicked in a computer or in a system on a chip via digital signal processing to provide not only filtering but also amplification, tuning across the IF band (e.g., 44.1 kHz to 192 kHz), phase shifting, multiplication (e.g., RF mixing) that includes image-reject mixing, and detecting [amplitude-modulated (AM) and frequency-modulated (FM) demodulators] via software.

One popular implementation of a software-defined radio (SDR) on a chip is a two-band 2-m and 70-cm transceiver made by Baofeng. This radio sells for about $40 and uses a dedicated SDR integrated circuit, the RDA 1846. Because this radio does not have too much RF filtering in the front end, it is susceptible to overload when used near a transmitter (e.g., amateur radio FM repeater). But otherwise it performs very well.

Another popular implementation of the SDR is via several do-it-yourself software-defined radio programs that run on a computer, such as the Winrad program. There are other SDR programs that will run on PCs. For example, once the Winrad program is installed on a PC (e.g., XP or Windows 7 operating system), all that is needed is a down-converted low-frequency IF signal to be provided to the audio inputs of the computer's sound card. The PC then takes over the IF filtering, tuning, and demodulation. Because the low-frequency IF signal has a wide bandwidth of at least half the sampling frequency of the sound card, radio stations can be tuned via the Winrad software program over a 20+-kHz range. Thus the Winrad program also provides the equivalent of a variable-frequency oscillator.

For many SDRs, switch-mode mixers will be used, and image rejection will be handled in the software via digital signal processing. So now let's look at two examples of SDR front-end systems. They both provide two channels of down-converted low-frequency IF signals to the two input channels of a computer's sound card.

Figure 12-1A provides a first example of an SDR front-end block diagram. An antenna is connected to generally a fixed-frequency RF band-pass filter, but a tunable filter can be used. This RF band-pass filter will be needed to remove signals whose frequencies are outside the band of interest so that out-of-band signals do not mix back down to the IF band. For example, in the 40-m amateur radio band, signals around 7.2 MHz should be passed through, but frequencies above 8 MHz should be attenuated. The RF band-pass filter in an SDR system generally is not used to remove image signals but to remove out-of-band signals because the mixer that is used is a harmonic mixer.

Briefly, almost every RF mixer is a harmonic mixer of some sort. So what is meant by harmonic mixing? A harmonic mixer is capable of providing a signal that

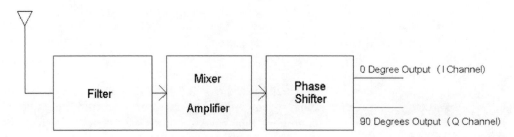

FIGURE 12-1A First example of an SDR front-end system using analog phase shifters.

falls into the IF band by way of out-of-band RF signals producing a difference signal with a harmonic of the local oscillator's frequency.

Let's take an example of a theoretical broadcast AM radio system in which there is no tuned RF filter prior to mixing. Recall from Chapter 8 that superheterodyne radios always use a tuned RF filter via variable capacitor VC1_RF to pass the radio station's signal to the mixer or converter circuit. And recall that the local oscillators or converter circuits in Chapter 8 have large oscillator signals that purposely cause distortion in order to provide a multiplying effect in the mixer or converter circuits. This distortion owing to the large amplitude of the oscillator signal also results in harmonics of the local oscillator frequency. Table 12-1 provides an example with a 455-kHz IF system (note that LO = local oscillator).

TABLE 12-1   Out-of-Band Radio-Frequencies Mixing Undesirably to the 455 kHz IF from Harmonics of the LO (Local Oscillator) Signal

	Oscillator Frequency Range	RF Frequency Range for 455-kHz IF
LO fundamental frequency	990 kHz–2.1 MHz	535 kHz–1,645 kHz
Second harmonic of LO	1.98 MHz–4.2 MHz	1.52 MHz–3.74 MHz
Third harmonic of LO	2.97 MHz–6.3 MHz	2.51 MHz–5.84 MHz

As can be seen from the table, without any filtering prior to RF mixing, unwanted signals of frequencies from 1.52 MHz to 5.84 MHz will mix into the 455-kHz IF band and thus cause the radio to receive extraneous signals. Note that noise from 1.52 MHz to 5.84 MHz also will mix into the IF band, resulting in extra or increased noise levels when the 455-kHz IF signal is demodulated.

So in Figure 12-1A a filter is used to ensure that out-of-band noise and signals are not down-converted to the IF band. The output of the mixer then is connected to a constant-amplitude phase-shift circuit that provides two outputs. Relative to one of the outputs of the phase-shifter network is a constant 90-degree shift over a range of frequencies. A first output of the phase-shifter circuit is normally defined as the *I channel*, or *in-phase channel*, whereas the second output that has the constant 90-degree phase shift relative to the first output is named the *Q channel*, or *quadrature channel*.

A computer's sound card then captures the signals from the I and Q channels of the phase-shifter circuit, which contain signals of a low IF (e.g., 3 kHz to 96 kHz). Now the question is, Why connect two types of signals to the sound card?

As stated earlier, the RF filter in Figure 12-1A does not filter out image signals. So there must be some other way to remove image signals. By using I and Q signals and connecting them to an image-signal-reject mixer circuit that is emulated by digital signal processing *in the computer*, image signals are attenuated. It will be shown in much more detail in Chapter 21 that signals that are 90 degrees relative to each other can be used to cancel out image signals while passing the desired signals, or vice versa, that is, passing image signals while canceling out the desired signals.

Figure 12-1B shows a more common way of generating I and Q signals at low IFs for coupling into the computer. The analog phase-shift network is replaced by an I and Q mixer system, which consists of two mixers. So in Figure 12-1A there is one mixer and an analog phase-shift network. And in Figure 12-1B there are two mixers instead to provide the phase-shifting function that replaces an analog phase-shifting network. A first mixer down-converts the RF signal to a low IF signal at reference 0 degree, whereas the second mixer down-converts the same RF signal to the same low IF but with a phase shift of 90 degrees. The low IF signals from both mixers are amplified by relatively low-bandwidth operational amplifiers, and the outputs of the amplifiers then are sent to the sound card input of the computer for digital signal processing.

The I and Q mixer shown in Figure 12-1B consists of two mixer circuits. The first mixer circuit receives a 0-degree phase signal from the local oscillator, whereas the second mixer circuit receives a 90-degree phase signal from the local oscillator. By multiplying at 0 and 90 degrees, two (low-frequency) IF signals are generated such that one of the IF signals is 90 degrees shifted from the other IF signal.

Figure 12-1C illustrates some of the SDR functions within a computer that is running the SDR software program. The computer's sound card channel 1 and channel 2 inputs digitize the I and Q signals to further filter them. Also, the software

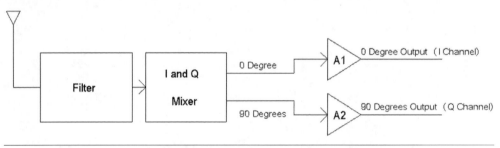

FIGURE 12-1B Generating I and Q signals via a mixing method.

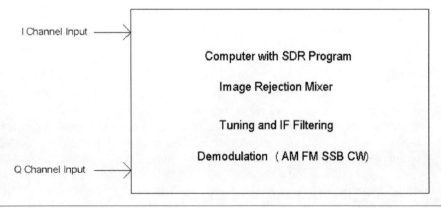

FIGURE 12-1C SDR functions within a computer or system on a chip.

program (e.g., Winrad) allows for fine-tuning of the phases and amplitudes of the I and Q signals for maximum image rejection through the image-reject filter in the computer.

With the I and Q signals, demodulation of broadcast AM signals is achieved by multiplying the I signal by itself and multiplying the Q signal by itself, summing the squares of the I and Q signals to provide a summed signal, and then taking the square root of that summed signal. Although this process may seem mysterious when compared with envelope detection, it is not.

But here is a hint on how this demodulation works: Use the trigonometric identity that $[\sin(Bt)]^2 + [\cos(Bt)]^2$ always is equal to 1 and that the I signal is $[1 + m(t)]\cos(Bt)$ and the Q signal is $[1 + m(t)]\sin(Bt)$, where $m(t)$ represents the audio information. Chapter 16 provides a detailed explanation on how AM demodulation works via I and Q signals.

Other functions included in an SDR are demodulation of single-sideband signals and frequency-modulation (FM) signals. And depending on the sampling rate of the sound card, the SDR software program allows for tuning to a bandwidth of half the sample rate or up to the sampling rate of the sound card. For example, if the sound card samples at a rate of 192 kHz, an SDR software program allows tuning over a range of at least 96 kHz or up to 192 kHz. Figure 12-1D shows a screen capture of the

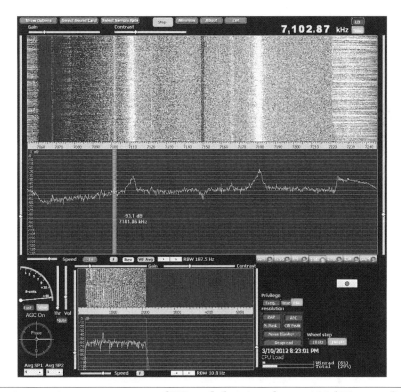

FIGURE 12-1D   A screen capture of Winrad showing that at 192-kHz sampling frequency, tuning for radio stations spans 192 kHz.

Winrad SDR software program running. As seen in the figure, the local oscillator is set for 7,150 kHz in this particular example to tune into the amateur-radio 40-meter band. The tuning spans from 7,054 kHz to 7,246 kHz for a range of 192 kHz. Since amateur radio traditionally has single-sideband transmission on the lower sideband for the 40-meter band, lower sideband (LSB) is chosen for the demodulation.

In this example, 192 kHz of tuning range is available because of the computer's sound card. Not all computers will come with this "extended" capability. For example, many laptop computers will only allow for 48 kHz or 96 kHz of tuning range. The user can first run Winrad to confirm the sampling rate and tuning range. If the tuning range seems restricted, go to the Winrad top menu to select the sampling rate (Figure 12-1E).

Should the sampling rate top out at a number lower than 192,000, upgrading the sound card to a 192-kHz sampling rate is recommended. However, in an SDR, it is still possible to have very usable tuning results with existing sound cards with sampling rates of 44.1 kHz, 48 kHz, or 96 kHz. One solution is to obtain a variable-frequency oscillator such as a direct-digital-synthesis (DDS) generator, which is available in kit or completed form. The DDS generator can tune to the stations instead of the Winrad software program.

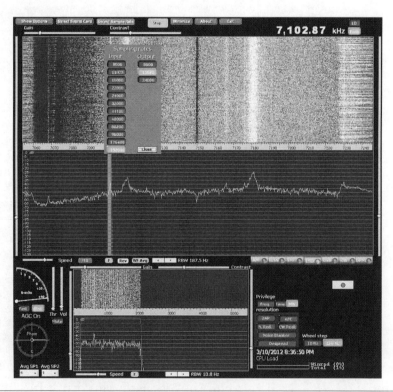

FIGURE 12-1E   Selecting sampling rate via Winrad.

SDR Front-End Circuits, Filters, and Mixers

As stated previously, the SDR front-end circuits usually require some type of filter prior to mixing or down-converting to the low-frequency IF signal for the computer to process. The filter generally must cut off or attenuate low-frequency noise in the IF band of about 100 kHz and below. Therefore, the simplest filter can consist of a high-pass filter with a cutoff frequency at about 200 kHz to 100 kHz. With a high-pass filter, low-frequency noise such as power-line hum can be attenuated effectively so that the demodulated signal does not include this type of low-frequency noise.

However, ideally, the filter should be a combination of a high-pass filter and a low-pass filter in which the low-pass filter prevents signals from out of the band of interest to be demodulated. The low pass filter is especially required if the SDR front-end circuit includes a switch mode RF mixer. A switch mode RF mixer is also a harmonic mixer which means that RF signals near a harmonic of the local oscillator or switching frequency will also mix down to the IF. Examples of switch mode mixers include commuting mixers and sampling circuits. It should be noted that a band-pass filter has a high-pass and a low-pass filtering characteristic. Thus, at a minimum, a single-inductor/capacitor (LC) band-pass circuit should be used for filtering. But a multipole band-pass filter may be used for wider bandwidth and sharper cutoff characteristics.

More elaborate filters usually result in better noise performance and overload characteristics, though. More filtering restricts the signals to a smaller portion of the RF spectrum and thus removes interfering signals for "jamming," or overloading, the mixer.

Figure 12-2 presents examples of simple LC and multistage band-pass filters. The circuits located at the top portion of the figure illustrate a simple LC band-pass network, whereas the circuit at the bottom shows a multistage LC band-pass filter for sharper cutoff to out-of-band signals and noise.

In terms of simplicity for mixing or down-converting the RF signal to a low-frequency IF signal, a commutating switch such as the one used in the coil-less radio in Chapter 11 works very well in terms of handling high-level RF signals as well as in terms of low noise performance.

In addition, a sampling circuit provides RF mixing action. For example a sample-and-hold circuit can serve as a mixing circuit. For those who have experimented with digitizing audio signals, one knows that aliasing can occur when the input signal's frequency exceeds half the sampling frequency. For example, if a digital audio system is sampling at 44.1 kHz, then an input signal of 25 kHz connected into the analog-to-digital converter will cause a 19.1-kHz aliasing signal to appear at the output. But the 19.1-kHz signal is just a signal that has a difference frequency of the sampling frequency and the input frequency or, alternatively stated, 44.1 kHz – 25 kHz = 19.1 kHz . So the sampling circuit can be used as a mixer.

Figure 12-3A shows a commutating switching mixer. The commutating mixer has been used since at least around 1930. This mixer circuit switches between a noninverting output RF signal and an inverting output RF signal. So a push-pull or complementary analog (RF) signal is needed at the inputs of this mixer. Under ideal conditions, this particular mixer is balanced in that neither the input RF signal nor the

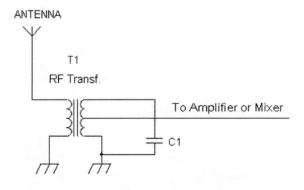

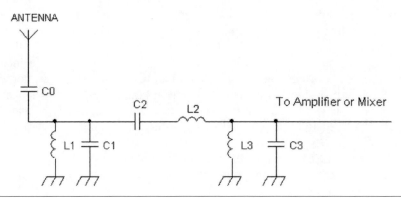

FIGURE 12-2   Simple LC and multistage band-pass filter for an SDR front-end circuit.

oscillator signal appears at its output terminal. However, in practice, there is a very small leak-through of the RF and oscillator signals at the output.

As mentioned previously, sampling a signal is a form of mixing or multiplication. And in particular, a sample-and-hold circuit allows for the equivalent of peak detection such that the sampled output of the mixer generates an IF signal that has almost the same amplitude as the input RF signal.

Figure 12-3B shows a single sampling switch with a hold capacitor C. Two or more sample-and-hold circuits may be used to generate I and Q signals. In its simplest form, two of these circuits will provide 0- and 90-degree phase-shifted signals.

In another configuration using a four-pole sample-and-hold circuit designed by Dan Tayloe, I and Q signals also can be generated. A variant of the four-pole sample-and-hold circuit can be implemented by using four single-sampling switches and capacitors. One should note that the sampling mixer is very popular with SDR do-it-yourself (DYI) kits.

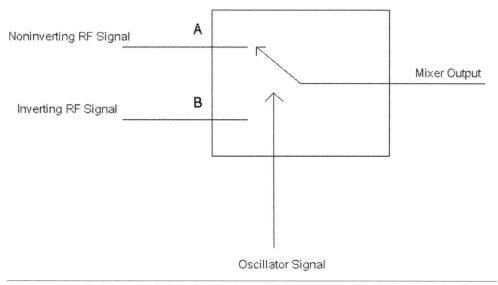

FIGURE 12-3A   Commutating switch-mode mixer.

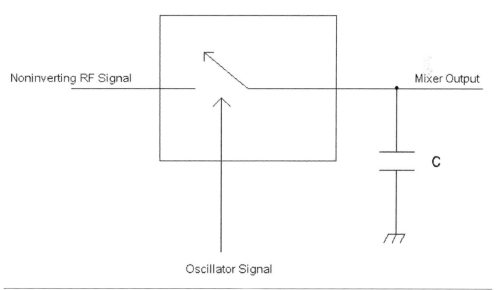

FIGURE 12-3B   A sample-and-hold circuit that can be used for RF mixing.

Phasing Circuits for 0- and 90-Degree Outputs for I and Q Signals

To generate I and Q signals, a circuit or series of circuits must be used to generate two outputs for a signal whose frequency range is defined. For SDR software programs, ideally, a phasing circuit should provide 0- and 90-degree outputs up to one-half the sampling rate of the sound card. For example, a sound card that samples up to 192 kHz will require that the phasing circuit provide a constant difference of 90 degrees from the I and Q outputs over a range of up to 192 kHz/2 = 96 kHz.

Most practical analog phasing circuits work over a range of frequencies that are defined as the lower and higher limit frequencies. For example, the lower limit frequency cannot be at 0 Hz or near DC (direct currect) but must be some positive number frequency such as 10 Hz or 300 Hz. The upper frequency limit theoretically can be as high as the designer wants, and this upper limit depends on the bandwidth of the amplifiers or components used.

Figure 12-4 presents a basic analog 0- and 90-degree phase-shift system. This phasing system must have two characteristics. Over the frequency range from the lower to the higher frequency limits,

1. The I and Q outputs must maintain constant amplitude relative to each other.
2. The I and Q outputs must maintain a constant 90-degree difference in phase.

Any amplitude variation or deviation from the 90-degree phase difference between the I and Q outputs over the frequency range will cause the image-reject mixer in the computer to not completely reject the image signal.

Figure 12-5 provides an example phasing circuit that generates 0- and 90-degree output.

Before I explain operation of this phasing circuit, I would like to recommend a terrific book on synthesizing filters and phase-shift networks by Arthur B. Williams. It is called the *Electronic Filter Design Handbook* and is published by McGraw-Hill. Any edition (e.g., the 1980 First Edition, 1989 Second Edition, 1995 Third Edition, or

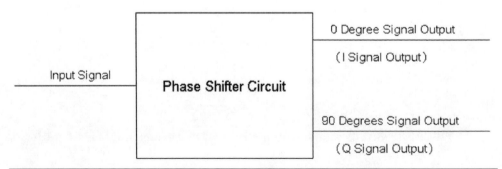

FIGURE 12-4   A phasing system with a single input terminal and two outputs for I and Q signals.

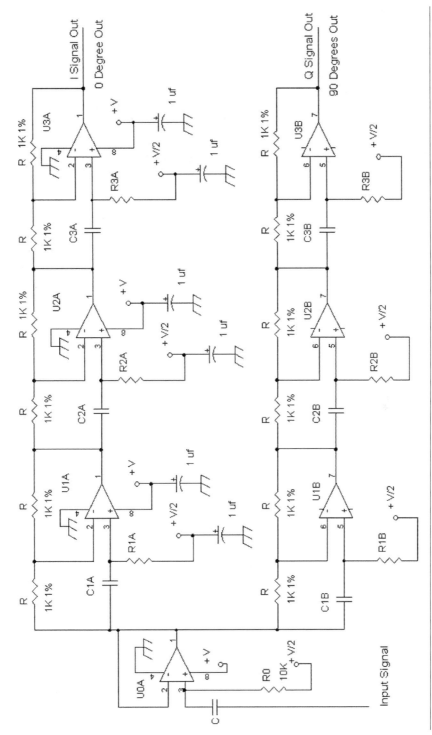

FIGURE 12-5 An analog phase-shifting circuit that provides I and Q signals.

the 2006 Fourth Edition) of this book will be helpful. And it is from this book that the phasing circuit in the figure is taken. I cannot say enough about how great this book is in terms of showing how to design all types of filters with coils or active components. The gyrator circuit in Chapter 11 came from this book. Any hobbyist or engineer will find that this book contains practical designs without long mathematical derivations on filter design. In short, this is a great filter book for designers.

Figure 12-5 shows a phase-shifting circuit with six amplifiers: U1A, U1B, U2A, U2B, U3A, and U3B.

 Note A low-output-impedance amplifier (e.g., op amp U0A) should drive the resistors and capacitors associated with U1A and U1B.

Each amplifier section constitutes an all-pass network. So the output of any of the amplifiers will provide constant amplitude over frequency while phase shifting the input signal. The actual phase shift varies with the frequency of the input signal. So the output of the A section of amplifiers at U3A is the summation of phase shifts from U1A, U2A, and U3A. Each amplifier has a different RC time constant, such as C1A R1A, C2A R2A, and C3A R3A. These time constants cause the phase (as compared with the input signal) at the output of U3A to vary quite "wildly" as the frequency is varied. At first glance, this may look all wrong. But there is the B section that also generates phase shifts from U1B, U2B, and U3B in a similarly "wild" manner but with a constant difference of 90 degrees from the A section.

The amplifier gain resistors R are identical 1 percent resistors and can be in the range of 1,000 Ω to 10,000 Ω as long as the resistors are the same value for each amplifier. So, for example, all the resistors R can be 1,000 Ω 1 percent (or better, such as 0.25 percent tolerance).

So we have one set of phasing circuits in the A section and another set in the B section, both generating lots of phase shift when compared with the input. But by carefully designing the RC time constants for each amplifier, a 90-degree phase difference is achieved with constant amplitude for an input signal having a lower and higher frequency limit.

The phasing circuit in Figure 12-5 maintains a constant 90-degree phase difference over a frequency range. This range can be expressed as the ratio of the higher frequency limit to the lower frequency limit. For example, from the Arthur B. Williams book, the following values are calculated for a ratio of the higher frequency limit to the lowest frequency limit of about 10:1 for a frequency range of 300 Hz to 3,000 Hz with an error of about 0.1 degree. And it should be noted that although 300 Hz to 3,000 Hz typically is a range for a single-sideband audio phase shifter, for the SDR, a range of about 3,000 Hz to 30 kHz is more useful.

In order to scale the original range of 300 Hz to 3,000 Hz to a new range, just scale *either* the resistors or the capacitors coupled to the positive (+) terminal of the amplifier to the reciprocal of the scaling factor. For example, if 3,000 to 30,000 Hz is needed, then this is a scaling factor of 10. So just divide all the resistor values at the

(+) terminal of the amplifiers by 10, or alternatively, divide all capacitor value at the (+) terminal of the amplifiers by 10.

Do not divide both the values of the resistors and capacitors.

Table 12-2 illustrates a phase-shifting circuit that holds a 90-degree phase difference within about ±0.1 degree.

TABLE 12-2 Resistor and Capacitor Values for the Phase-Shifting Circuit that Maintains a 90-Degree Phase Difference between the I and Q Signal Outputs within 0.1 Degree

Frequency Range 300 Hz to 3 kHz	Frequency Range 3 kHz to 30 kHz	Frequency Range 9 kHz to 90 kHz
R1A: 16.2 kΩ	R1A: 1.62 kΩ	R1A: 1.62 kΩ
R1B: 54.9 kΩ	R1B: 5.49 kΩ	R1B: 5.49 kΩ
R2A: 118 kΩ	R2A: 11.8 kΩ	R2A: 11.8 kΩ
R2B: 237 kΩ	R2B: 23.7 kΩ	R2B: 23.7 kΩ
R3A: 511 kΩ	R3A: 51.1 kΩ	R3A: 51.1 kΩ
R3B: 1.74 MΩ	R3B: 174 kΩ	R3B: 174 kΩ
All capacitors: 1,000 pF 1%	All capacitors: 1,000 pF 1%	All capacitors: 330 pF 1%

The first column of the table shows the original resistor and capacitor values for a frequency range of 300 to 3,000 Hz. To scale the frequency range 10-fold for 3 to 30 kHz, the resistors are divided by 10, but the capacitors remain the same. The third column shows how to scale the frequency range of 3 kHz to 30 kHz to 9 kHz to 90 kHz simply by taking the (same) resistor values of the second column and dividing the capacitor values by 3 (1,000 pF/3 = ~ 330 pF).

Op amps for these circuits (in Table 12-2) should have at least a 10-MHz gain bandwidth product. For example, an AD823 op amp may be used. For the values shown for the frequency ranges of 3 kHz to 30 kHz and 9 kHz to 90 kHz, one can use an NE5532 or LM833 op amp.

And see Table 12-3 for a phase-shifting circuit that holds the 90-degree phase differential between I and Q channels within ±0.44 degree but over a frequency ratio of about 25:1 instead of the 10:1, as seen in Table 12-2. For the values shown in the table, the op amps used may be NE5532 or LM833.

TABLE 12-3   Resistor and Capacitor Values for the Phase-Shifting Circuit that Maintains a 90-Degree Phase Difference between the I and Q Signal Outputs within 0.44 Degree

Frequency Range 3 kHz to 75 kHz	Frequency Range 3.8 kHz to 96 kHz	Frequency Range 4.5 kHz to 114 kHz
R1A: 1.62 kΩ	R1A: 1.62 kΩ	R1A: 1.62 kΩ
R1B: 5.89 kΩ	R1B: 5.90 kΩ	R1B: 5.90 kΩ
R2A: 14.0 kΩ	R2A: 14.0 kΩ	R2A: 14.0 kΩ
R2B: 32.4 kΩ	R2B: 32.4 kΩ	R2B: 32.4 kΩ
R3A: 76.8 kΩ	R3A: 76.8 kΩ	R3A: 76.8 kΩ
R3B: 280 kΩ	R3B: 280 kΩ	R3B: 280 kΩ
All capacitors: 500 pF 1%	All capacitors: 390 pF 1%	All capacitors: 330 pF 1%

A question one may ask is, What is the consequence when the 90-degree difference in phase between the I and Q channels is off by a degree or two? That is, if the I and Q channels are 89 or 88 degrees in phase difference, what happens? It turns out that an image-reject mixer relies on a precise 90-degree phase differential in order to perfectly cancel out the image signal. An error of a degree or two causes an imprecise cancellation of the image signal, and thus some of the image signal will appear at the output of the image-reject mixer.

For a rough approximation, the residual image signal output can be expressed by the following:

$$\text{Residual image signal output} = 0.5 \sin(x)$$

where x is the error phase angle measured in radians. Note that 1 radian = 57.3 degree, so 0.1 radian = 5.73 degrees, and 0.01 radian = 0.573 degree.

For small values of x measured in radians that are much smaller than 1,

$$\sin(x) = x$$

For example, if $x = 0.1$ radian, then $\sin(0.1) \sim 0.1$. Again, recall that x is measured in radians. And for all practical purposes, if $x = 0.01$, then $\sin(0.01) = 0.01$. And again, to reiterate, x is measured in radians, and x [in the equation of $\sin(x) = x$ for small values of x] is *not* measured in degrees.

Also, 1 degree equals 0.0174 radian. So a 1-degree error will result in a residual image signal of about

$$\begin{aligned} 0.5 \sin(0.0174) &= (0.5)(0.0174) \\ &= (0.5)(1.74\%) \\ &= 0.87\% \end{aligned}$$

whereas a 0.5-degree error from the 90-degree difference phase between the I and Q channels will result in a 0.435 percent leak-through of the image signal.

For an appropriate approximation of the residual image signal output with a phase error or deviation from the 90 degree difference between the I and Q channels is 1 percent per 1 degree of phase error, see Chapter 21, which presents an analysis of an image-reject mixer and provides further details pertaining to phase error.

Multipliers for Generating 0- and 90-Degree Phases

While the phase-shifting circuit in Figure 12-5 provides an accurate phase difference of 90 degrees between the I and Q channels, its accuracy depends on precision parts. Also, to increase the range of frequencies, more amplifier stages must be added. For example, if a frequency ratio of 500:1 is needed with less than 0.5 degree of error, 10 amplifiers instead of the 6 shown in Figure 12-5 will be required.

So an alternative to phase-shifting networks is to use multiplier circuits. Chapters 8 and 9 showed that a mixing circuit creates a multiplying effect on two signals. And one of the outputs of the mixing circuit is a signal that has a frequency equal to the difference of the two input signals—the local oscillator signal and the RF signal.

But what if we connect the RF signal to two multipliers, and each multiplier is connected to an oscillator of the same frequency, and one of the multiplier has a phase-shifted oscillator signal? See Figure 12-6.

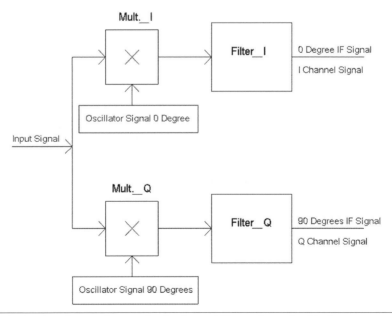

FIGURE 12-6 An alternative method of providing a 0- and a 90-degree output for I and Q signal generation.

It turns out that when two signals are multiplied to provide a difference-frequency signal, the phase of one or both of the two signals carries over to add a phase shift to the difference-frequency signal. So, if a mixer generates a signal that has a difference frequency (F1 – F2), then it also carries over the phase angle. That is, if signal 1 = $\cos(F1t + \phi)$ and signal 2 = $\cos(F2t)$, then the resulting difference-frequency signal by multiplication is

$$A \cos[(F1t + \phi - F2t)] = A \cos[(F1 - F2)t + \phi]$$

where A is a scaling factor–based conversion gain of the multiplier, and ϕ is the phase angle, such as 90 degrees.

So in Figure 12-6 the top multiplier has the RF signal multiplied by a cosine waveform, and the bottom multiplier has the RF signal multiplied by a 90-degree-shifted cosine waveform or a sine waveform. The output of both top and bottom mixers will provide a signal with the exact same difference frequency (frequency of the RF signal minus the frequency of the cosine waveform). However, the bottom mixer will provide a difference-frequency signal with a 90-degree phase shift in relation to the difference-frequency signal from the top mixer.

The output of each mixer then is connected to a filter to pass the difference-frequency signal while removing other signals, such as a signal whose frequency is the sum of the frequencies of the RF and cosine waveform. Thus the outputs of the (identical) filters provide I (0-degree) and Q (90-degree) signals, which typically are low IF signals (e.g., <100 kHz) for the SDR.

One advantage of multiplying mixers to generate I and Q signals is that there is essentially no lower- or higher-limit frequency to worry about. So, if the frequency ratio changes from 10:1 to 1,000:1 in maintaining 90 degrees of phase shift between the two channels, the multiplier circuit in Figure 12-6 does not change, whereas the phase-shifting circuit shown in Figure 12-5 increases in complexity.

One small downside to using switching mixers versus a phase-shifting network is that the signal generator driving the switching mixer initially has to run at four times the local oscillator frequency. If an oscillator has a perfect square wave with 50 percent duty cycle, then the signal generator can run at twice the local oscillator frequency. In order to synthesize a 90-degree phase-shifted version from the generator, at least twice the local oscillator frequency is needed for a digital divider circuit.

Example Radio Circuits for Software-Defined Radios

For a first experiment in an SDR, an AM radio will be shown. It will contain some circuits that were shown in previous chapters but will have a lower IF signal for inputting to a computer's sound card.

Figure 12-7 presents a block diagram of an AM radio front-end circuit for an SDR. The figure shows an AM radio with a built-in loop antenna that couples RF signals

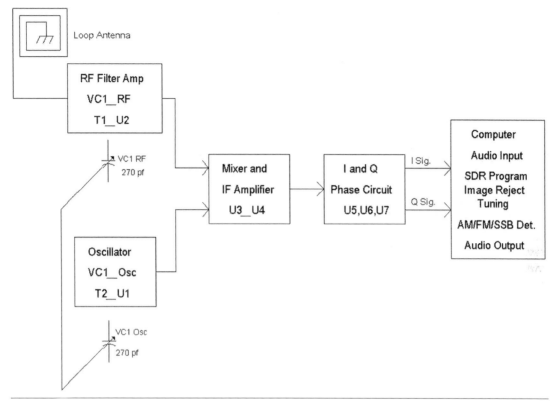

FIGURE 12-7   A front-end block diagram for an SDR that tunes the standard AM band.

into an RF transformer. The RF transformer tunes with a variable capacitor to pass frequencies of the tuned stations while removing signals outside the standard AM band. Since a switch-mode mixer will be used, removal of signals outside the AM band is essential. However, the variable tuned circuit, variable capacitor VC1_RF and an inductor T1, does not have the narrow bandwidth to reject image signals because of the low IF of less than 20 kHz. Fortunately, via the I and Q signals fed to the computer, the image signals will be rejected (or selected) via the digital signal-processing algorithms in the software.

The switch-mode mixer also is connected to a local oscillator tuned via VC1 Osc that is set to about 15 kHz above the tuned RF signal such that the output of the mixer circuit provides a low IF of about 15 kHz. The 15-kHz IF signal from the mixer is connected to a phase-shifting circuit similar to the one in Figure 12-5. And the I and Q outputs of the phase-shifting circuit then are coupled to the audio inputs of a sound card in a computer. As stated previously, tuning is accomplished mostly by the variable capacitor, so the entire 1-MHz range of the broadcast AM band can be accessed. Thus the 192-kHz tuning range of the computer via Winrad and the sound card is insufficient for tuning across the entire AM band.

The SDR program Winrad then is used to detect the I and Q signals, which then provide audio signals to the speakers in the computer.

Parts List

- **C1, C2, C3, C4, C6, C8, C24:** 0.01 uF
- **C5, C11, C12, C18, C22, C25:** 33 µF, 16 volts
- **C7, C9:** 0.001 µF, 5% film
- **C10, C23:** 0.01 µF, 5% film
- **C13:** 5 pF for R22 = 220 kΩ; or 47 pF for R22 = 22 kΩ
- **C14, C16, C17, C19, C20, C21:** 500 pF 1%
- **C15, C26, C27, C28, C29, C30:** 1 µF, 35 V
- **R1, R2, R4, R5, R7, R8:** 12 kΩ
- **R3:** 2 MΩ
- **R6:** 4.7 Ω
- **R9:** 1,800 Ω
- **R10:** 1,000 Ω
- **R11, R13, R14, R16, R17, R19, R23, R25, R26, R28, R29, R31:** 1,000 Ω 1%, 0.5%, or 0.25%
- **R12:** 1,620 Ω 1%
- **R15:** 14.0 kΩ 1%
- **R18:** 76.8 kΩ 1%
- **R20, R32:** 100 Ω
- **R21:** 22 kΩ
- **R22:** 220 kΩ, or 22 kΩ for lower gain so as to reduce overloading the sound card
- **R24:** 5.90 kΩ 1%
- **R27:** 32.4 kΩ 1%
- **R30:** 280 kΩ 1%
- **T1:** 42IF100, 42IF110, or 42IF300
- **T2:** 42IF100, 42IF110, or 42IF300 primary winding or 330-µH variable inductor
- **VC1:** twin-gang variable capacitor 270 pF and 270 pF
- **Loop antenna** to low-side tap of T1
- **U1:** 74HC04
- **U2:** UA733 or LM733 video differential amplifier
- **U3:** 74HC4053 or 74HCT4053
- **U4, U5, U6, U7:** NE5532 or LM833
- **U8:** LM78L05

Note Before building the circuit, with an inductance meter, set T1's primary winding inductance value across the main winding (e.g., *do not* measure at the tap of T1 or T2) to 322 µH, and set T2's inductance value to 312 µH. This will ensure that the oscillator's frequency is at least 10 kHz higher than the tuned RF frequency with T1.

Figures 12-8A and 12-8B show circuit diagrams for experimental AM front circuits for an SDR.

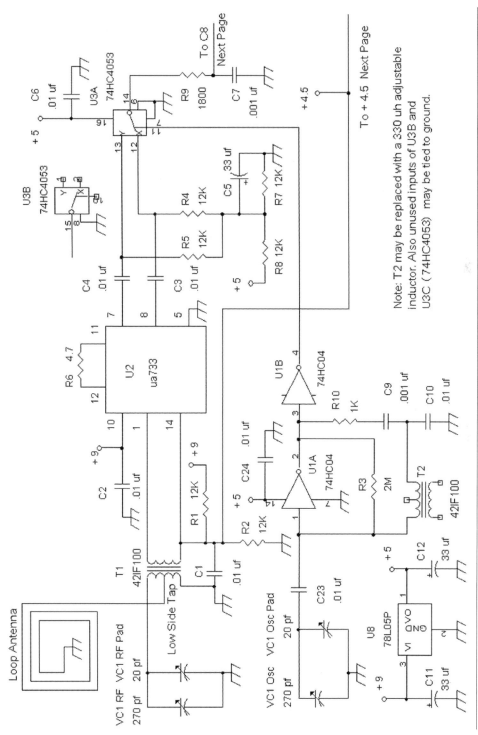

FIGURE 12-8A   Schematic diagram of an AM radio front-end circuit for an SDR that includes a commutating mixer.

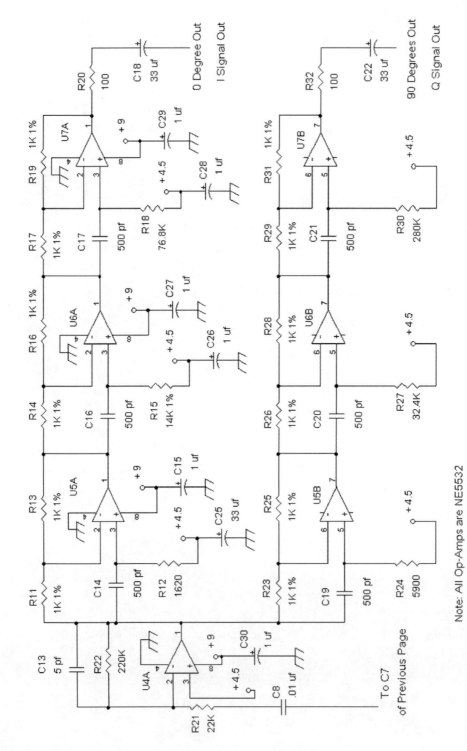

FIGURE 12-8B Schematic diagram of an AM radio front-end circuit with the analog phase-shifting circuit.

Before proceeding with building this circuit, with an inductance meter, adjust the inductance value of T1's primary winding to 322 μH and T2 to 310 μH. Radio signals are tuned via VC1 RF and T1 to provide the RF signals to amplifier U2, which has noninverting and inverting analog signal outputs to the commutating mixer, U3. The mixer U3 is switched via the local oscillator signal from U1B, whose frequency is between 10 kHz to 20 kHz above the frequency of the tuned RF signal. The output of mixer U3 then is low-pass-filtered via R9 and C7 to provide a low-frequency IF signal in the range of 10 kHz to 20 kHz. This low-frequency IF signal is coupled to an IF amplifier, U4A, to further amplify the IF signal. The amplified IF signal then drives two banks of phase-shifter circuits to provide I and Q signal outputs. The I signal via U7A and the Q signal via U7B then are connected to the line input of the sound card of a computer with an SDR software program running. With the SDR program running, select "AM" for the demodulation and a demodulated bandwidth of 3 kHz to 5 kHz. Then, via the software program, tune for a radio station between 10 kHz and 20 kHz. To tune throughout the AM band range of 540 kHz to 1,600 kHz, adjust variable capacitor VC1.

The value of R22 may be selected to adjust the signal amplitude to match the requirements (e.g., input sensitivity) of individual sound cards. For some sound cards, R22 set to 220 kΩ may cause overloading. Thus R22 can be replaced with a 22-kΩ resistor to reduce the I and Q signals by 10. Note that C13 should be replaced with a 47-pF capacitor if R22 is 22 kΩ.

Alignment procedure: With a reference generator of a known frequency, such as the 537-kHz reference RF oscillator shown in Chapter 4, couple the generator's signal with a wire near the loop antenna. With an oscilloscope set to alternating-current (AC) coupling, probe either end of C3 or C4, and tune VC1 for maximum signal. Then connect a frequency counter to pin 4 of U2 and adjust T2 (or an adjustable inductor) to a frequency about 15 kHz above the RF oscillator's frequency. For example, if the RF oscillator is measured at 537 kHz, then adjust T2 such that pin 4 of U2 reads 552 kHz.

Figure 12-9A shows a screen capture of the Winrad program with the AM radio front-end circuit connected to a computer with a received radio station. In the figure, the tuned desired signal is highlighted by a wide vertical strip, whereas the image signal is highlighted by a vertical line. The image signal is about 40 dB lower than the desired signal (tuned signal). Or stated another way, the amplitude of the image signal is about 1 percent of the amplitude of the desired signal (tuned signal).

The Winrad program allows for fine adjustment to null out the image signal (Figure 12-9B). By clicking on the "Show Options" tab and scrolling down to "Channel Skew Calibration," fine adjustments can be made for both amplitude and phase to null out the image signal. See the vertical line in the screen capture where the image signal is attenuated into the noise floor.

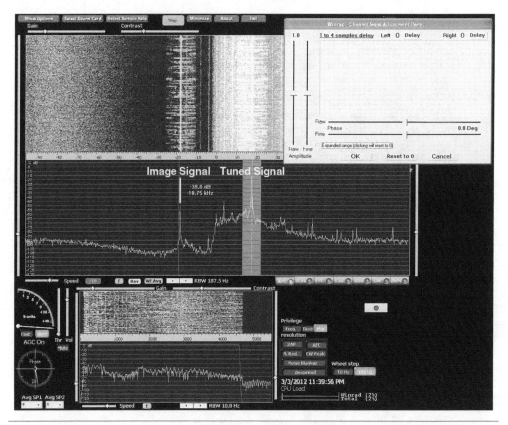

FIGURE 12-9A Screen capture of Winrad program with default adjustments for nulling out the image signal.

Second SDR Front-End Circuit for the 40-Meter Amateur Radio Band

Figure 12-10 shows a block diagram of an experimental front-end circuit for down-converting RF signals around 7.150 MHz to a low-frequency IF signal with I and Q output signals. Because the local oscillator will provide a fixed frequency, the I and Q output signals preferably are digitized at the highest sampling rate to provide the widest tuning range. For example, a 192-kHz rate allows tuning over a frequency range of 192 kHz via a software program such as Winrad. Fortunately, the 40-meter amateur radio band for voice transmission occupies approximately a 150 kHz of bandwidth for the frequency range 7.150 MHz to 7.300 MHz.

RF signals from an antenna are coupled to a 7.150-MHz band-pass filter. The output of the band-pass filter is buffered by an amplifier to drive the inputs of four analog switches. The analog switches are controlled by a four-phase oscillator signal that is used to make the four analog switches sample at 0, 90, 180, and 270 degrees.

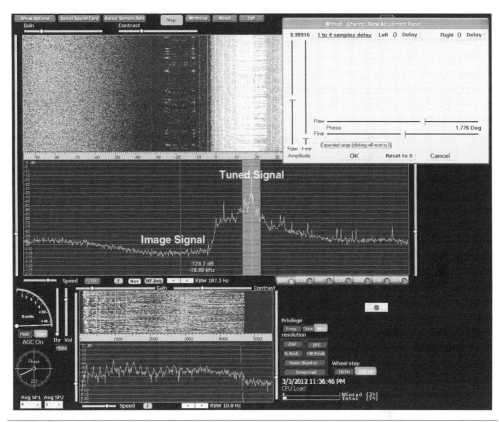

FIGURE 12-9B Image signal reduced further by adjusting amplitude and phase within the Winrad software program.

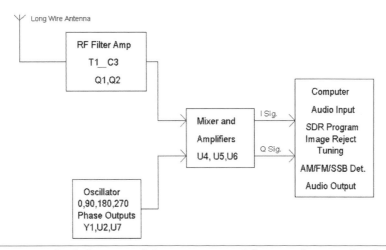

FIGURE 12-10 Block diagram of the experimental front-end circuit for the 40-meter band.

By sampling at 0 and 180 degrees, an I channel is provided. Similarly, by sampling at 90 and 270 degrees, a Q channel is provided. The output signals from the mixer are further amplified to provide I and Q signals of suitable amplitude for the computer's sound-card input circuits.

Parts List

- **C1, C8, C9:** 1 µF, 5% film, note: C1 can be 0.01 µF
- **C2:** 47 pF
- **C3:** 22 pF or 27 pF
- **C4, C5, C6, C7:** 0.0047 µF, 5% film
- **C10, C12, C15, C16, C17, C18, C24:** 0.01 µF
- **C11, C13, C14, C22, C23:** 33 µF, 25 volts
- **C19, C20:** 22 pF or 33 pF
- **C21:** 1 µF, 35 volts
- **R1:** 11 kΩ
- **R2:** 39 kΩ
- **R3:** 47 Ω
- **R4:** 470 Ω
- **R5:** 22 Ω
- **R6, R12, R13, R16:** 100 Ω
- **R7:** 3,300 Ω
- **R8, R10:** 1 kΩ 1%
- **R9, R11:** 100 kΩ 1%
- **R14:** 1 MΩ
- **R15:** 330 Ω
- **T1:** 10.7-MHz IF transformer 42IF129
- **Y1 crystal:** 28.636 or 28.322 MHz fundamental frequency crystal
- **U1:** 74HC4066
- **U2, U3:** NE5532 or LM833
- **U4:** 74AC04
- **U5:** 78L05 or 7805
- **U6, U7:** 74AC74
- **U8:** 74AC08
- **Q1:** MPHS10
- **Q2:** 2N3906

Figures 12-11A and 12-11B present schematic diagrams of the front-end circuit for receiving 40-meter RF signals. Note that this circuit can operate from a regulated 9-volt to 12-volt DC supply or a battery.

An antenna is coupled to a low-impedance winding of an RF transformer T1, which is really a 10.7-MHz IF transformer with an added capacitor C3 to lower the resonating frequency to 7.22 MHz. T1 as configured in this circuit has an input impedance of about 50 Ω for matching to an antenna. The other side of T1 at the low-side tap has about a 1:10 turns ratio, which then provides about a 10-fold increase

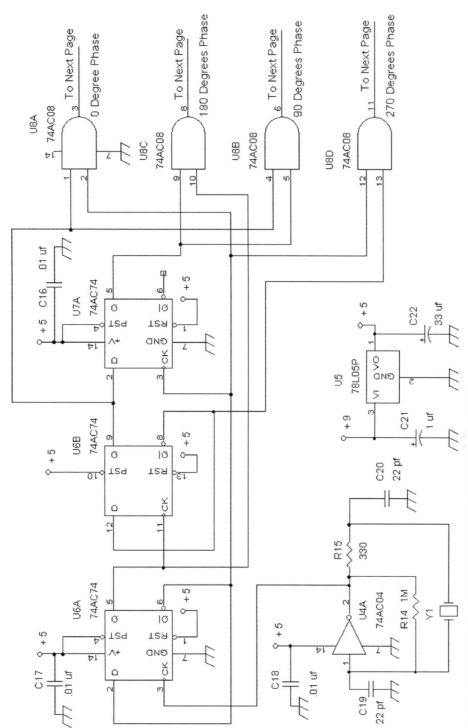

FIGURE 12-11A Schematic diagram of an experimental 40-meter RF circuit for an SDR local oscillator circuit with four-phase oscillator signals.

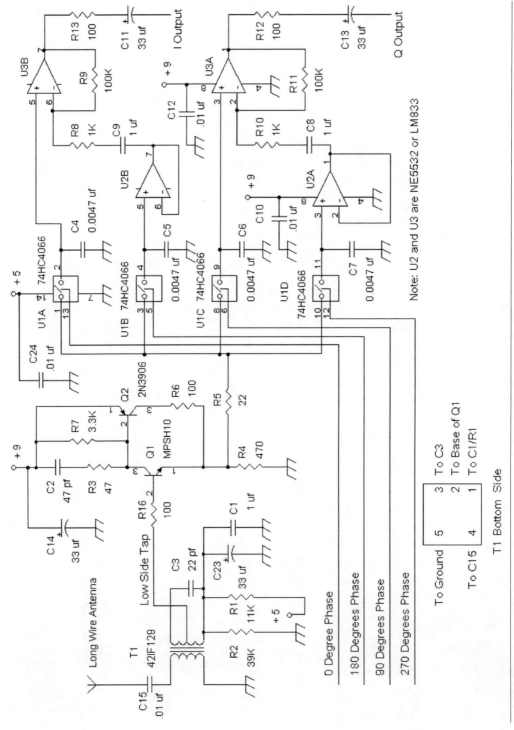

FIGURE 12-11B   Schematic diagram of the front-end circuit including front-end input filter and mixing circuit.

in RF signal when compared with the signal at the antenna input terminal. To drive the analog switch mixer, the signal from the low-side tap of T1 is buffered by a compound complementary transistor amplifier Q1 and Q2. This amplifier provides a high-impedance input resistance so that the Q of T1 is not affected and so that the amplifier also provides a low output resistance capable of driving the mixing circuit consisting of four sample-and-hold switches in U1.

 To adjust T1, connect a 7.22-MHz signal to C15, and measure with an oscilloscope the signal output at the emitter of Q1 [MPSH10, and note the pin-out for this transistor (BEC), which is different from a 2N3906 (EBC)]. Then adjust T1 for maximum signal level.

The control pulses that turn on the switches in U1A–U1D are nonoverlapping pulses, being a quarter cycle in duration at 0-, 90-, 180-, and 270-degree relative phases. The sampling frequency of each of these nonoverlapping pulses is the crystal oscillator's frequency divided by 4. The sample-and-hold signals for 0 and 180 degrees appear at capacitors C4 and C5. An "almost" differential amplifier consisting of U3B and U2B amplifies roughly the difference in voltage across C4 and C5.

The reason for saying that U3B and U2B almost form a differential amplifier is because normally the (+) input of U3B would include a resistive divider circuit. But since there is none, there is an error of about 1/gain. Since the feedback resistors in U6A (and U6B) result in a gain of about 100, the error is about 1/100, or 1 percent. Similarly, the 90- and 270-degree sampled signals are amplified almost differentially by a factor of 100 via amplifier U3A and U2A.

 Taking the difference between the two signals that are sampled 180 degrees apart in Figure 12-11B amounts to essentially the same as the circuit shown in Figure 12-8A, consisting of differential output amplifier (Ua733) U2 coupled to the two input terminals of the single-pole double-throw switch U3A. The output signals in Figure 12-11B of U3B and U3A then provide I and Q signals to the computer's sound card.

It should be noted that because the gain bandwidth product of the op amps in U6 is 10 MHz, it is operating near the limit to provide a 100-kHz bandwidth for a closed-loop gain of 100. The NE5532 is configured to provide a gain of 100, so the frequency response is 100 kHz, which means that there is very little negative feedback at 100 kHz. Normally, this can be a problem because the gain and frequency response of amplifiers are not tightly specified. That is, only a minimum specification of the gain bandwidth product of the op amp is guaranteed. So, if U6, a dual op amp, were replaced by two single op amps of similar specifications, chances are that the I and Q channels would suffer a mismatch in frequency response and ability to maintain a 90-degree phase difference over the range of frequencies (e.g., 100 kHz). But because in a dual op amp the characteristics in the die are matched (e.g., equally bad or equally good on both op amps in the package), it is often possible to still have good matching on the I and Q channels. Of course, higher-frequency dual op amps can be used in place of the NE5532.

In reference to Figure 12-11A, the crystal oscillator at about 28.6 MHz runs at four times the mixing frequency but gets frequency divided by 2 via U1A to ensure a 50 percent duty-cycle square wave at 14.3 MHz. This 14.3-MHz square-wave signal will be used later to generate a 90-degree phase-shifted version of the 7.150 MHz. A second frequency-divider circuit U2A further frequency divides the 14.3-MHz signal from U1A to provide a 7.150-MHz signal at a reference phase of 0 degree. By feeding this reference phase 7.150-MHz signal to a latch flip-flop circuit and clocking the latch with an inverted-phase 14.3-MHz signal, the output of the latch circuit provides a 90-degree phase-shifted signal at 7.150 MHz. The actual frequency of the crystal is 28.636 MHz, which results in 7.159 MHz instead of 7.150 MHz.

The 0- and 90-degree signals from U2A and U2B, along with output signals from U1A, are fed to a decoder circuit consisting of four AND gates, U3. The outputs of the AND gate provide the 0-, 90-, 180-, and 270-degree nonoverlapping signals for the analog switch mixer U4.

This front-end circuit was able to receive single-sideband transmissions on the 40-meter band. However, this circuit is just a starting point. Improvements can be made, and if time permits, another front-end circuit for receiving the 40-meter band will be presented.

It is suggested that the reader may just want to purchase DIY SDR front-end kits. One vendor is SoftRock, which sells receivers and transceivers for the amateur radio enthusiast.

 To transmit below the 10-meter band (i.e., transmit below 28 MHz), the reader must hold at least a general class amateur radio license. And to transmit on any amateur radio band, a license is required. Consult the ARRL, the Amercian Radio Relay League, at www.arrl.net for more information on licensing.

If the reader is just interested in operating receivers, no license is required. And SoftRock sells many receiver kits.

Well, this ends the first part of this book, which is geared more for the hobbyist to just build or experiment with radio circuits. It has been an interesting ride for the last 12 weeks for me in designing, building, testing, and writing about radio circuits.

The second part of this book will concentrate on tutorials of various subjects related to electronics, so I hope that readers will stick around for that. I will do my best to explain some of the principles of electronics with no more than high school math. So hang on, there is more to come.

Chapter 13

Oscillator Circuits

In Chapters 8, 9, and 10, oscillator circuits were used as local oscillators and converters (mixer and oscillator combination). In this chapter we will explore some "simple" analyses of a one-transistor oscillator circuit and a differential-pair oscillator. Both of these types of inductor/capacitor (LC) circuits have been used in some of the superheterodyne radio designs, and in a sense, this chapter is a "tutorial" on how these circuits work in more detail.

Although the oscillator circuits presented in this chapter are "simplified" for analysis sake and for ease of building and experimentation, the basic oscillator principles still apply. For example, the one-transistor circuit will use capacitors for "stepping down the voltage" instead of an oscillator coil/transformer.

The objectives for the analysis of oscillator circuits are as follows:

1. Introduce the reader to transconductance, which is related to the voltage-to-current gain.
2. Show how transconductance determines whether a circuit will oscillate reliably.
3. Illustrate that there are two types of transconductances—small-signal transconductance and large-signal transconductance.
4. Explain the role of the inductor and capacitors used in the oscillator circuit to determine oscillator frequency and equivalent turns ratio for a transformer.

But first let's take a systems approach to how oscillators work in general.

A condition for oscillation includes the following characteristics:

1. The total gain from the amplifier to the LC circuit and back to the input of the amplifier must exceed 1; typically, the gain is at least 2.
2. The total phase shift around the system must total to 0 or 360 degrees. So, if the amplifier has an inverting gain (e.g., common-emitter amplifier), the other components must deliver 180 degrees of phase shift. And if the amplifier has a noninverting gain (e.g., common or grounded-base amplifier), the other components typically must have a net phase shift of 0 degree.

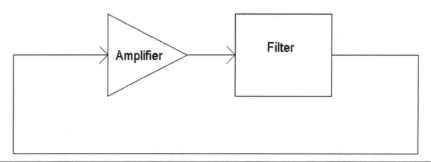

FIGURE 13-1 Basic oscillator system.

Figure 13-1 shows an oscillator system. A gain amplifier's output terminal is connected to a resonant filter, and the output of the resonant filter is fed back to the input terminal of the gain amplifier. The filter is commonly implemented as a parallel LC circuit.

The filter in the figure is usually a parallel LC circuit of some kind that has 0-degree phase shift at its resonant frequency of $1/2\pi\sqrt{LC}$. At a frequency above the resonant frequency, the phase shift usually lags or creates a negative phase shift. Above the resonant frequency, the LC network has more capacitive characteristics than inductive or resistive. And at frequencies below the resonant frequency, the LC network has more inductive characteristics that cause a phase lead or positive phase shift.

In an oscillator system, the positive-feedback mechanism responsible for causing the oscillation will readjust itself to shift the frequency of the oscillator to maintain the 0-degree phase shift. In a sense, the oscillator "servos" itself to correct for 0 degrees by readjusting its frequency.

So, for example, what if the gain amplifier in Figure 13-1 has some phase lag owing to a roll-off in frequency response, which causes a negative phase shift? The oscillator system is pretty "smart"; it lowers the frequency of the oscillation signal so that the LC filter is below the original resonance frequency, which, in turn, causes a phase lead to cancel out the phase lag of the amplifier. As a result, the oscillator will run at a lower frequency. This makes sense because a slower amplifier (see Figure 13-1) causes an oscillator to run slower in the form of a lower frequency.

The filter is commonly implemented as a parallel LC circuit (Figure 13-2). A voltage-driven LRC circuit can be used, such as the oscillator used in the SDR front-end circuit in Figure 13-3, U1A. Alternatively, and more commonly, the LRC circuit

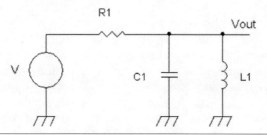

FIGURE 13-2 An LRC circuit that is voltage driven.

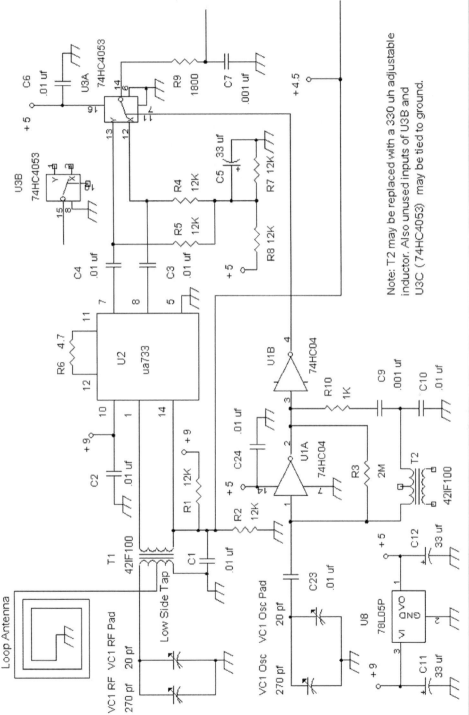

FIGURE 13-3   Schematic diagram of an AM radio front-end circuit for an SDR that includes a commutating mixer.

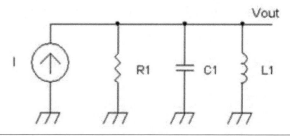

FIGURE 13-4   A current-driven LRC circuit.

is driven with a current source via the collector of a transistor, although there are exceptions where the emitter drives an LC or LRC circuit to form an oscillator (Figure 13-4).

Many active devices have current-source outputs such as a bipolar transistor with a collector terminal or a field-effect transistor with a drain terminal. The concept of a current source may not be familiar to some people because a voltage source is more easily visualized as a battery or a signal generator. In an ideal voltage source, the generated signal delivers a voltage signal independent of the current flowing into or out of the voltage source. In an ideal current source however, the generated current is independent of the voltage across the ideal current source.

One-Transistor Oscillator

Figure 13-5 shows an oscillator system with a voltage-dependent current source. The current source is controlled by the amount of voltage applied to the input of the system. Signal current is fed to a parallel LC circuit where two capacitors form a voltage divider network.

The resonant frequency is still $1/2\pi\sqrt{LC}$, where $C = C1 \times C2/(C1 + C2)$. Moreover, since C1 and C2 form a voltage divider, the capacitive voltage divider acts pretty much like a step-down transformer in terms of a turns ratio of $1/n = C1/(C1 + C2)$. The voltage-controlled current source is a transconductance device, which means that a change in input voltage results in a change in output current. Or, equivalently, the transconductance of a device $= \Delta I_{out}/\Delta V_{in}$, which can be construed as the ratio of

$$\frac{\text{Output AC current}}{\text{Input AC voltage}}$$

So now let's take a look at a particular transconductance device—the bipolar transistor. This transistor can be approximated by a part of the Ebers-Moll equation:

$$I_c = I_s e^{VBE/\left(\frac{KT}{q}\right)} = \text{collector current} \tag{13-1}$$

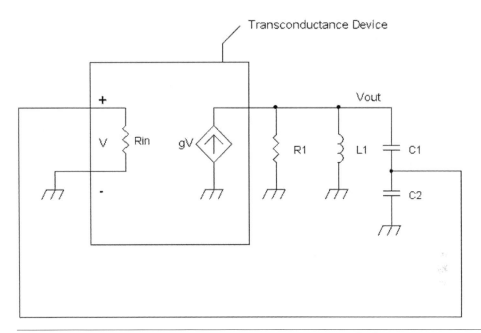

FIGURE 13-5   **An oscillator system using a voltage-controlled current source.**

where I_s is the transistor's reverse bias saturation current or leakage current (e.g., 0.01 pA to 0.0001 pA), VBE is the base-to-emitter voltage, e = 2.71828 . . . , and $(KT/q) = V_T = 0.026$ volt direct current (DC) at room temperature. V_T is also known as the *thermal voltage*.

The small signal transconductance of the bipolar transistor at a particular DC collector current I_{CQ} is given by

$$\text{gm\_small signal} = \frac{I_{CQ}}{\left(\frac{KT}{q}\right)} = \frac{I_{CQ}}{0.026 \text{ V}} = \frac{\Delta \text{collector current}}{\Delta \text{base emitter voltage}} = \frac{\Delta I_C}{\Delta \text{VBE}} = g_m \qquad (13\text{-}2)$$

If we did not know about Equation (13-2), there is another way to indirectly come up with the transistor's transconductance by plugging values into Equation (13-1). For Table 13-1, $I_s = 0.01$ pA. Note that the transconductance is independent of the value of I_s.

From this table, one can see that Equation (13-2) is valid and that the small-signal transconductance is linearly proportional to the DC collector current. And, as seen from Equation (13-2), it is valid for calculating the small-signal transconductance of a bipolar transistor. And by small signal, we are limiting the alternating-current (AC) signal into the base emitter of the transistor to about 5 mV to 15 mV peak to peak.

TABLE 13-1   Calculated Small Signal Transconductances of a Bipolar Transistor

VBE	I_c	$\frac{\Delta IC}{\Delta VBE}$	$\frac{I_{CQ}}{0.026 \text{ V}}$
0.59867	0.1000 mA		
0.59967	0.1039 mA	0.0039 mA/1 mV	0.00384 mA/ 1 mV @ 0.1 mA
0.65854	1.0000 mA		
0.65954	1.0392 mA	0.0392 mA/1 mV	0.0384 mA/ 1 mV @ 1.0 mA

If the base-emitter AC signal is larger than 15 mV peak to peak, gross distortion will occur at the output of the collector, and a large-signal transconductance model will have to be used. For example, Figure 13-6 shows an input signal at the base and emitter of a transistor of about 5 mV peak (10 mV peak to peak) and 39 mV peak (78 mV peak to peak). As can be seen in the figure, for small input signals, the collector current output looks pretty much like a sine wave, but for large input signals, the collector current is larger in amplitude but highly distorted.

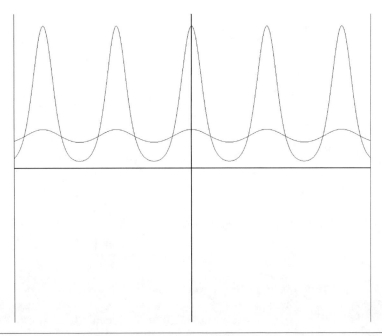

FIGURE 13-6   Output collector current for small and large input sinusoidal signals at the base-emitter junction of a transistor. The X axis is time, and the Y axis is the output current.

When an output signal gets distorted, harmonics are generated, and as the input signal level is increased, the fundamental frequency signal at the output is no longer proportional to the input signal's amplitude. Thus the output level of the fundamental frequency signal gets compressed (Figure 13-7). In the figure, the straight line shows the input signal increasing linearly, whereas the output starts to compress.

So let's take a look at how the large-signal transconductance G_m is different from the small-signal transconductance g_m measured in amperes per volt for a DC collector current of 1 mA = I_{CQ} for various sine-wave amplitudes measured at a peak voltage V_p or as a function of $\sin(\omega t)$ into the base-emitter junction of the transistor (Table 13-2). And the output collector current is the amplitude of the signal current at *only* the fundamental frequency. Note that I_{CQ} is the quiescent (DC) collector current.

See Figure 13-8 for input signals of 26 mV peak and 52 mV peak. Note that as the input signal is increased with a sinusoid, the collector output gets narrower, approaching a delta pulse as the input keeps on increasing. The spectrum of the narrow pulse contains fundamental and harmonic frequencies, with the amplitude of the harmonic signals being close to the amplitude of the fundamental frequency signal (Figure 13-8).

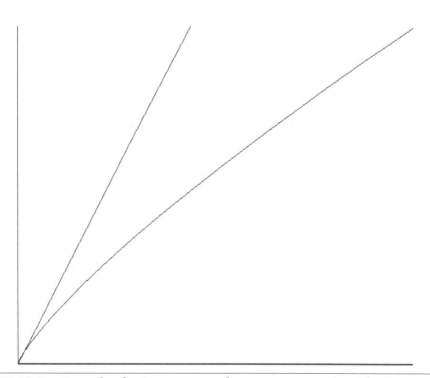

FIGURE 13-7   Example of gain or transconductance compression. The X axis is the input, and the Y axis is the output.

TABLE 13-2 Comparisons of Large and Small Signal Transconductances as a Function of Base-Emitter Peak Sine Wave Driving Voltages

Input Signal Level V_p	g_m @ 1-mA I_{CQ}	G_m @ 1-mA I_{CQ}	G_m/g_m for Any I_{CQ}
0.001 is small signal	0.038 mho	0.038 mho	1.0
0.013 is large signal		0.037 mho	0.97
0.026 is large signal		0.034 mho	0.893
0.052 is large signal		0.026 mho	0.698
0.104 is large signal		0.016 mho	0.432
0.156 is large signal		0.012 mho	0.304

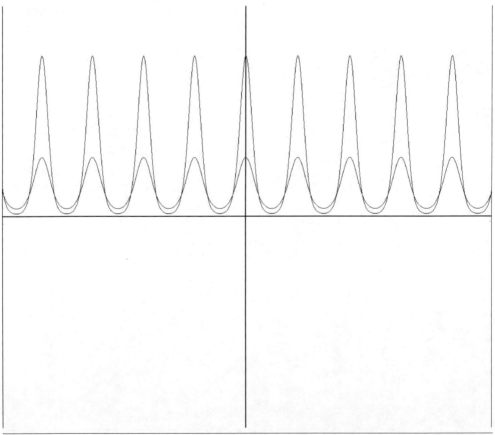

FIGURE 13-8 Output current of a transistor with 26-mV peak and 52-mV peak AC signals at the input. The X axis is time, and the Y axis is the output current.

From Figures 13-6 through 13-8 we see that when the input of a bipolar transistor amplifier is overdriven with a large signal, there is a lowering of effective transconductance pertaining to signals with the fundamental frequency. So how does this lowering of the effective transconductance of the transistor owing to overloading affect other characteristics such as input resistance in a bipolar transistor amplifier?

In a grounded base amplifier, the AC input small signal resistance at a quiescent collector current is just $\Delta VBE/\Delta I_E$, where ΔI_E is the change in emitter current. But we know that in a transistor with current gain $\beta > 10$, for practical purposes, $\Delta I_E = \Delta I_C$, the input resistance is $\Delta VBE/\Delta I_C$, which is just the reciprocal of g_m, where $g_m = \Delta I_C/\Delta VBE = I_{CQ}/0.026$ volt. So the small-signal input resistance of a common-base amplifier at a quiescent collector current, I_{CQ} is $1/g_m = 1/(I_{CQ}/0.026$ volt).

Also note for a common-emitter amplifier with the emitter AC grounded the input resistance is $\Delta VBE/\Delta I_B$, where ΔI_B is the change in base current. But $\beta I_B = I_C$, where β is the current gain, and $\Delta I_B = \Delta I_C/\beta$. Because of β, the small-signal input resistance of a common-emitter amplifier with emitter AC grounded is $\beta(\Delta VBE/\Delta I_C) = \beta/g_m$.

Now that the input resistances of common-base and common-emitter amplifiers have been established, let's take a look at the effect of input resistances for small and large signals at the input *where the input resistances pertain to the fundamental frequency signal (i.e., harmonics are ignored)*. Effectively, the large-signal resistance is the small-signal resistance multiplied by a factor of g_m/G_m. Intuitively, the reason why the input resistance across the base emitter junction (or emitter-to-base junction) increases as the input signal is increased. With large signals at the input, the input currents are no longer sinusoidal but instead are distorted in such a manner that the fundamental frequency input current decreases while the input currents of the harmonics increase. For example, when the sinusoid input voltage signal is > 26 mV peak, the input current has a periodic narrowed pulse waveform. The frequency spectrum of this periodic narrowed pulse waveform consists of signals of the fundamental frequency and at least the first few harmonics that are almost equal in amplitude. For example, a 1 mA peak-to-peak periodic narrowed pulse waveform (see Figure 13-8 with 52 mV peak AC signal) does not have a 1 mA peak to peak fundamental frequency sinusoidal component. Instead, the fundamental frequency signal component of the narrowed pulse is smaller (< 1 mA peak-to-peak) because the narrowed pulse contains harmonics as well. And because of the smaller fundamental signal amplitude current at the input, the input resistance increases as the input voltage signal is increased to cause distortion in the form of a narrowed pulsed input current waveform.

As can be seen in Table 13-3, driving a common-base amplifier with a sine wave of 100 mV peak (i.e., 200 mV peak to peak) actually will increase the input resistance by about two-fold. For example, with a DC collector current of 1 mA $= I_{CQ}$, the input resistance of a common base amplifier is 26 Ω for small signals such as a 1 mV peak sine wave AC signal across the emitter-base junction. If the input signal is increased to a 104 mV peak sine wave, the input resistance is increased from 26 Ω to 60 Ω pertaining to the fundamental frequency of the sine wave. In another example, if the $I_{CQ} = 0.1$ mA, the small signal input resistance into the common base amplifier is 260 Ω, and the large-signal input resistance with a 104 mV peak sine wave is 600 Ω.

TABLE 13-3 Muliplying Factor g_m/G_m to Determine the Large-Signal Input Resistance

Input Signal Level V_p	$1/g_m$ @ 1-mA I_{CQ}	$1/G_m$ @ 1-mA I_{CQ}	g_m/G_m for Any I_{CQ}
0.001 is a small signal	26 Ω	26 Ω	1.0
0.013 is a large signal		27 Ω	1.03
0.026 is a large signal		29 Ω	1.12
0.052 is a large signal		37 Ω	1.43
0.104 is a large signal		60 Ω	2.31
0.156 is a large signal		86 Ω	3.29

This table is also applicable to common-emitter amplifiers with the emitter AC grounded just by multiplying the resistance values ($1/g_m$ for small signals or $1/G_m$ for large signals) by β, the current gain. For a common emitter amplifier with $I_{CQ} = 1$ mA, the small-signal input resistance is $\beta/g_m = \beta(26\ \Omega)$ and the large-signal input resistance with a 104 mV peak sine wave is $\beta/G_m = \beta(60\ \Omega)$. For example, if $\beta = 100$, the small and large signal resistances for 1 mV and 104 mV peak sine wave signals are 2,600 Ω and 6,000 Ω, respectively.

So, after a rather lengthy discussion on small- and large-signal effects on transconductance and input resistance, we are now ready to analyze a one-transistor Colpitts oscillator circuit (Figure 13-9A) and its alternative diagram (Figure 13-9B).

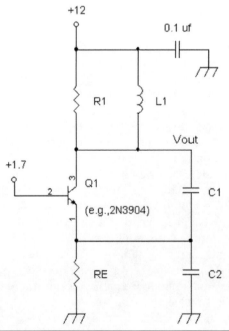

FIGURE 13-9A Colpitts oscillator circuit.

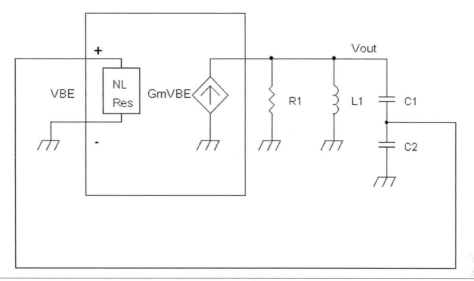

FIGURE 13-9B   Alternative diagram for a Colpitts oscillator.

For the values given where R1 = 2,700 Ω, R_E = 1,000 Ω, C1 = 0.0018 μF, C2 = 0.056 μF, and L1 = 15 μH, the resonant frequency is about 1 MHz, thus providing a 1-MHz signal at the emitter or collector of Q1. With 1.7 volts at the base of Q1, the emitter DC voltage is about 1 volt, which sets up a collector current of 1 mA. The current gain β is greater than 10 (e.g., typically 100 or more). The 1-mA collector current provides a small-signal transconductance g_m of 0.038 mho (amperes per volt).

Since at 1 MHz C2's impedance almost grounds the emitter or Q1, the initial gain of the system is $(g_m RL)(1/n)$, where $1/n$ is the step-down ratio C1/(C1 + C2) = 1/31 for this example. Therefore, the initial gain of the system A_v = 0.038 × 2,700/31 = 3.2. Normally, the initial small-signal gain of this type of oscillator is between 2.2 and 4. A higher initial small-signal gain can be set, but at the expense of generating more distortion with a larger-amplitude output voltage.

From the initial small-signal gain of A_v = 3.2, a reasonable prediction can be made as to what the input amplitude will be. It turns out that $1/A_v$ = 1/3.2 = 0.312 is also approximately G_m/g_m that is, in other words,

$$A_v = g_m/G_m$$

From Table 13-2, when G_m/g_m = 0.304, the input voltage is about 156 mV. So we can expect with G_m/g_m = 0.312 that the input signal voltage across C2 will be on the order of about 156 mV.

The output voltage then will be about 156 mV × 0.312 × 0.038 × 2,700 = 5.0 volts peak. Note that 0.312 × 0.038 = G_m, the large-signal transconductance, which is smaller than the small-signal transconductance by a multiplying factor of about 0.312.

The Q of the tank circuit is determined primarily by the load resistor R_L = 2,700; that is,

$$Q \approx 2\pi f(R_L)C1 = 30$$

The actual capacitance is really smaller because C1 is in series with C2. But because C2 >> C1, C1 can be used in the calculation. So the Q of the circuit is about 30 (give or take 10 percent).

In general, the actual output will be smaller than the predicted output voltage value because the inductor has losses and the transistor includes series emitter and base resistors that reduce its transconductance. For example, with the one-transistor oscillator with the values R1 = 2,700 Ω, R_E = 1,000 Ω, C1 = 0.0018 μF, C2 = 0.056 μF, and L1 = 15 μH, the actual output was smaller by at least 30 percent. That is, instead of the 5.0 volts peak, fewer than 3.5 volts was provided with a standard fixed inductor and a 2N3904 transistor. It turns out that at 1 MHz, the 15-μH inductor is very lossy and thus has a low Q that lowers the output voltage. By maintaining A_v = 3.2 by using the same values for R1 = 2,700 Ω and R_E = 1,000 Ω and raising the frequency to about 3 MHz by dividing the capacitor values by 10 (i.e., C1 = 180 pF and C2 = 0.0056 μF) and replacing the 2N3904 transistor with a 2N4401, the output increased to about 4.25 volts peak, which is within about 15 percent of the calculated 5.0 volts peak.

It should be noted that the input impedance at the emitter of the transistor is often low enough to cause additional signal loss, which thereby lowers the loop gain of the oscillator. This extra lowering of the loop gain results in a lower oscillator output signal. Normally, C2 is made sufficiently large to load into the emitter with minimal signal loss.

Differential Pair Oscillator

Figure 13-10 shows a differential-pair oscillator circuit that is similar to the ones shown in Figures 9-2 and 9-4 of Chapter 9. Essentially, this circuit has the same elements as the one-transistor oscillator circuit with the exception of an extra transistor to "buffer" the signal from the step-down capacitive voltage divider circuit that results in $1/n$ step-down ratio = C1/(C1 + C2).

The advantage of the second transistor Q2 is that the input resistance is at least a factor of β higher than the input resistance to a common-base amplifier, which has an input resistance of $1/g_m$ for small signals or $1/G_m$ for large signals.

The actual small- and large-signal input resistance at the input of Q2 is = $2\beta/g_m$ and $2\beta/G_m$, respectively. The reason for the factor of 2 is that the input resistance of Q1 at its emitter acts as a local feedback (degeneration) resistor for Q2. Normally, the input resistance to a common-emitter amplifier is β/g_m. Intuitively, the reason for having twice the input resistance is that only half the V_{in} signal is across the base-emitter junction of Q2, thereby resulting in half the current drain into the base of Q2, which translates into twice the input resistance.

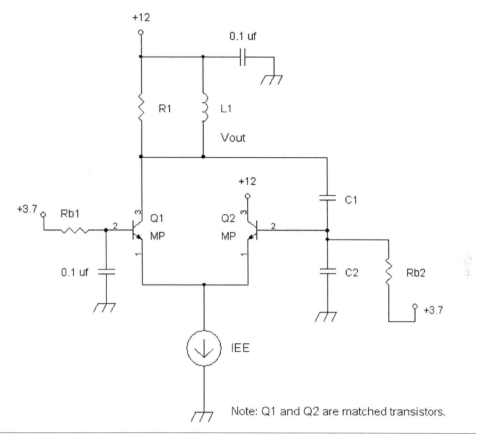

FIGURE 13-10   Differential-pair oscillator circuit.

In Figure 13-10, the quiescent currents of Q1 and Q2 are equal; thus the small-signal input resistance at the emitter of Q1 is $1/g_m$, and the output resistance of emitter follower Q2 is also $1/g_m$. From Figure 13-11A, Q2 is modeled as a unity-gain amplifier with an output resistance of R_out = $1/g_{mQ2}$.

But the input resistance to the emitter of Q1 is also $1/g_{mQ1}$. So the signal voltage at the emitter of Q1 is just one-half V_{in}, where V_{in} is the signal from the capacitive divider circuit C1 and C2. Therefore, the small-signal transconductance from V_{in} at the base of Q2 to the collector output of Q1 is the following:

$$\Delta I_{C1}/\Delta V_{in} = \frac{1}{2}g_{mQ1}$$

where $g_{mQ1} = I_{CQ1}/0.026$ volt and I_{CQ1} is the DC collector current of Q1. Because the current gain β is generally high, the emitter and collector currents of each of the transistors are equal. However, the summed equal emitter or collector currents of Q1 and Q2 equal the emitter tail current IEE, that is,

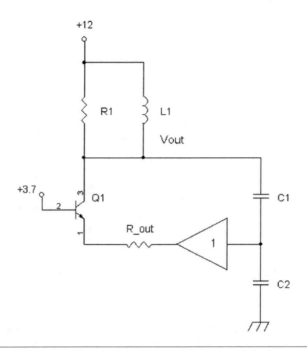

FIGURE 13-11A Differential-pair oscillator with Q2 modeled as a unity-gain amplifier with output resistance R_out.

$$I_{CQ1} = I_{CQ2} = (\tfrac{1}{2})IEE$$

by various substitutions,

$$\Delta\, I_{C1}/\Delta V_{in} = \tfrac{1}{2} I_{CQ1}/0.026 \text{ volt} = \tfrac{1}{4}IEE/0.026 \text{ volt} = \text{small-signal transconductance of the differential-pair amplifier}$$

And the small-signal loop gain for the oscillator circuit is

$$R1 \times \frac{C1}{C1 + C2} \times \frac{1}{2}(I_{CQ1}/0.026 \text{ volt}) = R1 \times \frac{C1}{C1 + C2} \times \frac{1}{4}(IEE/0.026 \text{ volt})$$

Usually the small-signal loop gain is set for 2.5 or more to ensure reliable oscillation.

Now let's take a look at the large-signal characteristics of a differential-pair circuit (Figure 13-11B). The large-signal equation for the collector current of Q1 is as follows for large values of β (e.g., >100):

$$I_{C1\_large signal} = IEE[e^{(VB1 - VB2)}] / [1 + e^{(VB1 - VB2)}]$$

With sinusoidal signals of various amplitudes at the base of Q1 in Figure 13-11B, the output current at the collector of Q1 is shown in Figure 13-12. Note in Figure 13-11B that the base of Q2 is grounded AC signal-wise.

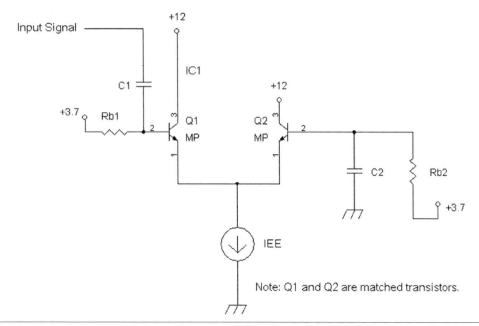

FIGURE 13-11B   A differential-pair amplifier that can be used for the oscillator.

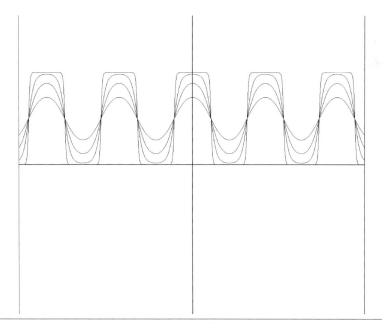

FIGURE 13-12   Q1 collector output current for various amplitude inputs at the base of Q1. The X axis is time, and the Y axis is the output current.

In Figure 13-12, the input levels at the base of Q1 are peak sinusoidal input levels of 26 mV, 52 mV, 104 mV, and 416 mV, respectively. At 26 mV peak, the output current still looks low in distortion, with about 2 percent third-order harmonic distortion. But as the amplitude input level is increased from 52 mV to 416 mV peak, we see that the output current increases in distortion. Distortion is seen at 104 mV, with a soft clipping waveform changing to hard clipping resembling a square wave at 416 mV at the input.

As a result of this distortion, the fundamental frequency signal current at the collector of Q1 will compress as the input signal is increased. So, although the distortion is symmetric compared with the distortion of a single-ended transistor amplifier, as shown in Figure 13-10, there is also a drop in transconductance in a differential-pair transistor amplifier as the input signal is increased.

Table 13-4 shows the approximate* drop-off in large-signal transconductance as a function on input voltage across the bases of the differential amplifier. Note that the large-signal transconductance is output-signal current of the fundamental frequency divided by the input-signal voltage of the fundamental frequency.

TABLE 13-4   Ratio of Large- to Small-Signal Transconductance as a Function of Sinusoidal Input Level

Input Signal (Peak Sinusoidal Voltage)	G_m/g_m for Any IEE
0.001	1.000
0.026	0.950
0.052	0.825
0.078	0.675
0.104	0.560
0.130	0.475
0.156	0.410
0.182	0.360
0.208	0.310
0.234	0.280
0.260	0.250
0.286	0.230
0.312	0.210
0.338	0.195
0.364	0.185

*As read off a graph from *Communication Circuits: Analysis and Design,* by Kenneth K. Clarke and Donald T. Hess (Addison-Wesley, 1971).

For designing a 3-MHz differential-pair oscillator as seen in Figure 13-10, consider the following: Given that L1 = 15 µH, C1 = 180 pF, C2 = 1,620 pF, and Q = 10, find R1, the emitter tail current IEE and voltage output at the collector of Q1. Note that C1/(C1 + C2) = 1/10.

The series capacitance of C1 and C2 results in 162 pF. And 162 pF with the 15-µH inductor L results in a resonant frequency of 3.23 MHz. Therefore, Q = 2π(3.23 MHz) R_L (162 pF) = 10. Solving for R1 yields R1 = 3,000 Ω for the closest 5 percent–valued resistor.

First, pick an initial small-signal loop gain of 3 to solve for the tail current IEE:

$$3 = R1 \times \frac{C1}{C1 + C2} \times \frac{1}{4}(IEE/0.026 \text{ volt}) = 3,000 \times \frac{1}{10} \times (IEE/0.104)$$

where IEE = 1.04 mA ≈ 1 mA. Because the initial small-signal loop gain = g_m/G_m = 3,

$$G_m/g_m = 0.33 \approx 0.310$$

From Table 13-4, this results in about 208 mV across the base of Q2. And because the capacitive voltage divider network C1 and C2 attenuates the collector signal voltage from Q1 by 10, the output voltage at Q1's collector must be 10 × 208 mV peak, or about a 2-V peak sinusoidal waveform, and thus

$$V_{out} = 2 \text{ volts peak (at the collector of Q2)}$$

Actually, the output voltage can be calculated via Table 13-4 without directly knowing the tail current IEE when the small-signal loop gain is chosen or established. Since Q2 has a much higher impedance input resistance at the base of Q2 compared to the input resistance at the emitter of the Colpitts oscillator, loading effects from Q2 are ignored because the equivalent large-signal input resistance is $2\beta/G_m$ and this large signal resistance is further multiplied by n^2 or 100 in this example. Note that $1/n$ = C1/(C1 + C2) = (180 pF)/(180 pF + 1620 pF) = 1/10, or n = 10, and n^2 = 100. Just for "fun," if β = 100, then the small-signal input resistance is $2\beta/g_{mQ1}$. Then

$$g_{mQ1} = \frac{1}{2}IEE/0.026 \text{ volt}$$

where IEE = 1 mA. Thus g_{mQ1} = 0.0192 mho.

So $2\beta/g_{mQ1}$ = 10,000 = Ω small signal–wise. Large signal–wise because g_m/G_m = 3,

$$2\beta/G_{mQ1} = 3 \times 10,000 \text{ Ω} = 30,000 \text{ Ω}$$

The equivalent large-signal resistance loading the tank circuit across L1 then is

$$n^2 \times 30 \text{ kΩ} = 100 \times 30 \text{ kΩ} = 3 \text{ MΩ}$$

which is insignificant to R1 = 3 kΩ in terms of loading the LC tank circuit.

Unless inductor L1 is a lossless (ideal) inductor, when the oscillator circuit is built based on the equations, a smaller-amplitude oscillation signal will be generated in the range of 10 to 25 percent lower than the calculated amplitude. For example, if the 15-μH L1 inductor is not ideal or lossless and has an unloaded Q of 100, that would be like paralleling a 30-kΩ resistor with R1 = 3,000 Ω, which results in a 2,700-Ω resistor in parallel with L1. Also, the transistors Q1 and Q2 each have internal base series resistors and internal emitter series resistors that cause a lowering of transconductance from the small-signal equation such as (½) IEE/0.026. A circuit was constructed, and the output voltage was about 1.65 volts peak or low by about 18 percent.

References

1. Class notes EE140, Robert G. Meyer, UC Berkeley, Fall 1975.
2. Class notes EE240, Robert G. Meyer, UC Berkeley, Spring 1976.
3. Paul R. Gray and Robert G. Meyer, *Analysis and Design of Analog Integrated Circuits*. New York: John Wiley & Sons (any edition, e.g., Third Edition 1993)
4. Kenneth K. Clarke and Donald T. Hess, *Communication Circuits: Analysis and Design*. Reading: Addison-Wesley, 1971.

Chapter 14

Mixer Circuits and Harmonic Mixers

An idealized mixer is a multiplier circuit that takes two signals and literally multiplies the values of each signal to provide an output. In one sense, this concept is easier to understand with multiplication of numbers. For example, there are two analog signals, each connected to an analog-to-digital converter (ADC). The digital outputs from the two ADCs then provide numbers from each signal, and these numbers are updated over time. If the outputs of the ADCs are fed to a digital multiplier, then the output of the digital multiplier will provide a stream of numbers that are the product of the two streams of numbers from the outputs of the ADCs. The output of the digital multiplier then is connected to a digital-to-analog converter (DAC) to provide an analog signal that represents the product of the two analog signals.

But back in the days of the superheterodyne radio in the late teens of the twentieth century, multiplication, or an approximation of it, had to be done all in the analog domain. Therefore, vacuum tubes such as triodes and pentodes were used commonly to provide an analog multiplication effect of two signals such as the radio-frequency (RF) signal and the local oscillator signal. For the triode, which is a very linear amplification device, a very large local oscillator voltage was supplied to its cathode or grid to essentially turn the triode on and off to cause a chopping effect on the RF signal such that the output signal at the plate of the triode provided an intermediate-frequency (IF) signal.

Pentodes were handled similarly to triodes in terms of RF mixing, with the oscillator signal and RF signal combined in the control grid. The pentode had a screen grid, and transconductance depended on the voltage supplied to the screen grid. Usually, the screen grid was connected to a direct-current (DC) source to set a fixed transconductance. However, some mixer designs connected the RF signal to the control grid, whereas the local oscillator's signal was connected to the screen grid to modulate the transconductance of the pentode. The modulated transconductance then provided a multiplying effect of the RF signal.

But other analog multiplying methods and circuits also were used in the early years of radio. For example, diodes could be used as mixers or as part of a more complex mixer circuit such as a double-balanced modulator. And in particular, in the late 1920s or by 1930, switch-mode mixers using solid-state copper oxide diodes were used in commutating switch-mode mixers or balanced modulators.

Mixers can be viewed in two ways:

1. Two (or more) signals are applied to a nonlinear device to generate distortion and thus provide an IF signal.
2. Or more commonly, mixers are symbolized as multipliers that generate an output signal that is the product of two (or more) input signals.

Thus the objective for this chapter is to cover analog mixer circuits. Switch-mode mixers will be explained in Chapter 15.

Adding Circuits Versus Mixing Circuits

Analog circuits that are adders will sum or combine two signals without distortion signals at the output. The output is just the summation of the two input signals. Figure 14-1 shows various examples of summing circuits. The top circuit is a simple two-resistor signal-combining network. And the bottom circuit shows an operational amplifier (op amp) summing circuit that has an output of

$$\text{Vout} = -[\text{Vin\_1}\,\frac{\text{Rfeedback}}{\text{Rin\_1}} + \text{Vin\_2}\,\frac{\text{Rfeedback}}{\text{Rin\_2}}]$$

Mixing circuits, however, are nonlinear in nature. When two signals are combined into a mixer circuit, either by modulation and/or by nonlinearities of the mixer circuit, signals at the output of the mixer contain distortion products related to harmonics and intermodulation of the two signals (Figure 14-2).

First, let's look at harmonic distortion. Harmonic distortion simply means that the output of the mixer generates a signal that is at least one multiple (2, 3, 4, . . .) of the input frequency. So, for example, a junction field-effect transistor (FET) has approximately a square-law relationship of the following:

$$I_{out} = I_D = I_{DSS}(1 - V_{gs}/V_p)^2$$

where V_{gs} is the gate-to-source voltage, V_p is the cutoff voltage of the FET, and is a constant based on DC current when the gate-to-source voltage $V_{gs} = 0$. From expanding the equation, we get

$$I_{out} = I_{DSS}[1 - 2(V_{gs}/V_p) + (V_{gs}/V_p)^2] \qquad (14\text{-}1)$$

Note that the first term in the brackets is 1, a DC current term, whereas the second term in the brackets is a linear-amplification term, $-2(V_{gs}/V_p)$ where V_{gs}

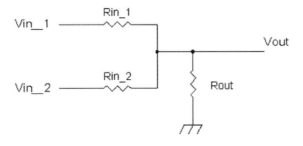

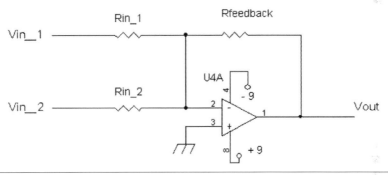

FIGURE 14-1   Examples of summing circuits.

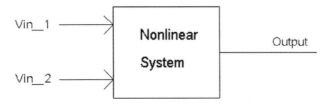

FIGURE 14-2   A mixer system with nonlinearities that generate distortions.

contributes to the output current I_{out}. However, the third term, $+(V_{gs}/V_p)^2$, squares the input signal across the gate and source of the FET.

Now let's take a look graphically at what squaring a signal means (Figure 14-3). The input signal is an alternating-current (AC) signal that has a range of positive and negative voltages, but the squared input signal is nonnegative, which includes 0 and positive voltages only and at twice the frequency of the input. As one can see, the squaring of the input signal results in a second harmonic signal.

So what happens if the input signal is a combination of two signals of different frequencies? Figure 14-4 shows two "high" frequency signals with frequencies F1 and F2 in the lower portion that are just slightly different in frequency. The smaller-amplitude high-frequency signal (with a frequency of F1) is lower in frequency than the higher-amplitude high-frequency signal (with a frequency of F2). When these two signals are added (summed) and squared, the resulting waveform is seen at the top of the figure and resembles an amplitude-modulated (AM) waveform. The envelope of the signal at the top of the figure has a frequency that is the difference of the two high-frequency signals with a frequency of F2 – F1.

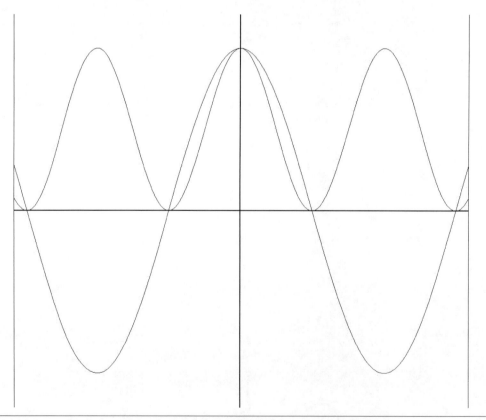

FIGURE 14-3   Squaring a sinusoidal signal produces a twice-frequency component. The X axis shows time, whereas the Y axis shows the amplitude.

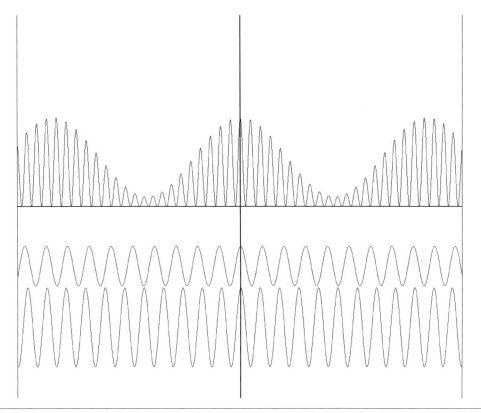

FIGURE 14-4   Two "high" frequency signals that are combined and squared, resulting in a modulating waveform. The X axis shows time, whereas the Y axis shows the amplitude.

So, clearly, a nonlinear device causes a modulation effect, which, in turn, causes a mixing effect. If we just substitute two sinusoidal signals of frequencies F1 and F2 into a squaring function, we get the following:

$$[\cos(2\pi F1t) + \cos(2\pi F2t)]^2 = [\cos(2\pi F1t)]^2 + [2\cos(2\pi F1t)\cos(2\pi F2t)] + [\cos(2\pi F2t)]^2$$
$$(14\text{-}2)$$

Note that the first and third terms, $[\cos(2\pi F1t)]^2$ and $[\cos(2\pi F2t)]^2$, will generate second harmonics of each of the two input signals, *and more important, the middle term shows multiplication of the two input signals.*

From a trigonometric identity, $2[\cos(\alpha)][\cos(\beta)] = \cos(\alpha + \beta) + \cos(\alpha - \beta)$ and in particular, that middle term

$$[2\cos(2\pi F1t)\cos(2\pi F2t)] = \cos(2\pi(F1 - F2)t] + \cos[2\pi(F1 + F2)t]   (14\text{-}3)$$

Note that the squaring function provides both a difference-frequency term (F1 – F2) and a summing-frequency term (F1 + F2). And it should be noted that signals that carry the difference- and summing-frequency terms are intermodulation distortion products or signals.

And in superheterodyne receivers, the IF signal from the mixer has a difference frequency such as (F1 – F2), where F1 is the frequency of the local oscillator, and F2 is the frequency of the input RF signal.

Distortion Can Be a Good Thing (for Mixing)

We have seen how an FET with an approximate square-law characteristic can act as a linear summer and as a mixer/multiplier. This square-law characteristic generates distortion in terms of harmonic and intermodulation distortion products at the drain of the FET.

FETs are used as mixers, and often to provide good mixing action. Typically the local oscillator's voltage (i.e., combined with the RF signal) is in the amplitude level of volts peak to peak. The reason for this is that the FET is such a linear device that huge amounts of input voltage are required to drive the FET into "usable" distortion that provides adequate levels of the IF signal.

Single-Bipolar-Transistor Distortion

Now suppose that we would like to use a bipolar transistor for mixing. As mentioned previously, a bipolar-transistor amplifier with its emitter grounded AC-wise has an exponential function:

$$I_c = I_s e^{VBE/(\frac{KT}{q})} = \text{collector current}$$

where VBE is the total voltage across the base emitter junction of the transistor, which includes a DC bias voltage V_{BEQ} plus the AC signal voltage V_{sig}. Thus

$$VBE = V_{BEQ} + V_{sig}$$

The collector current can be expressed equivalently as

$$I_c = I_s e^{(V_{BEQ} + V_{sig})/(\frac{KT}{q})} = I_c = (I_s e^{V_{BEQ}/(\frac{KT}{q})})(e^{V_{sig}/(\frac{KT}{q})}) \qquad (14\text{-}4)$$

The DC or quiescent collector current is just

$$(I_s e^{(V_{BEQ}/(\frac{KT}{q}))}) = I_{CQ} \quad \text{or} \quad I_c = I_{CQ}(e^{(V_{sig})/(\frac{KT}{q})}) \qquad (14\text{-}5)$$

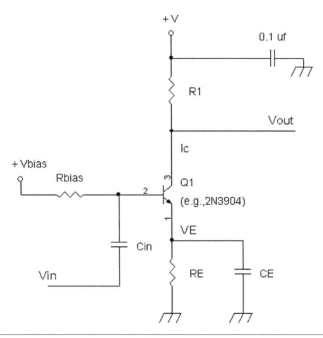

FIGURE 14-5 **A common emitter amplifier.**

Figure 14-5 presents an example of a common emitter amplifier. In the figure, a common emitter amplifier with $\beta > 100$, the DC emitter current is essentially equal to the DC collector current. Thus $V_E/R_E = I_{CQ}$.

Note that $(e^{V_{sig}/(KT/q)})$ is a modulating factor of the DC quiescent collector current and represents the AC signal or time-varying signal in the collector current.

So what about the "mysterious" term $(e^{V_{sig}/(KT/q)})$? What is that equal to? In Chapter 13 with the introduction to e, it was stated that $e = 2.71828\ldots = e^1$. And one way to calculate e is

$$e^1 = 1 + 1^1/1 + \{1^2/[(2)(1)]\} + \{1^3/[(3)(2)(1)]\} + \cdots + [1^n/(n!)] = 2.71828\ldots$$

And

$$e^x = 1 + x^1/1 + \{x^2/[(2)(1)]\} + \{x^3/[(3)(2)(1)]\} + \cdots + [x^n/(n!)] \qquad (14\text{-}6)$$

where $n!$ is not yelling out n with great loudness (a high school joke), but $n!$ is n factorial:

$$n! = (n)(n-1)(n-2)\ldots(1)$$

For example,

$$4! = (4)(3)(2)(1) = 24$$

Also, in general,

$$e^{x/a} = 1 + x^1/[(a^1)(1)] + \{x^2/[(a^2)(2)(1)]\} + \{x^3/[(a^3)(3)(2)(1)]\} + \cdots + \{x^n/[(a^n)(n!)]\}$$
(14-7)

The preceding general equation is rather daunting. But generally, we are concerned only with a few or at most the first four terms of this equation. If more terms are to be calculated, there are computers to do that!

Now let's use the preceding general equation to figure out what happens when

$$x = V_{sig}$$

and

$$a = \frac{KT}{q} = KT/q = 0.026v$$

with v in units of volts.

And we will just calculate the first three or four terms:

$$e^{V_{sig}/0.026v} = 1 + (38/(v)V_{sig} + (740/(v^2)]V_{sig}^2 + (9467/(v^3)]V_{sig}^3 + \cdots$$
(14-8)

By using Equation (14-5),

$$I_C = I_{CQ}[e^{V_{sig}/(KT/q)}]$$

this then leads to the collector current being equal to

$$I_C = I_{CQ}[1 + (38/v)V_{sig} + (740/v^2)V_{sig}^2 + (9,467/v^3)V_{sig}^3 + \cdots]$$
(14-9)

The equation for the collector current shows that there is a squared term and a cubic term that will (at least) provide distortion of the second and third orders. From Equation (14-1) on the FET's square-law characteristic, a modulation effect on two signals occurs with the squared term. Since the bipolar transistor also includes a squared term, we can expect that the bipolar transistor will generate a modulation effect with two input signals as well. And this modulation effect is what we are looking for in a single-bipolar-transistor mixer.

It should be noted that the generalized Equation (14-9) is really only valid for peak sinusoid waveforms of about 26 mV or less. At input signals of 26 mV peak, the equation holds up very well in determining harmonic distortion. Higher-amplitude sinusoidal waveforms of 50 mV or more peak will cause gain compression, and predicting the harmonic distortion will become inaccurate. Therefore, for the higher amplitude signals, modified Bessel functions are used instead of the power series expansion in equation (14-9).

Before moving on to how harmonic and intermodulation distortion is calculated, we should review Equation (14-9):

$$I_C = I_{CQ}[1 + (38/v)V_{sig} + (740/v^2)V_{sig}^2 + (9{,}467/v^3)V_{sig}^3 + \cdots] \qquad (14\text{-}9)$$

Spreading out the terms, we get

$$I_C = I_{CQ}1 + I_{CQ}(38/v)V_{sig} + I_{CQ}(740/v^2)V_{sig}^2 + I_{CQ}(9{,}467/v^3)V_{sig}^3 + \cdots$$

$$(14\text{-}10)$$

Now let's look at the first four terms and see what they represent.

1. $I_{CQ}1 = I_{CQ}$ is DC collector current or bias current.
2. $I_{CQ}(38/v)V_{sig} =$ is the linear amplification term. The small-signal transconductance $g_m = I_{CQ}(38)$. For example, at 1 mA of I_{CQ}, the DC collector current results in 38 mA/V for the small-signal transconductance. And 100 µA of I_{CQ} results in 3.8 mA/V of small-signal transconductance.
3. $I_{CQ}(740/v^2)V_{sig}^2$ is the square-law term that generates a DC offset plus second-order harmonic and intermodulation distortion. Second-order intermodulation distortion from two signals of frequencies F1 and F2 results in output signals with frequencies of (F1 + F2) and (F1 – F2).
4. $I_{CQ}(9{,}467/v^3)V_{sig}^3$ is the cubic term, which produces third-order harmonic and intermodulation distortion. For two signals at the input of F1 and F2, the third-order intermodulation signals are more complicated than the second-order intermodulation components. The third-order intermodulation distortion products for the two input signals will generate signals that have frequencies of (2F1 – F2), (2F2 – F1), (2F1 + F2), and (2F2 + F1).

To calculate distortion, such as second- or third-order distortion, V_{sig} is set to a single or multiple sinusoidal signals. For mixing purposes, there are generally two signals at the input of the mixer, the RF signal and the local oscillator signal. Thus, for now, we will not concern ourselves with harmonic distortion but just concentrate on second-order intermodulation distortion because it is this distortion product that generates the difference-frequency signal (F1 – F2).

Consider the input two sinusoidal waveforms from Equation (14-2):

$$[\cos(2\pi F1t) + \cos(2\pi F2t)]^2 = V_{sig}^2 = [\cos(2\pi F1t)]^2 +$$
$$[2\cos(2\pi F1t)\cos(2\pi F2t)] + [\cos(2\pi F2t)]^2 \qquad (14\text{-}2)$$

Let's generalize the amplitudes of each input signal with A_1 and A_2 such that

$$V_{sig} = A_1\cos(2\pi F1t) + A_2\cos(2\pi F2t) \qquad (14\text{-}11)$$

which then yields the square-law term:

$$[A_1 \cos(2\pi F1t) + A_2 \cos(2\pi F2t)]^2 = V_{sig}^2 = [A_1 \cos(2\pi F1t)]^2 +$$
$$[2A_1 A_2 \cos(2\pi F1t) \cos(2\pi F2t)] + A_2 \cos(2\pi F2t)]^2 \quad (14\text{-}12)$$

Because we are interested only in the "modulated" or multiplying term between the two signals, the second term is of interest, which then reduces to

$$[2A_1 A_2 \cos(2\pi F1t) \cos(2\pi F2t)] = A_1 A_2 \cos[2\pi(F1 - F2)t] +$$
$$A_1 A_2 \cos[2\pi(F1 + F2)t] \quad (14\text{-}13)$$

So now, if we return to Equation (14-10) for just the first three terms, we have

$$I_C = I_{CQ}1 + I_{CQ}(38/v)V_{sig} + I_{CQ}(740/v^2)V_{sig}^2$$

and if we substitute for the two signals at the input for V_{sig} and refer to equations (14-11) and (14-12), we have

$$I_C = I_{CQ}1 + I_{CQ}(38/v)[A_1 \cos(2\pi F1t) + A_2 \cos(2\pi F2t)] + I_{CQ}(740/v^2)\{[A_1 \cos(2\pi F1t)]^2$$
$$+ [2A_1 A_2 \cos(2\pi F1t) \cos(2\pi F2t)] + [A_2 \cos(2\pi F2t)]^2\}$$

Removing or ignoring the second-order harmonic distortion terms $[A_1 \cos(2\pi F1t)]^2$ and $[A_2 \cos(2\pi F2t)]^2$, we get

$$I_C = I_{CQ}1 + I_{CQ}(38/v) [A_1 \cos(2\pi F1t) + A_2 \cos(2\pi F2t)] +$$
$$I_{CQ}(740/v^2)[2A_1 A_2 \cos(2\pi F1t) \cos(2\pi F2t)]$$

And using Equation (14-13) for substitution, we arrive at

$$I_C = I_{CQ}1 + I_{CQ}(38/v) [A_1 \cos(2\pi F1t) + A_2 \cos(2\pi F2t)] +$$
$$I_{CQ}(740/v^2)\{A_1 A_2 \cos[2\pi(F1 - F2)t] + A_1 A_2 \cos[2\pi(F1 + F2)t]\} \quad (14\text{-}14)$$

So what does Equation (14-14) really mean? For determining the intermodulation distortion products and the difference-frequency term, this equation is really accurate for signals of less than about 13 mV peak. Any larger signals will start to include errors, and these errors will increase as the input signal's amplitude rises above 26 mV peak.

Let's try the following scenario: The signal $A_1 \cos(2\pi F1t)$ is the oscillator signal with $A_1 = 0.013$. A_2 is the RF input signal with $A_2 < 0.013$ volt (e.g., typically 5 mV or less). The second-order intermodulation distortion IM_2 as a function of one of the input signals is

$$IM_2 = (I_{CQ}740/I_{CQ}38)A_1 A_2/A_2 = 19A_1 = 19(0.013) = 0.25 = 25 \text{ percent} = 25\%$$
$$(14\text{-}15)$$

Note: $(I_{CQ}740/I_{CQ}38) = (I_{CQ}/I_{CQ})(740/38) = 1(740/38) = 19$

But what does this 25 percent of IM_2 distortion mean? This means that the IM_2 distortion components have an equivalent transconductance of 25 percent of the small-signal transconductance. So, if the transistor is biased at 1 mA, the transconductance for generating the sum- or difference-frequency signal is $25\%(g_m = 25\%(0.0384)$ A/V, or 9.5 mA/V.

So, for example, for an oscillator signal of 0.013 volt peak sinusoidal waveform added or combined with a small-signal RF signal V_{RF} into the base and emitter junction of a bipolar transistor amplifier operating at 1 mA, the IF signal current I_{sig_IF} with the difference frequency is

$$I_{sig\_IF} = V_{RF} \times 25\%(g_m) = V_{RF} \times 9.5 \text{ mA/V}$$

Note that the conversion transconductance is I_{sig_IF}/V_{RF}. And for a 0.013-volt peak sinusoidal oscillator signal into the mixer, the conversion transconductance is $25\%(g_m)$.

Equation (14-15) shows that the conversion gain or IF signal transconductance is controlled by the oscillator's signal amplitude. But it should be noted that increasing the oscillator's peak amplitude A_1 beyond 0.013-volt will raise that 25 percent number to something higher that is no longer linear or proportional. By using other mathematical techniques that are more accurate than the power-series expansion [e.g., Equation (14-8) or (14-9)], it is found that if the oscillator signal is raised to 26 mV peak, or 2×13 mV peak, we get $45\%(g_m)$ instead of $50\%(g_m)$ for the conversion gain.

In previous chapters concerning the superheterodyne radio, it was mentioned that the oscillator voltage into the mixer or converter is in the range of about 200 mV to 300 mV peak to peak or 100 mV to 150 mV peak. The next section will explore a more accurate model for bipolar mixer circuits with large oscillator voltages that are beyond 13 mV peak.

Another technique to more accurately characterize the behavior of a simple transistor mixer is to view the transistor as having a time-varying transconductance that depends on the oscillator signal. Recall that the small-signal transconductance $g_m = I_{CQ}/(0.026 \text{ volt})$. Usually we define a small-signal transconductance as $g_{mQ} = I_{CQ}/(0.026 \text{ volt})$, where I_{CQ} is a constant DC current, which then makes g_{mQ} a constant small-signal transconductance.

But when an oscillator voltage is added with the bias voltage, the collector current is actually time-varying, and thus the transconductance is changing with time in relation to the amplitude and frequency of the oscillator signal. Therefore, the output collector current as a function of time can be thought of as

$$I_c(t) = g_m(t) \times V_{RF}$$

where $g_m(t)$ is a time-varying transconductance with its shape depending on the oscillator amplitude. This shape resembles the collector current of a bipolar transistor with a large overdrive signal, as seen in Figure 14-6. For large signals such as 100 mV

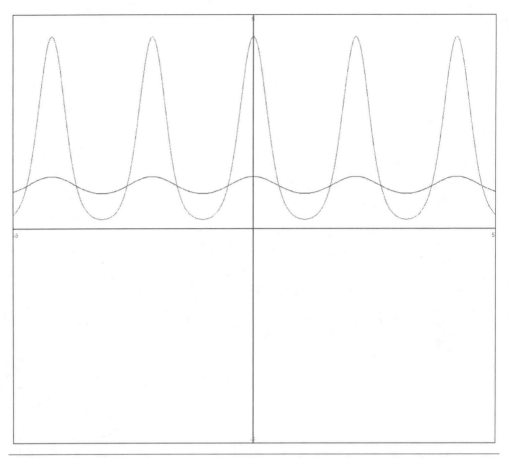

FIGURE 14-6 Collector current of the transistor for various input levels into the base-emitter junction. The *X* axis shows time, whereas the *Y* axis shows the amplitude.

to 200 mV peak to peak, the time-varying transconductance has a shape approaching a pulse. And the narrowing nature of this pulse as larger and larger amplitudes of oscillator signals are combined into a transistor mixer will limit the conversion transconductance.

As seen in the figure, because transconductance is directly related to the collector current, the transconductance is time-varying and has the same pulse "shape" as the oscillator amplitude is increased into the transistor mixer. A topic of mathematics known as *modified Bessel functions* explains the time-varying transconductance of a bipolar transistor mixer. Modified Bessel functions are rather involved, and they generally are taught in senior year or graduate school university classes. I will take a simplified approach instead, just showing the results of the conversion transconductance as a function of oscillator signal injection voltage.

Simple Transistor Mixer and Its Conversion Transconductance

Let me say at this point that I have probably overloaded quite a few readers with equations and trigonometric identities to explain how distortion in amplifiers is used in mixers or modulators. So this next section will detour back to a tabulated approach. After all, the math can be interesting but also can sidetrack the reader.

See Figure 14-7, a simple one-transistor mixer. Figure 14-7 is a simple transistor mixer circuit whereby the oscillator signal is combined with the RF signal to provide typically an IF signal. Table 14-1 lists the tabulated ratios of conversion transconductance to small-signal transconductance as a function of the oscillator signal's peak amplitude.

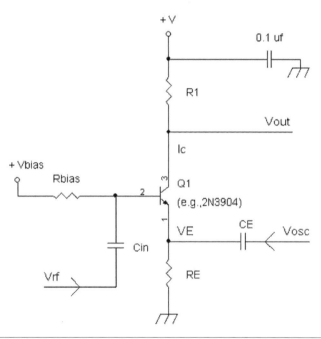

FIGURE 14-7   A simple transistor mixer circuit.

TABLE 14-1 Ratio of Conversion Transconductance to Small-Signal Transconductance for the Sum- or Difference-Frequency (F1 ± F2) Signals as a Function of the Oscillator Signal's Peak Amplitude

Oscillator Signal (Peak Sinusoidal Voltage)	$g_{m\_conversion}/g_{mQ}$ for Any I_{CQ}
0.013	0.242
0.026	0.446
0.052	0.698
0.078	0.810
0.130	0.893
0.182	0.926
0.260	0.949

As can be seen in Table 14-1, there is a law of diminishing returns, and generally, an oscillator signal of at least 100 mV peak to peak will provide sufficient drive. Table 14-2 shows the conversion transconductance at 1 mA DC collector current.

TABLE 14-2 Conversion Transconductance at 1 mA as a Function of Oscillator Drive Voltage

Oscillator Signal (Peak Sinusoidal Voltage)	$g_{m\_conversion}$ @ 1 mA I_{CQ} in mho or S
0.013	0.0093
0.026	0.0171
0.052	0.0268
0.078	0.0311
0.130	0.0343
0.182	0.0356
0.260	0.0364

Since the conversion transconductance $g_{m\_conversion}$ is proportional to the DC collector current, the numbers in the right column of Table 14-2 can be scaled appropriately. For example, to find $g_{m\_conversion}$ at 2 mA, just multiply the numbers by 2, and to find $g_{m\_conversion}$ at 100 μA, just divide the numbers by 10. Again, the values for $g_{m\_conversion}$ are only for output signals that provide frequencies of F1 – F2 or F1 + F2.

For a quick example, suppose that the oscillator injects a 78-mV peak sine wave into the mixer. What would be the conversion gain for an IF tank circuit with an equivalent parallel resistance $R = 150$ kΩ when the DC collector current is 100 μA?

The conversion gain is $g_{m\_conversion} \times R$. At 1 mA for 78 mV peak,

$$g_{m\_conversion} = 0.0311 \text{ mho or } 0.0311 \text{ S}$$

so 100 µA is one-tenth of 1 mA, which leads to

$$g_{m\_conversion} = 0.00311 \text{ mho or } 0.00311 \text{ S}$$

Therefore, the conversion gain is 0.00311 mho $\times$ 150,000 Ω = 466. This would mean that an RF signal is amplified 466 times and converted to an IF signal of generally lower frequency. For example, if the RF signal is 1 mV peak to peak at 600 kHz and the local oscillator provides a 1,055-kHz signal at 78 mV peak with a 100-µA DC collector current for the one-transistor mixer and a 150-Ω load at 455 kHz, the output IF signal at 455 kHz will be 466 mV peak to peak.

Also, it should be noted that because the RF signal is generally small, the worst case or minimum input resistance into the base of the mixer is β/g_{mQ}, which is the small-signal input resistance at the (quiescent) DC collector current.

Differential-Pair Mixer

The differential-pair transistor amplifier is a building block to a true multiplying circuit. A double-balanced multiplier performs the operation of literally multiplying two signals without leaking through any signals from the inputs (e.g., RF signal and/or oscillator signal) to the output. Normally, this "true" multiplying circuit will require at least three or four differential-pair amplifiers. For example, see the data sheets and schematic diagrams to the MC1494 and the MC1495.

For this chapter, we will just look at a simple differential-pair mixer, which will provide a multiplying action but also will leak through the input signals, the RF input signal, and the oscillator signal at its output. However, the input signals generally can be filtered out from the output of a differential-pair mixer.

A differential-pair mixer usually consists of at least three transistors. See Figure 14-8. In Figure 14-8, the oscillator signal will be connected to the base of transistor Q1, and Q3's collector will be providing the modulating current that controls the signal current at the output of Q1 (or Q2). Generally, the top transistors are in a "limiting" mode, which means that the oscillator voltage across the bases of Q1 and Q2 is sufficiently high in amplitude (e.g., >300 mV peak) to cause Q1 and Q2 to generate a square-wave output current. At an amplitude level of about 1 volt peak to peak sine wave to the base of Q1, the output current of Q1 or Q2 does indeed resemble a square-wave signal. Figure 14-9 shows the output waveform of Q1 for various levels at the base of Q1. Note: In practice R1 is replaced with an IF filter.

With the base of Q2 AC grounded, the collector current from Q1 is

$$I_{C1} = (1EE + I_{sig})[e^{(VB1)}/(1 + e^{(VB1)})]$$

where IEE is the DC emitter "tail" current, and I_{sig} is the AC signal current for modulating, which means that $I_{sig} = g_{m3} \times V_{RF} = I_{RF} \sin(2\pi F2)t$.

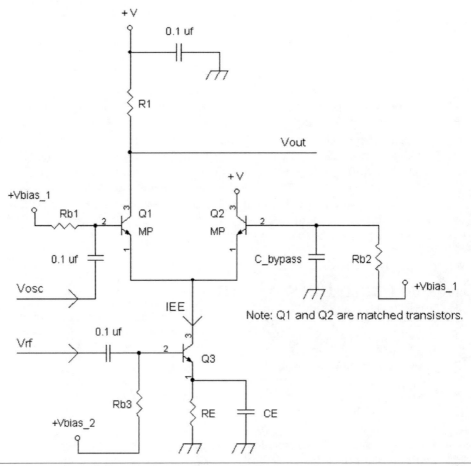

FIGURE 14-8 A differential-pair mixer with three transistors.

If VB1 is a sinusoid signal at a frequency of F1 at about 1 volt peak to peak, then the expression

$$[e^{(VB1)}/(1 + e^{(VB1)})] = \{0.5 + \tfrac{2}{\pi}[\sin(2\pi F1)t + \tfrac{1}{3}\sin(2\pi 3F1)t + \tfrac{1}{5}\sin(2\pi 5F1)t$$
$$+ \tfrac{1}{7}\sin(2\pi 7F1)t + \cdots]\}$$

which approximately represents a square-wave signal. And if $I_{sig} = I_{RF}[\sin(2\pi F2)t]$, then

$$I_{C1} = [IEE + I_{RF}\sin(2\pi F2)t]\{0.5 + \tfrac{2}{\pi}[\sin(2\pi F1)t + \tfrac{1}{3}\sin(2\pi 3F1)t + \tfrac{1}{5}\sin(2\pi 5F1)t$$
$$+ \tfrac{1}{7}\sin(2\pi 7F1)t + \cdots]\} \tag{14-16}$$

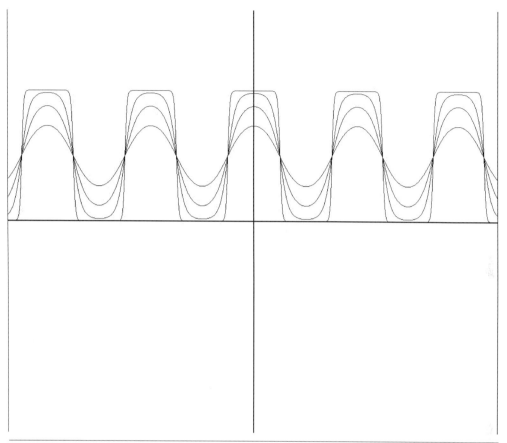

FIGURE 14-9   Q1 collector output for various input sine-wave input levels at 0.026, 0.052, 0.104, and 0.416 V peak. The *X* axis is time, whereas the *Y* axis is the amplitude.

If the harmonics are ignored, we have

$$I_{C1} = [IEE + I_{RF} \sin(2\pi F2)t] \, [0.5 + \tfrac{2}{\pi}\sin(2\pi F1)t] \tag{14-17}$$

Since the RF signal at frequency F2 and oscillator signal at frequency F1 can be filtered out, let's just look at the relevant product term related to the IF signal, which is a signal at frequency F1 – F2 or F1 + F2. Therefore,

$$\begin{aligned}
I_{C1\_IF} &= [I_{RF} \sin(2\pi F2)t][\tfrac{2}{\pi}\sin(2\pi F1)t] = I_{RF}\tfrac{2}{\pi}[\sin(2\pi F1)t][\sin(2\pi F2)t] \\
&= I_{RF}\tfrac{2}{\pi}\tfrac{1}{2}\{\cos 2\pi\,[(F1 - F2)t] - \cos(2\pi)[(F1 + F2)t]\} \tag{14-18} \\
&= I_{RF}\tfrac{1}{\pi}\{\cos 2\pi[(F1 - F2)t] - \cos(2\pi)[(F1 + F2)t]\} \tag{14-19}
\end{aligned}$$

And if the RF signal voltage is small, such as less than 10 mV peak,

$$I_{RF} = g_{m3} \times V_{in}$$

where g_{m3} = IEE/0.026 volt and V_{in} is the amplitude of the RF signal, then

$$I_{C1\_IF} = V_{in} \times g_{m3}\frac{1}{\pi}\{\cos(2\pi)[(F1 - F2)t] - \cos(2\pi)[(F1 + F2)t]\}$$

The conversion transconductance is $g_{m3}(1/\pi)$. So, if IEE = 1 mA, then the conversion transconductance is $0.0384/\pi = 0.0122$ mho.

One should note that the square-wave signal from the differential-pair mixer consists of odd harmonics that are attenuated by a factor of $1/N$, where N is the odd harmonic. So, for example, the third harmonic is one-third the amplitude of the fundamental frequency and the eleventh harmonic is one-eleventh the amplitude of the fundamental frequency. Therefore, the conversion transconductance for mixing with the harmonic of frequency F1 is just multiplied by a factor of $1/N$.

Thus the differential-pair mixer equation (14-16) shows that the RF signal current also multiplies with signals whose frequencies are odd multiples of F1. This then means that mixes with the harmonics of F1 and thus provides harmonic mixing.

Harmonic Mixer Circuits

In the previous two circuits, the one-transistor mixer and differential-pair mixer, we were concerned only with the mixers generating a sum- or difference-frequency signal, with frequency (F1 – F2) or (F1 + F2). But we also have seen that both the single-transistor and differential-pair mixer circuits generate signals with harmonics of the oscillator frequency.

If we define the frequency F1 as F_{OSC} and frequency F2 as F_{RF}, then we can show examples of simple mixing (sum and difference frequencies) and harmonic mixing in Table 14-3.

Now let's return to the one-transistor mixer. We can compare the ratio of conversion transconductance to small-signal transconductance, this time including mixing at harmonics of the oscillator frequency. Table 14-4 shows that the conversion gain for harmonic mixing is almost the same as the conversion gain for simple mixing when the oscillator injection voltage is about 500 mV peak sine wave into the base-emitter junction. However, one downside to having high-amplitude injection voltages

TABLE 14-3 Out-Frequencies for Simple and Harmonic Mixers

Frequencies to Mixer	Simple Mixing	Second-Harmonic Mixing	Third-Harmonic Mixing
F_{OSC} and F_{RF}	$F_{OSC} \pm F_{RF}$	$2F_{OSC} \pm F_{RF}$	$3 \pm F_{OSC} \pm F_{RF}$

TABLE 14-4   Transconductance Ratios for a Single-Transistor Mixer

Oscillator Signal (Peak Sinusoidal Voltage)	$g_{m\_conversion}/g_{mQ}$ Fundamental Frequency	$g_{m\_conversion}/g_{mQ}$ Second Harmonic	$g_{m\_conversion}/g_{mQ}$ Third Harmonic
0.013	0.242	0.030	0.0025
0.026	0.446	0.107	0.018
0.052	0.698	0.302	0.093
0.078	0.810	0.460	0.197
0.130	0.893	0.642	0.379
0.182	0.926	0.736	0.505
0.260	0.949	0.810	0.625
0.520	0.974	0.902	0.779

into the mixer is that more filtering may be required at the mixer's output to filter out all the signals of fundamental and harmonic frequencies related to the oscillator frequency.

So the bottom line is if harmonic mixing is needed in a one-transistor mixer, make sure that the oscillator voltage is at least 130 mV peak into the base-emitter junction. $g_{m\_conversion}/g_{mQ}$ is known from this table and $g_{mQ} = I_{CQ}/0.026$ volt, $g_{m\_conversion}$ can be found. For example, if the oscillator injection voltage is 0.052 V peak into the base-emitter junction, then for the second harmonic of the oscillator signal mixing with the RF signal, we have

$$g_{m\_conversion\_2nd}/g_{mQ} = 0.302$$

and thus

$$g_{m\_conversion\_2nd} = g_{mQ}(0.302)$$

If $I_{CQ} = 1$ mA, then $g_{mQ} = 0.0384$, and then

$$g_{m\_conversion\_2nd} = 0.0384(0.302) = 0.0116 \text{ mho}$$

Now let's take a look at the harmonic conversion transconductance of the differential-pair mixer. For mixing at the fundamental frequency, Equation (14-18) states

$$I_{C1\_IF} = [I_{RF} \sin(2\pi F2)t][\tfrac{2}{\pi}\sin(2\pi F1)t] = I_{RF}\tfrac{2}{\pi}[\sin(2\pi F1)t][\sin(2\pi F2)t] \quad (14\text{-}18)$$

But we can modify this equation for any Nth harmonic by factoring N, which leads to

$$I_{C1\_IF\_N} = [I_{RF} \sin(2\pi F2)t][\frac{2}{\pi}\frac{1}{N}\sin(2\pi NF1)t] =$$
$$IRF \frac{2}{\pi N}[\sin(2\pi NF1)t][\sin(2\pi F2)t] \qquad (14\text{-}19)$$

And this leads to a modification of Equation (14-19):

$$I_{C1\_IF} = I_{RF} \frac{1}{\pi}\{\cos(2\pi)[(F1 - F2)t] - \cos(2\pi)[(F1 + F2)t]\} \qquad (14\text{-}20)$$

And for an Nth harmonic, where N is an odd number, the output signal is

$$I_{C1\_IF\_N} = I_{RF} \frac{1}{\pi N}\{\cos(2\pi)[(NF1 - F2)t] - \cos(2\pi)[(NF1 + F2)t]\} \qquad (14\text{-}21)$$

And thus the conversion transconductance is

$$g_{m3}\frac{1}{\pi N} \qquad (14\text{-}22)$$

where g_{m3} = IEE/0.026 volt.

For example, for harmonic mixing based on the fifth harmonic to provide an output signal whose frequencies are $5_{FOSC} \pm F_{RF}$, the conversion transconductance is

$$g_{m3} \frac{1}{5\pi}$$

where g_{m3} = IEE/0.026 volt, the small-signal transconductance of Q3 in Figure 14-9.

Mixer Oscillator Circuits

At this point of writing this chapter I am glad to leave the world of complicated equations. So back to circuits!

In essence, some oscillator circuits can double as a mixer circuit as well. You have seen these circuits in Chapters 8 and 10, and they are called *converter circuits*. The conversion gain analysis is the same as in the preceding section for a one-transistor mixer. Typically, the self-oscillating voltage is usually set for about 100 mV to 200 mV peak into the base-emitter junction for the converter circuit, so the conversion transconductance for simple mixing will be about the same as the small-signal transconductance.

Figure 14-10 presents a schematic for an oscillator converter circuit. Figure 14-10 shows a common base oscillator that has a low-impedance oscillating voltage fed to the emitter of Q1 via a tap from the oscillator transformer. The base of Q1 is coupled to the RF input signal. It should be noted that the RF signal source should have a low-impedance drive so as to ensure that oscillation occurs reliably. This is why if one

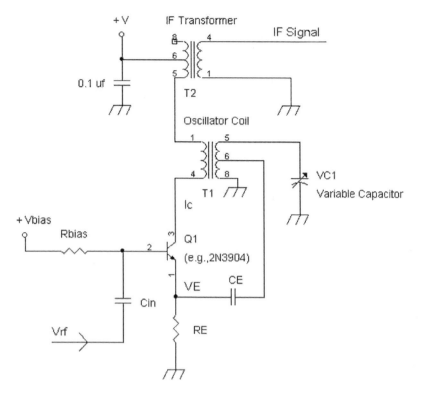

Note: The IF Transformer has an internal capacitor.

FIGURE 14-10   An oscillator mixer circuit or converter circuit.

inspects the converter circuits in Chapter 8 or Chapter 10, the RF signal is stepped down via the secondary winding of the antenna coil. The RF signal's amplitude should be much smaller than the oscillator's signal. For example, if the oscillator's amplitude is 250 mV peak into the base-emitter junction, the RF signal should be under 25 mV peak.

Also note that the RF signal's frequencies should be way beyond the pass band of the oscillator's tank circuit. For example, if a very low IF is desired, such as 10 kHz, the RF signal itself may cause the oscillator frequency to pull over to the RF frequency. Thus, if the oscillator is set at 1,010 kHz and the incoming RF signal is at 1,000 kHz, there can be a problem of oscillator-frequency stability. Therefore, the RF signals whose frequencies are close to the oscillator's frequency can cause the oscillator to lock onto or synchronize with the incoming RF signal via injection locking.

Figure 14-10 also can be used as a harmonic converter. If this is desired, the oscillator's signal should be increased by increasing the DC collector current to provide a "reasonable" conversion gain. This can be done by increasing the bias voltage at the base of Q1 by increasing V_{bias} and/or by decreasing the resistance of emitter bias resistor R_E.

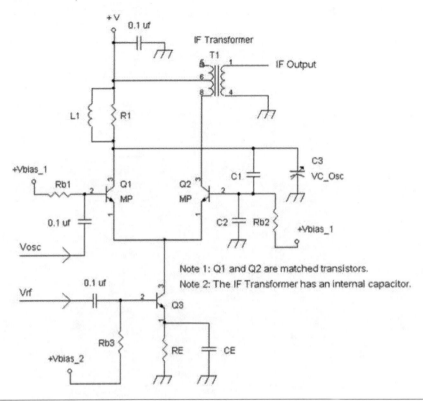

FIGURE 14-11   A differential-pair oscillator mixer circuit.

Figure 14-11 shows a differential-pair oscillator with an emitter tail current source that also amplifies the RF signal source. Thus the emitter current is a combination of a DC current and the amplified RF signal current. The IF signal is extracted via the collector of Q2. This is not as common a configuration for a mixer oscillator circuit compared with the one-transistor circuit in Figure 14-10 but nevertheless is workable. Care should be taken as to the RF level connected to the base of Q3.

If RF gain reduction is needed, an emitter capacitor C_E can be replaced with a series capacitor and resistor combination (RC), and the transconductance of Q3 is lowered.

Conversion Gain

Conversion gain is usually the ratio of the voltage amplitudes of the output IF signal to the input RF signal. When the conversion transconductance is known, the conversion gain is just

$$g_{m\_conversion} \times R1 = \text{conversion gain}$$

where R1 is the equivalent load resistance (usually) in a parallel LC tank circuit. If the mixer's output current is connected to a tapped transformer, then the turns ratio must be taken into consideration. For example, many IF transformers have a turns ratio of 3:1 in the primary winding. If the IF transformer has an equivalent parallel resistance of 200 kΩ, then the resistance at the low-side tap is $(1/3^2) \times 200$ kΩ, or 22.2 kΩ = R1. Also note that the output signal at the secondary winding of the IF transformer will be lower than at the primary winding owing to the step-down ratio.

Before I end this chapter, I offer Table 14-5, which lists various conversion transconductances of different devices for comparison. For the one-transistor mixer, assume that the driving voltage at the base-emitter junction is above 125 mV peak, which translates to a conversion transconductance of about 90 percent of the small-signal transconductance of a bipolar transistor.

TABLE 14-5   Conversion Transconductances of Transistors, Vacuum Tubes, and MOSFETs

Device	Conversion Transconductance
Transistor at 1 mA	0.0345 mho
Transistor at 100 μA	0.00345 mho
Transistor at 10 μA	0.00035 mho
1R5 pentagrid tube at 5 mA	0.00030 mho
12BE6 pentagrid tube at 10 mA	0.00047 mho
40604 dual-gate MOSFET at 10 mA	0.00280 mho

As can be seen in Table 14-5, the two vacuum tubes, 1R5 and 12BE6, which were used in superheterodyne radios, have conversion transconductances very close to that of a bipolar transistor running at a collector current of 10 μA. This fact will give you a hint as to how the ultralow-powered superheterodyne radios of Chapter 9 were designed.

References

1. Class notes EE140, Robert G. Meyer. UC Berkeley, Fall 1975.
2. Class notes EE240, Robert G. Meyer. UC Berkeley, Spring 1976.
3. Paul R. Gray and Robert G. Meyer. *Analysis and Design of Analog Integrated Circuits*. New York: John Wiley & Sons (any edition).
4. Kenneth K. Clarke and Donald T. Hess. *Communication Circuits: Analysis and Design*. Reading: Addison-Wesley, 1971.
5. *RCA Transistor Thyristor and Diode Manual* (Technical Series SC-15), Somerville: RCA Corporation, 1971.

6. Mary P. Dolciani, Simon L. Bergman, and William Wooton. *Modern Algebra and Trigonometry*. New York: Houghton Mifflin, 1963.

7. William E. Boyce and Richard C. DiPrima. *Elementary Differential Equations and Boundary Valued Problems*. New York: John Wiley & Sons, 1977.

Chapter 15

Sampling Theory and Sampling Mixers

After showing a whole host of equations from the last two chapters, I will try a graphical approach to the subject of sampling. However, some equations will have to be shown.

The objectives of this chapter will be the following:

1. To show what sampling is in terms of multiplying a signal
2. To show how sampling can be used for mixing, such as generating an intermediate-frequency (IF) signal
3. To investigate switch-mode mixers

Sampling Signals as a Form of Muliplication or Mixing

A typical sampling circuit is a switch that is turned on for a short duration that momentarily passes an input signal to the output terminal of the switch (Figure 15-1A). In Figure 15-1A, an input signal such as a direct-current (DC) signal or a sine-wave signal is "gated" through for a short duration via a sampling signal. Thus this sampling-switch circuit is sometimes called a *chopper modulator*, which is a single balanced mixer circuit. A single balanced mixer means that only one of the input signals is nulled or removed from the output, but the other input signal appears at the output. In Figure 15-1A, the sampling signal that drives the control signal to the switch does not appear at the output when the input signal is zero. However, the input signal does leak through to the output terminal. And if the input signal is a DC signal, the output of the sampling switch is just a series of pulses where the height or amplitude of each pulse is proportional to the input signal.

So, if the input signal is +1 volt DC, then the output of the sampling circuit will generate a series of pulses that are +1 volt in amplitude. And if the input signal is

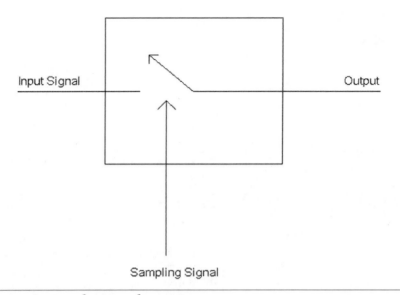

FIGURE 15-1A   Sampling-switch circuit.

–2 volts DC, then the output of the sampling switch will generate a series of negative pulses at –2 volts.

 Unless labeled, all the waveforms have the *X* axis denoting time and the *Y* axis denoting amplitude.

Figure 15-1B shows the output of the sampling-switch circuit when the input signal is +1 volt. Since the sampling switch seems to just convert the input signal into a series of pulses of the same amplitude, we can view the sampling switch in another way that is useful for analyzing the frequency components at the output.

Equivalently, the sampling-switch circuit looks like a 0-volt to +1-volt pulse train that ideally is multiplied by the input signal (Figure 15-1C). The 0-volt to +1-volt pulse train is also known as a *unit pulse train*.

So now let's take a look at a sampling system viewed as a multiplier of the input signal (Figure 15-2). From the figure, everywhere the pulse is +1 volt , that pulse is multiplied by the input sine-wave signal.

Figure 15-3 shows that the input sine wave that has a peak amplitude of 0.5 volt when multiplied by the unit-pulse-train signal results in a pulse train that has a sampled sinusoidal outline or envelope also with a 0.5-volt peak amplitude.

Figure 15-4A shows the outline of the multiplied or modulated pulse-train signal more clearly. For ideal sampling, the widths of the pulses are as narrow as possible and are supposedly close to zero in width. In practice, the pulses have some finite width. For analysis, we will use a finite pulse width for the pulse train.

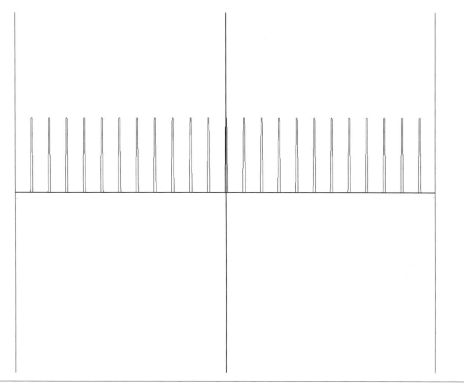

FIGURE 15-1B   Sampling-switch output with a +1-volt DC input signal.

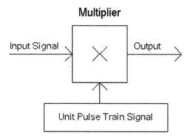

FIGURE 15-1C   An equivalent system that produces the same output signal as the sampling-switch circuit in part A.

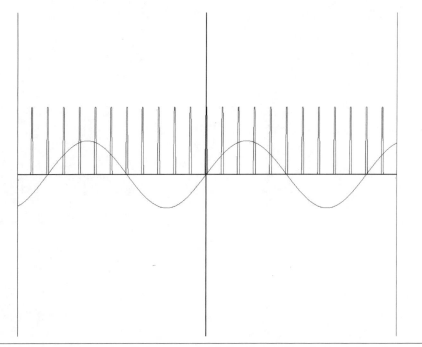

FIGURE 15-2　An input signal sine wave of 0.5 volt peak and the unit pulse train.

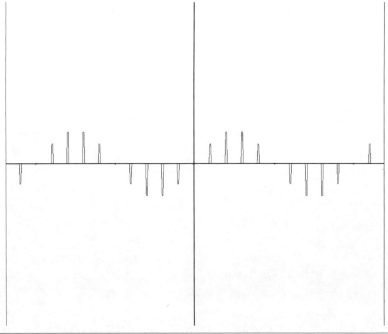

FIGURE 15-3　Unit pulse train multiplied by the input sine-wave signal.

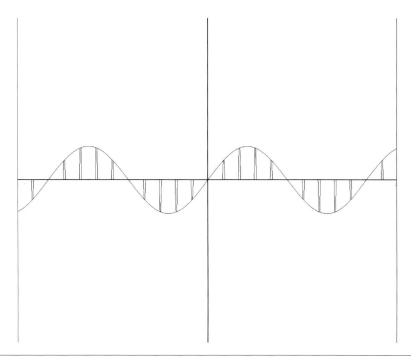

FIGURE 15-4A   Multiplied-unit-pulse-train signal with a "drawn" outline to denote the envelope of the resulting sampled signal.

As stated in Chapter 14 for a single-transistor mixer, we found that increasing the sinusoidal input into the transistor's base-emitter junction causes the collector current to output a narrower and narrower pulse. And we found that the narrower the pulse, the more harmonics were generated. For example, for a transistor mixer, increasing the oscillator's signal level to 0.52 volt peak into the base-emitter junction causes the second and third harmonic signal amplitudes to almost match the amplitude of the fundamental. See Table 14-4, where the conversion transconductances of the second and third harmonics are within 20 percent of the fundamental oscillator frequency for a 0.52-volt peak sinusoidal waveform.

Thus we can surmise that a pulse train of narrow pulses has frequency components consisting of harmonics that are almost the same amplitude as the fundamental. Also, a positive-going pulse train also must have an average DC term.

A periodic pulse-train signal, where one of the pulses is an even function that is centered at the origin at $t = 0$, can be written as

$$P(t) = DC + a_1 \cos(2\pi f_s t) + a_2 \cos(2 \times 2\pi f_s t) + a_3 \cos(3 \times 2\pi f_s t) + \cdots + a_N \cos(N \times 2\pi t f_s t) \qquad (15\text{-}1)$$

where N is a positive integer, f_s is the sampling frequency in hertz, t is time, and DC is the average value of the pulse train. The DC term is also found by taking the average value of one period P of the pulse. And a_N is a weighting factor based on the frequency coefficients of a Fourier series of the pulse-train signal.

We can define the input signal that will be sampled as

$$S(t) = b_1 \cos(2\pi f_{in}t) \tag{15-2}$$

And because the multiplication of the periodic pulse-train signal and the input signal amounts to sampling of the input signal, we have the sampled input signal as

$$S(t) \times P(t) = b_1 \cos(2\pi f_{in}t)[DC + a_1 \cos(2\pi f_s t) + a_2 \cos(2 \times 2\pi f_s t) + a_3 \cos(3 \times 2\pi f_s t) + \cdots + a_N \cos(N \times 2\pi f_s t)]$$

Expanding the terms, we get

$$S(t) \times P(t) = [b_1 \cos(2\pi f_{in}t)]DC + [b_1 \cos(2\pi f_{in}t)][a_1 \cos(2\pi f_s t)] + [b_1 \cos(2\pi f_{in}t)][a_2 \cos(2 \times 2\pi f_s t)] + [b_1 \cos(2\pi f_{in}t)][a_3 \cos(3 \times 2\pi f_s t)] + \cdots + [b_1 \cos(2\pi f_{in}t)][a_N \cos(N \times 2\pi t f_s t)] \tag{15-3}$$

Now let's take a look at some of the product terms:

1. $[b_1 \cos(2\pi f_{in}t)]DC$ is just the input signal that is scaled by the average DC voltage.
2. $[b_1 \cos(2\pi f_{in}t)][a_1 \cos(2\pi f_s t)]$ is the input signal multiplied by a fundamental sampling-frequency sinusoidal waveform.
3. $[b_1 \cos(2\pi f_{in}t)][a_2 \cos(2 \times 2\pi f_s t)]$ is the input signal multiplied by a second-harmonic sampling-frequency sinusoidal waveform.
4. $[b_1 \cos(2\pi f_{in}t)][a_3 \cos(3 \times 2\pi f_s t)]$ is the input signal multiplied by a third-harmonic sampling-frequency sinusoidal waveform.
5. $[b_1 \cos(2\pi f_{in}t)][a_N \cos(N \times 2\pi t f_s t)]$ is the input signal multiplied by an Nth harmonic sampling-frequency sinusoidal waveform.

So, from the preceding, we see for a sampling system that there is a linear term from the first product term, as well as multiplication (mixing) between the input signal and signals of the fundamental and harmonics of the sampling frequency from the subsequent terms. For sampling with a narrow pulse train or using an ideal sampling pulse train, the harmonic signal amplitudes are equal to that of the fundamental-frequency signal. That is, with ideal sampling, $a_1 = a_2 = \cdots = a_N$.

Finite Pulse-Width Signals

For a finite pulse-width signal, calculating the weighting factors can be challenging, so here is the result for any N for a pulse centered at the origin:

$$a_N = \frac{2}{N\pi}\sin(N\pi T/P) \tag{15-4}$$

where N is the Nth harmonic including the fundamental ($N = 1$), T is the duration of the rectangular pulse, and P is the period of the pulse.

For a quick verification, let's take a look at a square-wave pulse that has a level at 0 volt and 1 volt. The duration of the square-wave pulse is one-half the period for a 50 percent duty cycle. Thus

$$T/P = \frac{1}{2}$$

And

$$a_N = \frac{2}{N\pi} \sin(N\pi/2)$$

For the fundamental frequency, $N = 1$, so the coefficient

$$a_1 = \frac{2}{1\pi} \sin(1\pi/2) = \frac{2}{\pi}$$

because the sine of 90 degrees, or $\pi/2$, equals 1. This $2/\pi$ factor then is in agreement with the coefficient for the fundamental frequency of a square wave with levels of 0 and 1. Also note that the DC term or this square-wave signal from 0 to 1 is $\frac{1}{2} = DC$; thus the square-wave signal $SQ(t)$ has the following DC term and coefficients:

DC	a_1	a_2	a_3	a_4	a_5	a_6	a_7	a_8	a_9
$\frac{1}{2}$	$\frac{2}{\pi}$	0	$\frac{-2}{3\pi}$	0	$\frac{2}{5\pi}$	0	$\frac{-2}{7\pi}$	0	$\frac{2}{9\pi}$

$$SQ(t) = \frac{1}{2} + \frac{2}{\pi}\cos(2\pi f_s t) - \frac{2}{3\pi}\cos(6\pi f_s t) + \frac{2}{5\pi}\cos(10\pi f_s t) - \frac{2}{7\pi}\cos(14\pi f_s t) \cdots \quad (15\text{-}5)$$

Thus, sampling the input signal $S(t) = b_1 \cos(2\pi f_{in} t)$ with $SQ(t)$ results in

$$S(t) \times SQ(t) = [b_1 \cos(2\pi f_{in} t)][\frac{1}{2} + \frac{2}{\pi}\cos(2\pi t f_s t) - \frac{2}{3\pi}\cos(6\pi f_s t) + \frac{2}{5\pi}\cos(10\pi f_s t) - \frac{2}{7\pi}\cos(14\pi f_s t) \cdots]$$

For a quadrature pulse signal $QP(t)$, used in a Tayloe detector, which is a 25 percent duty-cycle pulse (from 0 to 1) that results in

$$T/P = \frac{1}{4}$$

the DC term, and the coefficients

$$a_N = \frac{1}{N\pi} \sin(N\pi/4)$$

where N is an integer > 0, are as follows:

DC	a_1	a_2	a_3	a_4	a_5	a_6	a_7	a_8	a_9
$\dfrac{1}{4}$	$\dfrac{\sqrt{2}}{\pi}$	$\dfrac{1}{\pi}$	$\dfrac{\sqrt{2}}{3\pi}$	0	$\dfrac{-\sqrt{2}}{5\pi}$	$\dfrac{-1}{3\pi}$	$\dfrac{-\sqrt{2}}{7\pi}$	0	$\dfrac{\sqrt{2}}{9\pi}$

Note that every fourth harmonic signal is zero or missing in a 25 percent duty-cycle pulse signal $QP(t)$:

$$QP(t) = \tfrac{1}{4} + \tfrac{\sqrt{2}}{\pi}\cos(2\pi f_s t) + \tfrac{1}{\pi}\cos(4\pi f_s t) + \tfrac{\sqrt{2}}{3\pi}\cos(6\pi f_s t) - \tfrac{\sqrt{2}}{5\pi}\cos(10\pi f_s t) \dots \tag{15-7}$$

Thus, if an input signal $S(t)$ is sampled by the 25 percent duty-cycle pulse-train signal, we have

$$S(t) \times QP(t) = b_1 \cos(2\pi f_{in} t)[\tfrac{1}{4} + \tfrac{\sqrt{2}}{\pi}\cos(2\pi f_s t) + \tfrac{1}{\pi}\cos(4\pi f_s t) +$$

$$\tfrac{\sqrt{2}}{3\pi}\cos(6\pi f_s t) - \tfrac{\sqrt{2}}{5\pi}\cos(10\pi f_s t) \dots]$$

From the preceding two example pulse-train signals, we find that sampling is indeed a way of multiplying the input signal with pulse-train signals that include a DC term and alternating-current (AC) signals, including the fundamental frequency and harmonics of the sampling signal. Multiplication of the input signal with the sampling signal then provides a form of harmonic mixing as well.

Figure 15-4B shows the relative spectrum associated with an ideal sampling signal that has a very narrow pulse width, and Figure 15-4C shows the spectrum associated with a square-wave-sampling waveform. It should be noted that when two signals are multiplied such as from $S(t) = b_1 \cos(2\pi f_{in} t)$ and the fundamental sampling frequency signal of the square-wave signal $SQ(t)$, $[b_1 \cos(2\pi f_s t)] [\tfrac{2}{\pi}\cos(2\pi f_s t)]$, the resulting signal is a double-sideband-suppressed carrier signal. That is, there are no signals from the multiplication operation that include a "carrier" frequency of f_s. We will find that double-sideband-suppressed carrier signals are the building blocks for I and Q signals in Chapter 16.

In Figures 15-1 through 15-4, examples of sampling are shown. And generally in a sampling system, the frequency range of the input signal f_{in} is less than half the sampling frequency f_s. That is, if f_s, the sampling frequency, is 96 kHz, then the maximum frequency of f_{in} is less than 48 kHz. However, in some sampling mixers, the frequency range of f_{in} is above one-half the sampling frequency, and as a result, aliasing signals occur. Is aliasing necessarily bad? We know that aliasing is bad in digital audio signal processing when the input frequency violates the Nyquist sampling rate and causes aliasing distortion signals within the audio band. But actually for radio-frequency (RF) mixing, the aliasing signals are signals that can serve as intermediate-frequency (IF) signals.

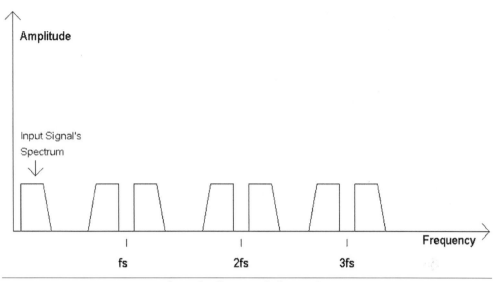

FIGURE 15-4B   Spectrum of an ideally sampled signal.

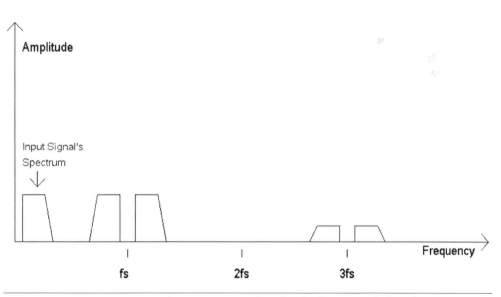

FIGURE 15-4C   Spectrum of an input signal multiplied or sampled with a square-wave signal.

Aliasing Is a Mixing Effect

For using a sampling system as a mixer to provide an IF signal, let's start by thinking about what happens when an input signal matches the exact frequency and phase of the sampling signal. Figure 15-5 presents an example pulse-train signal. For a sine-wave input signal of the same frequency, see Figure 15-6.

In Figure 15-6 we see that the sine-wave signal at the zero crossings lines up with the sampling pulses, which amounts to sampling a 0 volt DC signal. Figure 15-7 shows the result of sampling a signal of the same frequency at 0-degree phase or at the zero crossing. Note the finite negative and positive glitches that occur during the time the pulse is turned on. And if the sampling pulses are narrowed further, these glitches will trend toward zero in amplitude.

For an ideal sampling pulse-train signal, where the pulse duration is narrowed toward 0 µs, these glitches will trend toward zero.

Now let's take a look at sampling at a different portion of the input signal, that is, viewing the sampling at another phase of the input signal (Figure 15-8). In the figure, since the peak amplitude of the phase-shifted sine wave is about one-half the amplitude of the "unit" sampling pulses, the output should show a train of pulses of about one-half height (Figure 15-9).

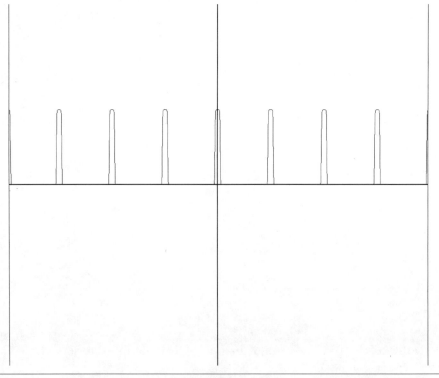

FIGURE 15-5 A pulse-train signal that will be sampling an input signal of the same phase and frequency.

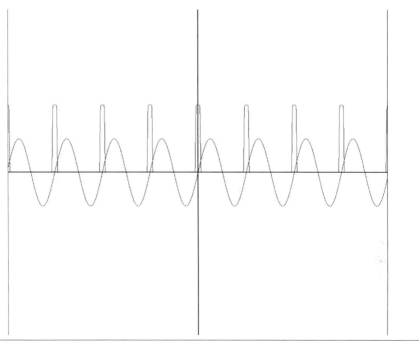

FIGURE 15-6   A sampling pulse train and an input signal with its zero crossing being sampled.

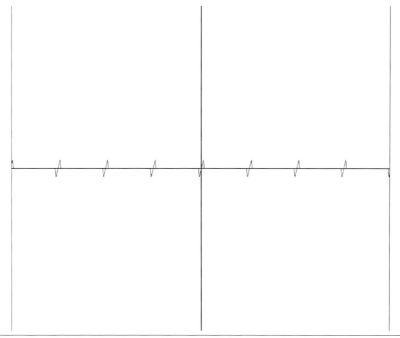

FIGURE 15-7   Sampling at the zero crossing of an input signal.

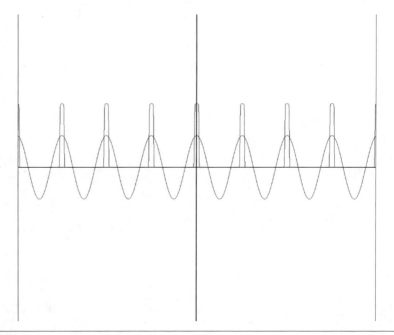

FIGURE 15-8 Sampling of the input signal that has been phase-shifted 90 degrees.

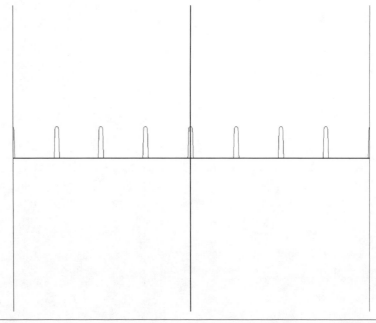

FIGURE 15-9 Result of sampling the peak amplitude of the phase-shifted input signal.

Figure 15-9 shows that sampling the input signal that has been phase-shifted to 90 degrees resulted in a train of pulses whose pulse amplitude is the peak amplitude of the input signal (e.g., one-half). So, even though the input signal's frequency is the same as the sampling frequency, the phase shift of 90 degrees in this example is equivalent to an input signal set to 0.5 volt DC.

Now let's see what happens when the input signal is phase-shifted further (Figure 15-10). As seen in the figure, the input signal is phase-shifted by 270 degrees, causing the sampling pulses to capture the negative peaks of the input sine-wave signal. Again, the peak amplitude of the input signal is one-half the pulse's amplitude. Figure 15-11 shows the result of sampling the 270-degree phase-shifted signal.

As can be seen, for an input signal that has been phase-shifted 270 degrees, the output of the sampling system provides a constant pulse train whose pulse amplitude is –0.5.

Therefore, a way to equivalently express sampling a signal that exceeds the Nyquist rate and that causes aliasing is the following: For signals whose frequencies are over one-half the sampling frequency, the sampled signals from the sampling system where $P(t)$ is the unit pulse-train signal can be characterized as

$$V_{out}(t) = V_p \sin[2\pi(f_{in}t + \theta - f_s t)] \times P(t) \tag{15-8}$$

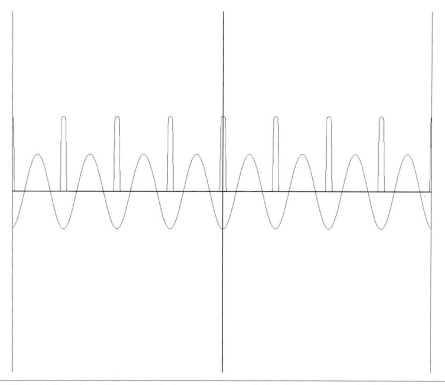

FIGURE 15-10   Sampling a 270-degree phase-shifted input signal of the same frequency as the sampling frequency.

where θ is the phase angle of the input signal relative to the sampling signal, and V_p is the peak voltage of the input sine-wave signal.

In a sense, when the input signal's frequency is close to the sampling frequency, the sampling system provides a "phase detector" output signal. For example, when the input signal has the same frequency as the sampling frequency, as shown in the preceding figures, we have

$$f_{in} = f_s$$

and using Equation (15-8), then

$$V_{out}(t) = [V_p \sin(\theta)] \times P(t)$$

which agrees with Figures 15-7, 15-9, and 15-11 for θ = 0, 90, and 270 degrees.

Now let's take a look at a situation where the input signal's frequency f_{in_hi} is just above the sampling frequency (Figure 15-12). This figure shows an input signal whose frequency is slightly above the sampling frequency that is sampled at various amplitude ranges of the input signal, which provides a varying signal at the output of the sampling system (Figure 15-13). This figure shows the result of sampling an

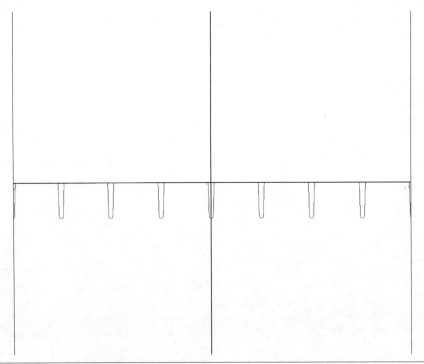

FIGURE 15-11 Sampling the 270-degree phase-shifted input signal of the same frequency as the sampling frequency.

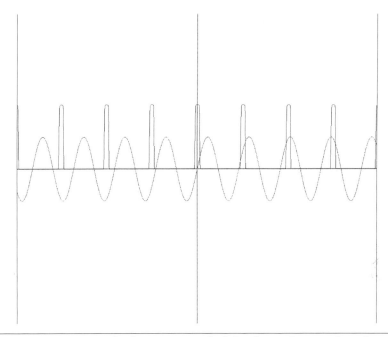

FIGURE 15-12   Input signal's frequency is slightly above the sampling frequency.

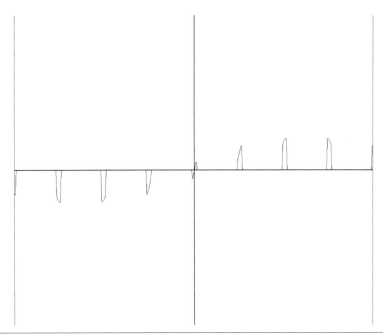

FIGURE 15-13   Resulting output from the sampler when the input signal is slightly higher than the sampling frequency.

input signal whose frequency f_{in_hi} is slightly higher than the sampling frequency f_s. Note that the input signal's peak sinusoidal amplitude is one-half the amplitude of the pulse-train signal. Notice that the resulting signal from the output of the sampler looks like a low-frequency signal at $(f_{in\_hi} - f_s)$, and it has the same amplitude as the input signal that is being sampled or multiplied by the pulse-train signal (Figures 15-14 and 15-15).

Figure 15-14 shows the pulse train signal and an input signal of frequency $(f_{in\_hi} - f_s)$, and Figure 15-15 thus shows that there is an equivalence to sampling at a low frequency of $(f_{in\_hi} - f_s)$ and sampling at a high frequency of f_{in_hi} that violates the Nyquist condition.

Now compare the Figures 15-15 and 15-13. Except for the small variance in the shape of the pulses, the output of the sampling system is basically the same for both figures. Note that if the pulse-train signal is narrowed sufficiently, then Figures 15-15 and 15-13 would be identical.

Similarly, now let's take a look at sampling an input signal whose frequency f_{in_low} is slightly below the sampling frequency (Figure 15-16). In this figure, the difference frequency of the sampling frequency and the input signal is $(f_s - f_{in\_low})$, which is equal to $(f_{in\_high} - f_s)$. So, for example, if the input signal with frequency f_{in_hi} is 10 kHz above

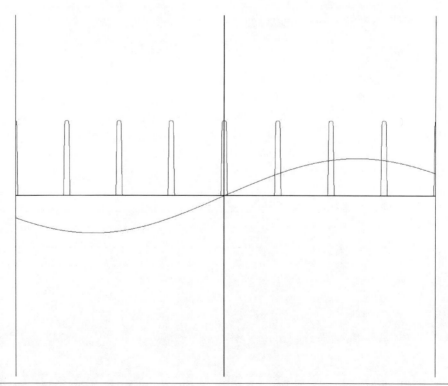

FIGURE 15-14   The pulse-train signal and an input signal of half pulse amplitude.

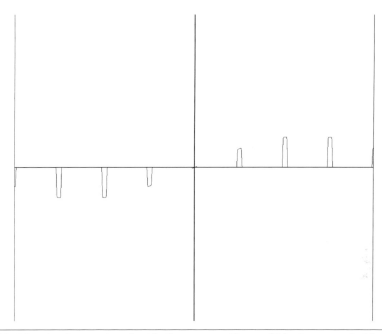

FIGURE 15-15   An input signal with frequency $(f_{in\_hi} - f_s)$ being sampled by the pulse-train signal.

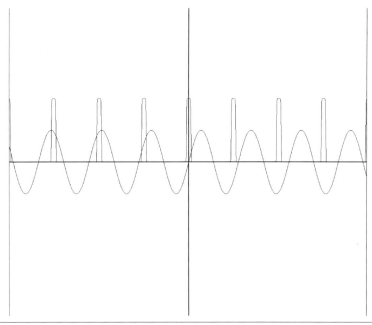

FIGURE 15-16   An input signal slightly below the sampling frequency.

the sampling frequency f_s, then the input signal with frequency f_{in_low} is 10 kHz below the sampling frequency f_s.

The output of the sampling system then produces a waveform such as that seen in Figure 15-17. This figure shows basically the same type of waveform as Figures 15-13 and 15-15 with an inversion in the shape of the modulating envelope. So let's see what happens when a low-frequency signal of $(f_s - f_{in\_low})$ is inverted and sampled (Figures 15-18 and 15-19).

In Figures 15-17 and 15-19 we see a similarity, very much like Figures 15-13 and 15-15, with a difference based on inversion of the waveforms. Thus the phase-detector "model" of the sampling system is confirmed as

$$V_{out}(t) = V_p \sin[2\pi t(f_{in}t + \theta - f_s t)] \times P(t) \tag{15-9}$$

When $f_{in} > f_s$, the pulse train is modulated with a positive sine-wave signal:

$$V_{out}(t) = V_p \sin[2\pi(f_{in} - f_s)t] \times P(t) \tag{15-10}$$

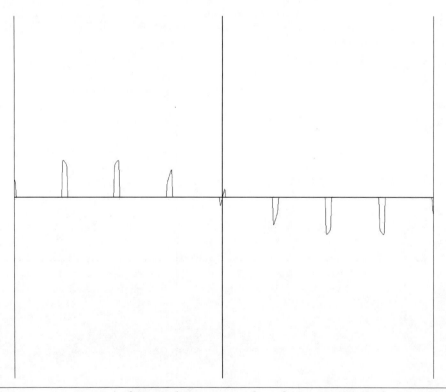

FIGURE 15-17   Output of a sampling system when the input signal's frequency is slightly lower than the sampling frequency.

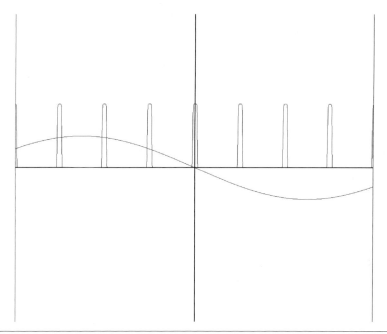

FIGURE 15-18   An inverted-sine-wave signal as the input signal that is sampled by the pulse-train signal.

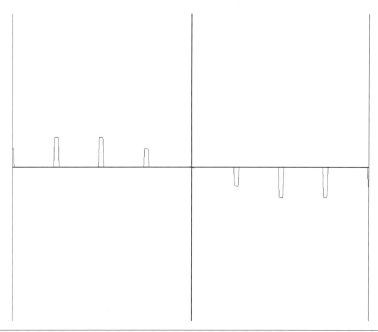

FIGURE 15-19   The output of the sampling system with an inverted sine wave at the input.

And if $f_{in} < f_s$, then the pulse-train signal is modulated by a negative sine-wave signal:

$$V_{out}(t) = -V_p \sin[2\pi(f_s - f_{in})t] \times P(t) \qquad (15\text{-}11)$$

Now let's take a look at the implications of aliasing by sampling a signal that is beyond the Nyquist frequency of $(\frac{1}{2})f_s$. Under ideal sampling conditions, where the pulse width of the pulse-train signal is sufficiently narrow, sampling such a signal whose frequency exceeds $(\frac{1}{2})f_s$ is the equivalent of sampling another signal of the same amplitude but with a frequency of $(f_s - f_{in})$ or $|f_s - f_{in}|$. This also implies that if the output of the sampling circuit includes a capacitor to form a sample-and-hold circuit, the actual conversion gain will be almost lossless when $|f_s - f_{in}| << f_s$. (See the sample-and-hold circuit in Figure 12-3B.)

For example, if a mixer is to translate amplitude-modulated (AM) radio signal at 990 kHz (f_{in}) to a low IF signal such as 10 kHz, then the mixing or sampling frequency can be 1,000 kHz (f_s), and thus

$$|f_s - f_{in}| = 10 \text{ kHz} << 1,000 \text{ kHz}$$

so the sampling mixer will have almost lossless conversion. That is, the 10-kHz IF signal will have almost the same amplitude as the incoming AM radio signal.

Generally, the phase of the IF signal is not that important. However, the fact that there is an identifying phase (noninverting or inverting) in the sampling or multiplying mixer when the incoming RF signal is above or below the mixing or sampling frequency provides added information. This added (phase) information is actually used in constructing an image-reject mixer (note that image-reject mixers will be covered in Chapter 21).

Before we move on to multiplexing mixers, I'd like to add a word or two about what is really inside the sine or cosine functions. That is, when we look at $\sin(x)$, what is x? The variable x has to be an angle. So, when you see $\sin(2\pi ft)$, the term $2\pi ft$ is an angle that varies with time. In addition,

$$2\pi f = \omega$$

where ω is the frequency in radians per second, and t is time measured in seconds. Therefore,

$$2\pi ft = \omega t = \text{some angle in radians}$$

and thus

$$2\pi ft = \omega t = \varphi(t) = \text{a phase angle that is time-dependent}$$

And it is known that

$$\Delta\varphi(t)/\Delta t = \text{the frequency in radians per second} \qquad (15\text{-}12)$$

Thus $\Delta\varphi(t)/\Delta t = \omega$ in radians per second, which is indeed the radial frequency of the sinusoidal waveform.

When studying frequency modulation (FM), $\Delta\varphi(t)/\Delta t$, the frequency, is handy to keep in mind because the change in phase divided by the change in time results in the (instantaneous) frequency of the signal.

Multiplexer Circuits as Balanced Mixers

In this section pertaining to multiplexer circuits such as an A-B switch, with A and B inputs, we will find that when the A-B switch is switched back and forth with equal duration (e.g., 50 percent duty cycle pulse), the output of the A-B switch will provide a signal that is equivalent to multiplying the input signal with a unit bipolar square-wave signal. A unit bipolar square-wave signal is defined to have two levels, one at –1 and the other at +1.

Figure 15-20 shows a multiplexer circuit for mixing. The figure also shows a simple switching circuit that provides mixing of two signals in a doubly balanced manner. A *double-balanced mixer* means that the output does not contain any leak-through from any of the AC input signals, such as the oscillator signal and the RF signal.

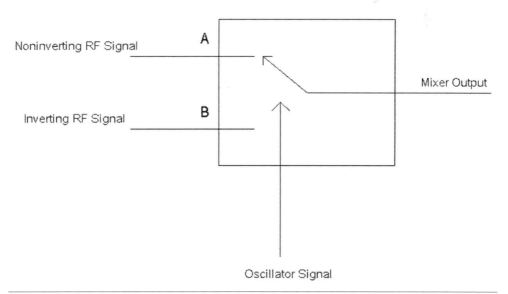

FIGURE 15-20   A double-throw single-pole multiplexer switch for mixing/multiplying RF signals.

The operation by multiplication also means that for DC input signals, the output must include one of the input signals. For example, consider multiplying an input signal S(t) with a unit-pulse-train signal P(t) for the oscillator signal:

$$S(t) \times P(t) = V_{out}$$

If $S(t)$ = 1 volt DC, then

$$1 \times P(t) = V_{out} = P(t)$$

In this example using the circuit in Figure 15-1A, by setting the input as a DC signal or constant voltage, the output of the multiplying function is the same as one of the input signals, P(t). Therefore, there are exceptions to the concept of balancing out or nulling out the input signal from the output of a multiplier or mixer.

Now let's return to Figure 15-20 and examine how it works as a multiplier. The RF inputs connected to each of the two inputs of the switch are complementary or push-pull. The oscillator signal that is fed to the switch control actuator is a 50 percent duty-cycle square-wave signal. Therefore, at the output of this mixer, half the time the output is the RF signal input and the other half of the time the output is the inverted RF input signal. For example, if the control signal is logic high, the output of the switch is connected to the noninverting RF signal. When the control signal is logic low, the output of the switch is connected to the inverting RF signal. Thus the output of the mixer is a signal that flips the sign or polarity of the incoming RF signal. In other words, this mixer commutes the RF signal in polarity as a function of the oscillator signal.

Therefore, the circuit in Figure 15-20 is sometimes called a *commutating mixer*. To understand the multiplying characteristic of this circuit, for the time being, consider setting the noninverting RF signal V_{RF} to +1 volt DC at the A input of the switch. If the A input is +1 volt DC, then the B input of the switch has to be –1 volt DC based on the fact that the B input is the inverting RF signal input.

Now, with +1 volt DC at the A terminal and –1 volt DC at the B terminal, and with the oscillator signal controlling the switch so as to send the voltages at the input terminals in an alternate manner, the output has to generate a square-wave signal of levels +1 volt and –1 volt at the same frequency as the oscillator signal. For example, if the oscillator signal is set to 10 MHz, with the input voltages at +1 volt and –1 volt, the output of the switch will generate a 10-MHz square-wave signal that is 2 volt peak to peak and centered around 0 volt.

If V_{RF} is increased to +2 volts DC at the A terminal, then the voltage at the B terminal of the switch has to be –2 volts DC, and the output of the mixer will provide a square-wave signal from –2 volts to +2 volts, or a 4-volt peak-to-peak signal at 10 MHz. Therefore, the output of the mixer is a square-wave signal that is scaled by the input signal, and in particular, this square-wave signal is bipolar, meaning that it outputs both positive and negative pulses. Also note that with positive voltages at input terminal A, the output square-wave signal matches the same phase as the oscillator signal.

Now what happens if V_{RF} at the A input is set to –1 volt DC? Then the B input will have to be +1 volt DC, and the output of the mixer still will generate a 2 volt peak-to-peak square-wave signal but with inverted phase in relationship to the oscillator's signal. Thus, at least with DC voltages, there is a multiplying effect from this circuit. And in general, if the input signal is an AC signal, there is still a multiplying effect from this mixer circuit. From the examples where the input signal is increased, decreased, or changed to a negative polarity, the output of the mixer maintains proportionality and phase.

And if $V_{RF} = 0$ volt, then both inputs of the multiplexer are 0 volt. The output of the multiplexer thus switches between 0 volt and 0 volt to provide a 0 volt (square-wave) signal or no signal. So, when the input signal drops to 0 volt, there is no carrier or oscillator signal at the output of the multiplexer, which balances out or removes the oscillator signal from the output.

We thus will characterize this dual-polarity square-wave signal and its Fourier series as

$$SQ_{bp}(t) = \tfrac{4}{\pi}\cos(2\pi f_s t) - \tfrac{4}{3\pi}\cos(6\pi f_s t) + \tfrac{4}{5\pi}\cos(10\pi f_s t) - \tfrac{4}{7\pi}\cos(14\pi f_s t)\ldots$$

$$(15\text{-}13)$$

Notice that the bipolar square-wave signal $SQ_{bp}(t)$ has no DC term. This makes sense because there is an equal number of positive and negative pulses from this signal, which averages to 0.

The output of the mixer now can be characterized as

$$V_{out} = V_{RF} \times SQ_{bp}(t) \qquad (15\text{-}14)$$

$$V_{out} = V_{RF} \times [\tfrac{4}{\pi}\cos(2\pi f_s t) - \tfrac{4}{3\pi}\cos(6\pi f_s t) + \tfrac{4}{5\pi}\cos(10\pi f_s t) - \tfrac{4}{7\pi}\cos(14\pi f_s t) \cdots]$$

$$(15\text{-}15)$$

Equations (15-14) and (15-15) thus show the multiplying effect of the input RF signal and the oscillator signal.

To this point, an intuitive explanation has been presented. In practice, if one builds the circuit shown in Figure 15-20, a multiplication of the RF and oscillator signals will be verified. However, there is another way to show more formally why the multiplexer circuit works as a mixer.

The signal at the A terminal can be thought of as being multiplied by a square-wave signal $SQ(t)$ from Equation (15-5) that has an amplitude range from 0 to 1. The signal at the B terminal can be thought at of as the inverted signal from the A terminal multiplied by a 180-degree phase-shifted version of $SQ(t)$. The mixer output then is the summation of these two multiplication operations.

A 180-degree phase-shifted version of $SQ(t) = [1 - SQ(t)]$. Figure 15-21 presents a graph of $SQ(t)$, and Figure 15-22 presents a graph of $[1 - SQ(t)]$. Thus, Figure 15-22 shows equivalent of the square-wave signal $SQ(t)$ shifted by 180 degrees.

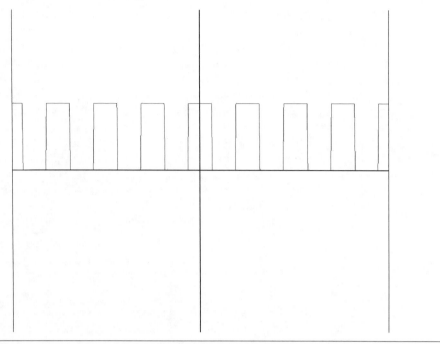

FIGURE 15-21   Square-wave signal SQ(t).

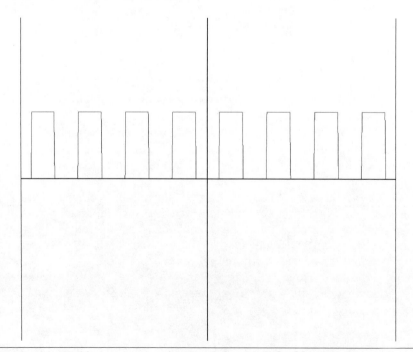

FIGURE 15-22   SQ(t), a square-wave signal that is shifted by 180 degrees.

For the derivation, then,

$$V_{out} = V_{RF} \times SQ(t) + -V_{RF} \times [1 - SQ(t)]$$
$$= 2V_{RF} \times SQ(t) - V_{RF} \tag{15-16}$$

However, from Equation (15-5),

$$SQ(t) = \tfrac{1}{2} + \tfrac{2}{\pi}\cos(2\pi f_s t) - \tfrac{2}{3\pi}\cos(6\pi f_s t) + \tfrac{2}{5\pi}\cos(10\pi f_s t) - \tfrac{2}{7\pi}\cos(14\pi f_s t) \ldots$$

and

$$V_{out} = 2V_{RF} \times [\tfrac{1}{2} + \tfrac{2}{\pi}\cos(2\pi f_s t) - \tfrac{2}{3\pi}\cos(6\pi f_s t) + \tfrac{2}{5\pi}\cos(10\pi f_s t) - \tfrac{2}{7\pi}\cos(14\pi f_s t) \ldots] - V_{RF} \tag{15-17}$$

$$V_{out} = 2V_{RF} \times \tfrac{1}{2} + 2V_{RF} \times [\tfrac{2}{\pi}\cos(2\pi f_s t) - \tfrac{2}{3\pi}\cos(6\pi f_s t) + \tfrac{2}{5\pi}\cos(10\pi f_s t) - \tfrac{2}{7\pi}\cos(14\pi f_s t) \ldots] - V_{RF}$$

and

$$V_{out} = V_{RF} + 2V_{RF} \times [\tfrac{2}{\pi}\cos(2\pi f_s t) - \tfrac{2}{3\pi}\cos(6\pi f_s t) + \tfrac{2}{5\pi}\cos(10\pi f_s t) - \tfrac{2}{7\pi}\cos(14\pi f_s t) \ldots] - V_{RF} \tag{15-18}$$

Note that the V_{RF} terms subtract each other out, leaving

$$V_{out} = 2V_{RF} \times [\tfrac{2}{\pi}\cos(2\pi f_s t) - \tfrac{2}{3\pi}\cos(6\pi f_s t) + \tfrac{2}{5\pi}\cos(10\pi f_s t) - \tfrac{2}{7\pi}\cos(14\pi f_s t) \cdots] \tag{15-19}$$

Equation (15-19) then leads to Equation (15-15):

$$V_{out} = V_{RF} \times [\tfrac{4}{\pi}\cos(2\pi f_s t) - \tfrac{4}{3\pi}\cos(6\pi f_s t) + \tfrac{4}{5\pi}\cos(10\pi f_s t) - \tfrac{4}{7\pi}\cos(14\pi f_s t) \cdots] \tag{15-15}$$

Figure 15-23 shows V_{RF} and $SQ(t)$, and Figure 15-24 shows $V_{RF} \times SQ(t)$. Figure 15-25 shows waveforms $-V_{RF}$ and $[1 - SQ(t)]$. Figure 15-26 shows the resulting waveform of $-V_{RF} \times [1 - SQ(t)]$. And Figure 15-27 shows the summation of the two products $V_{RF} \times SQ(t)$ and $-V_{RF} \times [1 - SQ(t)]$.

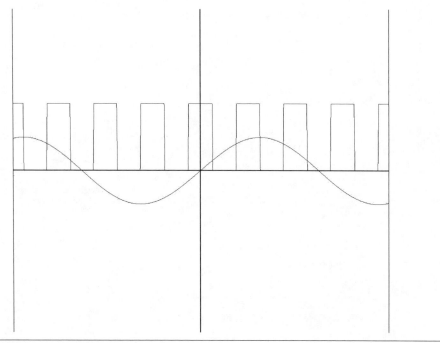

FIGURE 15-23 Waveforms V_{RF} and SQ(t).

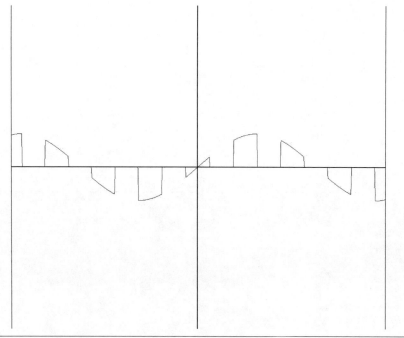

FIGURE 15-24 Waveform showing $V_{RF} \times$ SQ(t).

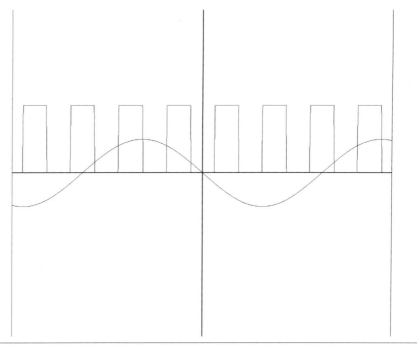

FIGURE 15-25   Waveforms $-V_{RF}$ and $[1 - SQ(t)]$.

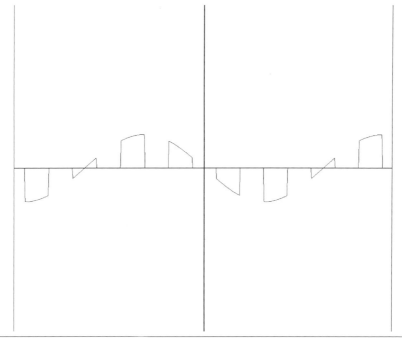

FIGURE 15-26   Waveform resulting from multiplying signals $-V_{RF}$ with $[1 - SQ(t)]$.

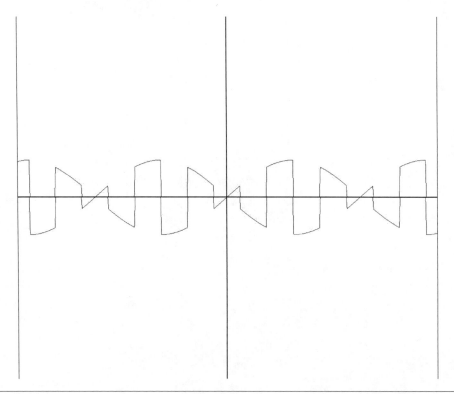

FIGURE 15-27   The summation of the two multiplied signals $V_{RF} \times SQ(t)$ and $-V_{RF} \times [1 - SQ(t)]$.

Adding or combining the waveforms from Figures 15-24 and 15-26 results in the waveform shown in Figure 15-27.

Now let's take a look at Figure 15-28, which shows the input signal V_{RF} multiplying the bipolar square-wave signal $SQ_{bp}(t)$. Note that Figures 15-27 and 15-28 are identical, which confirms that the multiplexer circuit in Figure 15-20 can be analyzed in terms to two unipolar square waves that have levels of 0 and +1. The first unipolar square wave at 0 degrees of phase is multiplied by an input signal, and the other unipolar square wave at 180 degrees of phase is multiplied by an inverse phase input signal, which when the two multiplied outputs are summed will provide a signal equivalent to the input signal, multiplying a bipolar square-wave signal that has levels of –1 and +1.

In terms of conversion gain, this mixer circuit has a conversion gain with the fundamental oscillator signal of $2/\pi$ and requires the generation of an inverting-phase RF signal. This inverting phase can be achieved via a transformer, a balanced-output amplifier, or an inverting-gain amplifier.

With $V_{RF} = b_1 \cos(2\pi f_{in} t)$ and using just the first term from Equation (15-15), we have

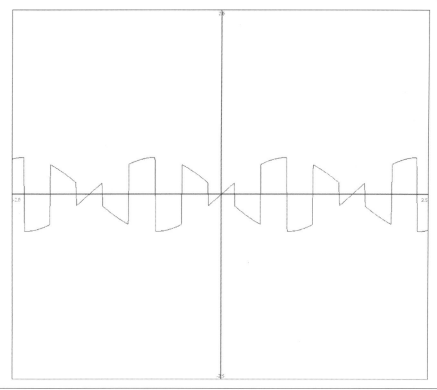

FIGURE 15-28   Square-wave form showing the product of input signal and bipolar square-wave signal $SQ_{bp}(t)$.

$$V_{out} = V_{RF} \times \tfrac{4}{\pi}\cos(2\pi f_s t) = [b_1 \cos(2\pi f_{in} t)][\tfrac{4}{\pi}\cos(2\pi f_s t)] =$$

$$b_1 \tfrac{4}{\pi}\tfrac{1}{2}\{\cos[2\pi(f_{in} + f_s)t] + \cos[2\pi(f_{in} - f_s)]\}$$

Thus the input signal's amplitude b_1 gets scaled by a factor of $(4/\pi)(\tfrac{1}{2}) = 2/\pi$, which is the conversion gain.

Care should be taken to avoid any DC voltage difference between the A and B terminals in Figure 15-20. Any potential difference at the inputs will cause a leak-through of the carrier signal. This makes sense: If we remove the input signal V_{RF} and still have a DC offset voltage between the two input terminals, the switch will toggle between the two different DC voltages at the A and B terminals, resulting in a square wave at the output. For example, if there is a +10-mV DC signal at the A input terminal and a +5-mV DC signal at the B input terminal, the multiplexer will generate a square-wave signal that has levels of +5 mV to +10 mV, which is a 5-mV peak-to-peak square-wave signal.

Also, the oscillator's duty cycle must be precisely 50 percent. If there is an asymmetry in the oscillator's waveform in terms of on-off duration, then there will be a leak-through of the RF input signal to the output. For example, if the oscillator's waveform is 99 percent in the on state and 1 percent in the off state, then this means that for all practical purposes virtually all the input signal from terminal A will be sent to the output of the switch.

Switch-mode and commutating mixers have been around since the late 1920s. Some of the pioneers of these types of mixers were

1. C. R. Keith, who in 1929 invented doubly balanced mixers with four diodes and eight adjusted transformer windings
2. R. S. Caruthers, who in 1934 invented switching-mode mixers with two, four, six, and eight diodes that included a singly balanced mixer
3. F. A. Cowan, who in 1934 invented an improved and simplified four-diode doubly balanced mixer design that is still used today

For curiosity's sake, let's take a look at the very familiar four-diode ring doubly balanced mixer developed by F. A. Cowan (Figure 15-29). The figure shows that even in 1934 or earlier, solid-state diodes were available that were made from copper oxide.

Figure 15-29 looks a little daunting with the two transformers and center tap windings. Therefore, let's simplify it in the following manner:

1. Replace the input transformer with two balanced or push-pull signal sources, a noninverting signal source, and an inverting signal source.
2. Replace the output transformer with two resistors, and take the output signal across the two resistors in a differential manner.

In Figure 15-30, the oscillator signal source is provided by Vosc, which normally is a large sinusoid signal that supplies sufficient voltage to switch on the diodes. When

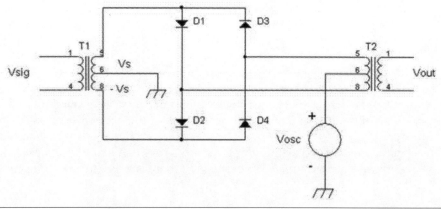

FIGURE 15-29   A four-diode ring mixer/modulator.

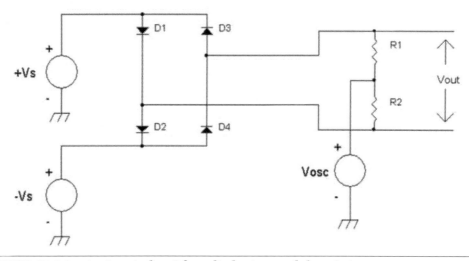

FIGURE 15-30   An "equivalent" four-diode ring modulator/mixer.

Vosc is a large positive voltage, diodes D2 and D3 are turned on, whereas diodes D1 and D4 are off. Thus, with Vosc being a large positive voltage, D2 and D3 are like wires, and D1 and D4 are like open circuits.

With D2 and D3 conducting, input signal +Vs is connected to the top of resistor R1, and input signal –Vs is connected to the bottom of R2. The output signal then is the potential difference across the top of R1 and the bottom of R2, which is +Vs – (Vs) = +2Vs when the oscillator signal Vosc is positive voltage.

However, when the oscillator signal Vosc is a large negative voltage, diodes D2 and D3 turn off and become open circuits, whereas diodes D1 and D4 conduct and become wires. Therefore, input signal –Vs is connected to the top of R1 and input signal +Vs is connected to the bottom of R2. The potential difference between the voltages from the top of R1 and bottom of R2 is then

$$-Vs - (+Vs) = -2Vs$$

In essence, the oscillator signal is commutating the input signal in the four-diode ring modulator/mixer very much like the A-B switch multiplexer shown in Figure 15-20. Therefore, the analysis of Figures 15-28 and 15-29 is the same as the analysis for multiplexer circuits and has the equation

$$V_{out} = K \times V_{sig} \times [\tfrac{4}{\pi}\cos(2\pi f_{osc}t) - \tfrac{4}{3\pi}\cos(6\pi f_{osc}t) + \tfrac{4}{5\pi}\cos(10\pi f_{osc}t) - \tfrac{4}{7\pi}\cos(14\pi f_{osc}t) \cdots]$$

where K is a gain factor, as determined by the transformers' turns ratios, diode losses, and loading.

Figures 15-31 and 15-32 are the front pages of the patents filed by C. R. Keith and F. A. Cowan that were published by the U.S. Patent Office. Note that patents can be

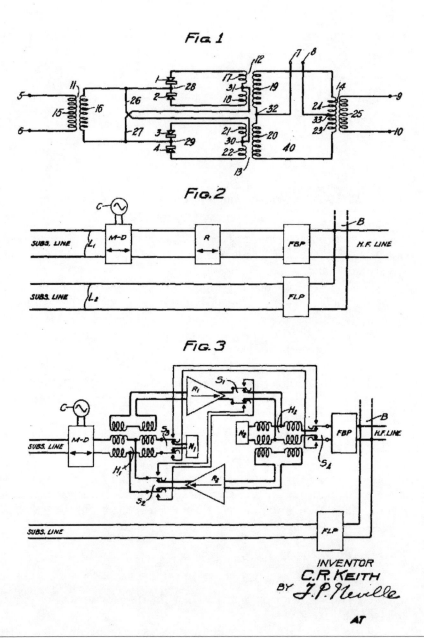

FIGURE 15-31   Mixer circuit by C. R. Keith.

Dec. 24, 1935.   F. A. COWAN   2,025,158

MODULATING SYSTEM

Filed June 7, 1934

FIG. 1

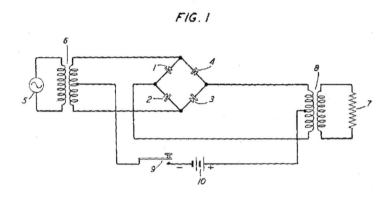

FIG. 2

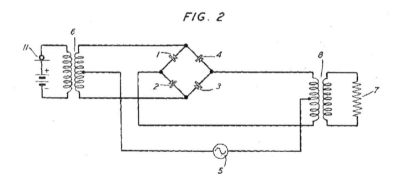

INVENTOR
F.A. COWAN

BY

ATTORNEY

FIGURE 15-32   A mixer by F. A. Cowan that is still used today.

accessed through www.google.com/patents. Note that Figure 15-30 is derived from Figure 15-32.

Tradeoffs in Performance of the Mixers

The sampling mixer in Figure 15-1A and the multiplexing mixer in Figure 15-20 have advantages and disadvantages. A sampling mixer with a hold capacitor has almost lossless conversion gain but generally requires a very low-impedance drive at its input. When the sampling switch is turned on, a significant load is presented to the input terminal because of the charging current into the capacitor. This loading effect usually requires an amplifier or a low-impedance transformer to drive the input. Also, the sampling mixer is not a doubly balanced mixer, which means that the input signal does find its way to the output. However, usually the input signal can be filtered out or canceled out via an extra sampling switch that is turned on later at one-half period of the oscillator's frequency.

Because there is no hold capacitor in the multiplexer mixer in Figure 15-20, driving the input does not require a very low-impedance output from the RF input signal source. However, an extra signal that is opposite in phase is required, which can be provided by a transformer with a center-tap winding or an extra amplifier. Conversion gain is less than that of the sample-and-hold mixer, but the noise performance is very good.

Both mixers are harmonic mixers, and a band-pass or low-pass filter is required on the path of the inputs. Otherwise, the output of these mixers will "map" noise and RF signals from out of the radio band of interest.

References

1. U.S. Patent 1,855,576, Clyde R. Keith, "Frequency Translating System," filed on April 9, 1929.
2. U.S. Patent 2,086,601, Robert S. Caruthers, "Modulating System," filed on May 3, 1934.
3. U.S. Patent 2,025,158, Frank Augustus Cowan, "Modulating System," filed on June 7, 1934.
4. U.S. Patent 5,471,531, Ronald Quan, "Method and Apparatus for Low-Cost Audio Scrambling and Descrambling," filed on December 14, 1993.
5. Kenneth K. Clarke and Donald T. Hess, *Communication Circuits: Analysis and Design*. Reading: Addison-Wesley, 1971.
6. Keith Henney, *Radio Engineering Handbook*, 3rd ed. New York: McGraw-Hill, 1941.
7. Allan R. Hambley, *Electrical Engineering Principles and Applications*, 2nd ed. Upper Saddle River: Prentice Hall, 2002.
8. Robert L. Shrader, *Electronic Communication*, 6th ed. New York: Glencoe/ McGraw-Hill, 1991.

9. Mischa Schwartz, *Information, Transmission, and Noise*. New York: McGraw-Hill, 1959.

10. Harold S. Black, *Modulation Theory*. Princeton: D. Van Nostrand Company, 1953.

11. B. P. Lathi, *Linear Systems and Signals*. Carmichael: Berkeley Cambridge Press, 2002.

12. Alan V. Oppenheim and Alan S. Wilsky, with S. Hamid Nawab, *Signals and Systems*, 2nd ed. Upper Saddle River: Prentice Hall, 1997.

13. Mary P. Dolciani, Simon L. Bergman, and William Wooton, *Modern Algebra and Trigonometry*. New York: Houghton Mifflin, 1963.

14. William E. Boyce and Richard C. DiPrima, *Elementary Differential Equations and Boundary Valued Problems*. New York: John Wiley & Sons, 1977.

15. Murray H. Protter and Charles B. Morrey, Jr., *Modern Mathematical Analysis*. Reading: Addison-Wesley, 1964.

16. E. B. Saff and A. D. Snider, *Fundamentals of Complex Analysis for Mathematics, Science, and Engineering*, 2nd ed. Upper Saddle River: Prentice Hall, 1993.

Chapter 16

In-Phase and Quadrature (IQ) Signals

This chapter will explore amplitude modulation in a little more detail. The objectives are

1. To take a brief look at broadcast amplitude-modulated (AM) signals
2. To examine double-sideband suppressed carrier (DSBSC) AM signals
3. To use the DSBSC signal to generate I and Q signals
4. To determine how I and Q signals are demodulated
5. To apply I and Q signals to software-defined radios (SDRs)

Broadcast AM signals are generally modulated with a carrier signal and an audio signal. There is only one phase of carrier signal, and in the standard envelope detection of the AM radio signal, this phase information is normally not used. The standard AM signal is known as an amplitude-modulated signal with carrier. This signal is illustrated in Figure 16-1 and characterized by Equation (16-1).

 Note All the waveforms have the *X* axis denoting time and the *Y* axis denoting amplitude.

$$\{1 + m[\cos(2\pi f_{mod})t]\}\ \cos(2\pi f_{carrier})t \qquad (16\text{-}1)$$

This equation describes the waveform in Figure 16-1, with the modulation m of the carrier equal to 0.5, or 50 percent. If $m = 100$ percent, the carrier momentarily drops to zero or gets "pinched" off (Figure 16-2).

Figure 16-2 shows an AM signal that has been modulated at a 100 percent level, which pinches off the carrier but also allows the amplitude of the carrier to increase to twice its level at 0 percent modulation. In general, the equation governing the standard AM signal is

$$[1 + m(t)]\cos(2\pi f_{carrier})\ t = \text{AM signal} \qquad (16\text{-}2)$$

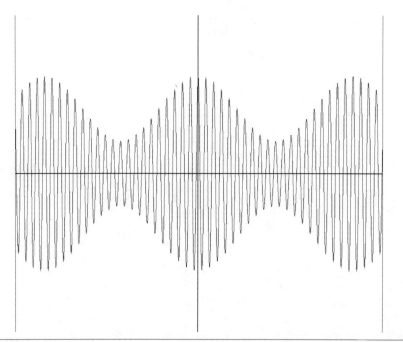

FIGURE 16-1 An example waveform of a broadcast AM signal with 50 percent modulation.

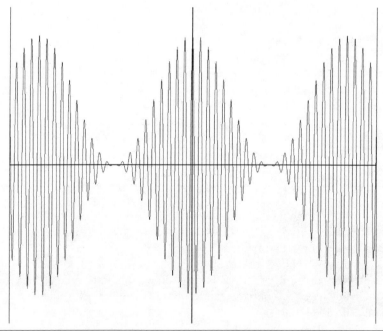

FIGURE 16-2 An AM signal with 100 percent modulation.

The modulating signal $m(t)$ is limited in range generally in the following manner to prevent distortion when demodulating the AM signal:

$$-1 < m(t) < 1$$

Given this limitation in range for $m(t)$, the value of $[1 + m(t)]$ is nonnegative, and the phase of the carrier signal $[\cos(2\pi f_{carrier})\, t]$ does not change as its carrier amplitude is being varied. The constant 1 in the term $[1 + m(t)]$ ensures that there is a carrier signal should the modulating signal drop to 0. This makes sense because when the music or voice signals drop in a broadcast AM radio program, the radio-frequency (RF) carrier signal is still present.

Introduction to Suppressed-Carrier Amplitude Modulation

For a suppressed-carrier amplitude-modulated signal, there is no carrier signal. That is, the sinusoidal signal $\cos(2\pi f_{carrier})t$ is not present in a suppressed-carrier AM signal. There are generally two types of suppressed-carrier AM signals, a double-sideband suppressed-carrier (DSBSC) AM signal and a signal-sideband suppressed-carrier (SSBSC) AM signal. By using a combination of DSBSC AM signals, one can also generate phase-modulated signals.

The basic signal to provide I and Q signals and single-sideband signals is the double-sideband signal $S(t)_{DSBSC}$. It is characterized by the following equation:

$$S(t)_{DSBSC} = [V_{sig}(t)]\cos(2\pi f_{carrier})t \qquad (16\text{-}3)$$

where $V_{sig}(t)$ is the modulating signal and generally is an alternating-current (AC) signal, and the range of $V_{sig}(t)$ is not necessarily restricted to –1 and +1. Figure 16-3 provides an illustration of a double-sideband suppressed-carrier signal $S(t)_{DSBSC}$. The figure looks rather strange in that one may expect to see that a sinusoidal modulating signal should result in a sinusoid envelope.

Figure 16-4 then overlays the modulating signal on top of the DSBSC AM signal for clarity. One should now note that the phase of the carrier signal $\cos(2\pi f_{carrier})\, t$ can change signs or phase from 0 to 180 degrees and vice versa. For example, if for a particular duration $V_{sig}(t) = +1$, then

$$S(t)_{DSBSC} = \cos(2\pi f_{carrier})t$$

and the phase of the carrier is 0 degrees. However, for another duration $V_{sig}(t) = -1$, then

$$S(t)_{DSBSC} = -\cos(2\pi f_{carrier})$$

which means that the carrier signal has inverted or changed in phase by 180 degrees.

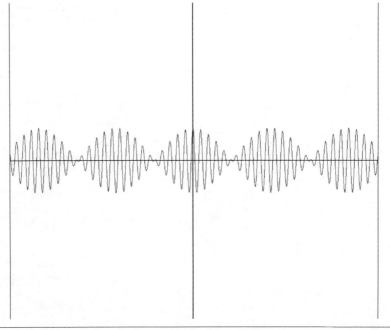

FIGURE 16-3 An example of a DSBSC (double-sideband suppressed-carrier) AM signal.

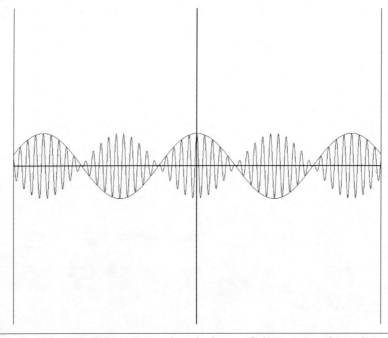

FIGURE 16-4 The DSBSC AM signal with the modulating signal overlayed.

Figure 16-5 shows a relationship between the modulating signal, the DSBSC AM signal, and the carrier signal. The figure shows a continuous-wave (CW) carrier signal below the DSBSC AM signal. Note that in the regions where the modulating signal is positive (above 0), the positive peaks of the CW carrier signal line up with the positive peaks of the DSBSC AM signal. However, when the modulating signal is negative (below 0), we see that the positive peaks of the CW carrier signal line up with the negative peaks of the DSBSC AM signal. Thus a DSBSC AM signal includes time-varying amplitude and phase modifications on the carrier signal.

As mentioned previously, the DSBSC (double-sideband suppressed-carrier) AM signal with an AC modulating signal never contains the carrier signal. For example, if

$$V_{sig}(t) = \cos(2\pi f_{mod})t$$

then

$$S(t)_{DSBSC} = [\cos(2\pi f_{mod})t][\cos(2\pi f_{carrier})t)]$$

$$= \tfrac{1}{2}\{\cos[2\pi(f_{carrier} + f_{mod})t] + \cos[2\pi(f_{carrier} - f_{mod})t]\}$$

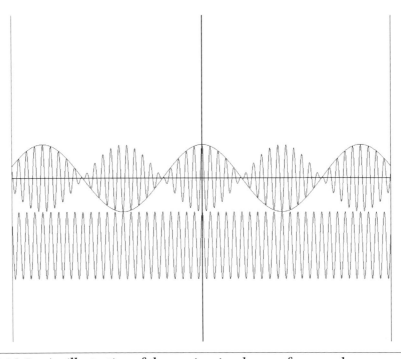

FIGURE 16-5   An illustration of the carrier signal as a reference phase compared with the DSBSC AM signal and modulating signal.

As expected, multiplying the modulation signal (e.g., audio) by the carrier signal results in two signals whose frequencies are above and below the carrier frequency $f_{carrier}$ by f_{mod}. The frequency ($f_{carrier} + f_{mod}$) is the upper-sideband frequency, and the ($f_{carrier} - f_{mod}$) frequency is the lower-sideband frequency.

In terms of demodulating the DSBSC (double sideband suppressed carrier) signal, simple envelope detection such as using a diode is not workable. A simple diode envelope detector will result in a demodulated signal that is distorted and full-wave rectified. For example, a sinusoidal modulating signal for the DSBSC signal will be detected with a diode, as shown in Figure 16-6.

From Figure 16-6 we see that simple envelope detection for the DSBSC AM signal results in distortion, and thus another way of detection is needed. Instead of a diode for demodulation, a synchronous detector is used.

A synchronous detector consists of a mixer or multiplier circuit and an oscillator that is precisely the same frequency as the carrier. The oscillator's frequency sometimes can be adjusted to be near the original carrier frequency, but often a separate signal is sent to lock the oscillator to the correct frequency. An example of such a DSBSC AM system is the one used in recovering the L-R channels in a stereo FM radio. The L-R audio signals are DSBSC AM modulated by a 38-kHz oscillator at

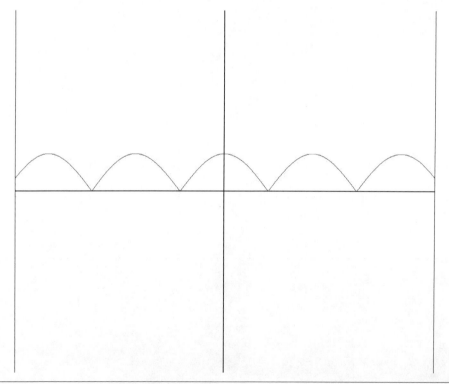

FIGURE 16-6 Using an envelope detector on a DSBSC AM signal results in full-wave rectification of the wanted signal.

the radio station, and the spectrum of this AM signal spans from 23 to 53 kHz, with the carrier missing at 38 kHz. At the receiver, to regenerate the 38-kHz carrier signal for synchronous detection, a 19-kHz "pilot" signal is sent from the radio station. By frequency multiplication of the 19-kHz pilot tone, a 38-kHz signal is generated and then used to recover the L-R audio signals (Figure 16-7).

The figure shows a demodulation system for a DSBSC AM signal by means of a mixer, oscillator, and low-pass filter. The received signal $S(t)_{DSBSC}$ is connected to one of the inputs of the mixer or multiplier, whereas the remaining input of the mixer is connected to the oscillator that "somehow" has the correct frequency. After mixing the two signals, the mixer is connected to a low-pass filter to extract the modulation signal such as audio. Is this all that is needed to demodulate the signal $S(t)_{DSBSC}$? No, to demodulate the signal correctly, the phase of the oscillator's signal is important.

Let's take at look at Equation (16-3) to see why the phase of the receiver's oscillator signal is important:

$$S(t)_{DSBSC} = [V_{sig}(t)] \cos(2\pi f_{carrier})t \tag{16-3}$$

To demodulate this signal, we will multiply $S(t)_{DSBSC}$ by a signal $2\cos[(2\pi f_{carrier}t + \theta]$ that has the same carrier frequency and includes an arbitrary phase-shift angle θ. Then

$$\begin{aligned} \text{Mixer output} &= \{2\cos[(2\pi f_{carrier})t + \theta]\}[V_{sig}(t) \cos(2\pi f_{carrier})t \\ &= V_{sig}(t)\{\cos[2\pi(2f_{carrier})t + \theta] + \cos(\theta)\} \end{aligned} \tag{16-4}$$

The first term in this equation is a high-frequency signal of frequency $2\pi f_{carrier}$ that will be removed by the low-pass filter, leaving only the second term related to the $\cos(\theta)$.

Thus the output via low-pass filtering is

$$\text{Filtered output} = [V_{sig}(t)]\cos(\theta) \tag{16-5}$$

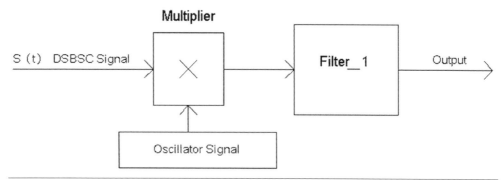

FIGURE 16-7   A demodulation system for a DSBSC AM signal.

If the phase angle θ from the receiver's oscillator is 0, then the $\cos(\theta) = \cos(0) = 1$, and we can see that the original modulating signal is recovered completely:

$$\text{Filtered output} = [V_{sig}(t)]\cos(0) = V_{sig}(t)$$

But if the oscillator's phase is 90 degrees, the filtered output is

$$[V_{sig}(t)]\cos\left(\tfrac{\pi}{2}\right) = 0$$

And if the oscillator's phase is 180 degrees off from the radio station's carrier signal phase, we get as the output

$$[V_{sig}(t)]\cos(\pi) = -V_{sig}(t) = \text{filtered output that inverts the phase of } V_{sig}(t)$$

What's interesting about using synchronous detection on a DSBSC AM signal is that with the "wrong" phase provided by the oscillator, the output of the detector can be zero. This result is almost as if the detector does not even "see" the input signal $[V_{sig}(t)]\cos(2\pi f_{carrier})t$.

At first look, this result of having the wrong phase for the detector may be a flaw in trying to demodulate a DSBSC (double-sideband suppressed-carrier) AM signal. But actually one can use this "flaw" to transmit two DSBSC AM signals (I and Q signals) and recover two channels of information, even though the two I and Q DSBSC AM signals occupy the same spectrum or bandwidth.

How I and Q Signals Are Generated

Before the structures of various I and Q modulators are shown, a table of trigonometric identities will be useful because generating various forms of I and Q signals involves multiplication of sine and cosine signals (Table 16-1).

Figure 16-8 shows a general I and Q modulator. The figure shows a first input signal $V_1(t)$ that is multiplied by an in-phase (I) carrier signal, generally a cosine signal such as $\cos(2\pi f_c t)$, where f_c is the carrier's frequency in hertz. Similarly, the second input signal $V_2(t)$ is multiplied by a quadrature-phase (Q) carrier signal that is 90 degrees phase-shifted from the I carrier signal, which is generally denoted as a sine signal such as $\sin(2\pi f_c t)$. Note that the carrier frequency f_c is the same for both I and Q carrier signals.

A combiner takes the output signals from each multiplier (M1 and M2) and can add or subtract them. The output of the combiner then provides a new signal.

I and Q signals that are combined have multiple uses. They are

1. Transmitting two separate channels of information within the same spectrum.
2. Frequency translating a signal of one frequency to another frequency, such as generating a single-sideband signal. The frequency shift can be upward or downward.

TABLE 16-1   Trigonometric Identities

$\cos(\alpha + \beta) = [\cos(\alpha)][\cos(\beta)] - [\sin(\alpha)][\sin(\beta)]$
$\cos(\alpha - \beta) = [\cos(\alpha)][\cos(\beta)] + [\sin(\alpha)][\sin(\beta)]$
$\sin(\alpha + \beta) = [\sin(\alpha)][\cos(\beta)] + [\cos(\alpha)][\sin(\beta)]$
$\sin(\alpha - \beta) = [\sin(\alpha)][\cos(\beta)] - [\cos(\alpha)][\sin(\beta)]$
$[\sin(\alpha)][\cos(\beta)] = (\frac{1}{2})\sin(\alpha + \beta) + (\frac{1}{2})\sin(\alpha - \beta)$
$[\cos(\alpha)][\cos(\beta)] = (\frac{1}{2})\cos(\alpha + \beta) + (\frac{1}{2})\cos(\alpha - \beta)$
$[\sin(\alpha)][\sin(\beta)] = -(\frac{1}{2})\cos(\alpha + \beta) + (\frac{1}{2})\cos(\alpha - \beta)$
$\cos(\alpha - \beta) = \cos(\beta - \alpha)$
$\sin(\alpha - \beta) = -\sin(\beta - \alpha)$
$\sin(\alpha) - \sin(\beta) = 2[\cos(\frac{\alpha + \beta}{2})][\sin(\frac{\alpha - \beta}{2})]$

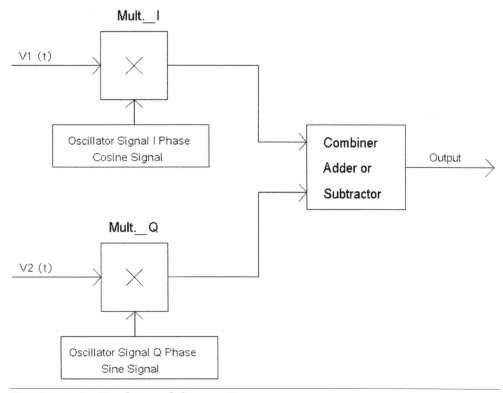

FIGURE 16-8   I and Q modulators using two mixers.

An example of transmitting two channels of information using I and Q signals or, equivalently, using quadrature modulation is the analog composite color television signals for NTSC (National Television System Committee) and PAL (Phase Alternate Line). Although television signals have been changed to a digital format, analog TV signals are still in use in many countries, even if the analog TV signals are not transmitted. DVD and Blu-ray disc players still have analog TV outputs, and many modern TV sets still have analog TV inputs. The use of quadrature modulation allows sending two channels of color video signals along with the luminance or black-and-white signal to provide decoding for red, green, and blue signals. When the outputs of the mixers are summed, the resulting signal is as follows:

$$\text{Summed output signal} = V_1(t) \cos(2\pi f_c t) + V_2(t) \sin(2\pi f_c t)) = \text{Sum}_{IQ}(t) \quad (16\text{-}6)$$

To generate a single-sideband signal, we provide input signal $V_2(t)$ as a 90-degree phase-shifted version of $V_1(t)$. For simplicity, let $V_1(t) = \cos(2\pi f_{mod} t)$ and $V_2(t) = \sin(2\pi f_{mod} t)$, and now take the summed output signal as

$$\text{Sum}_{IQ\_lsb}(t) = \cos(2\pi f_{mod} t) \cos(2\pi f_c t) + \sin(2\pi f_{mod} t) \sin(2\pi f_c t) \quad (16\text{-}7)$$

By using two trigonometric identities from Table 16-1, namely,

$$[\cos(\alpha)][\cos(\beta)] = \tfrac{1}{2}\cos(\alpha + \beta) + \tfrac{1}{2}\cos(\alpha - \beta) \quad (16\text{-}8)$$

and

$$[\sin(\alpha)][\sin(\beta)] = -\tfrac{1}{2}\cos(\alpha + \beta) + \tfrac{1}{2}\cos(\alpha - \beta) \quad (16\text{-}9)$$

let $\alpha = 2\pi f_{mod} t$ and $\beta = 2\pi f_c t$ and then

$$\text{Sum}_{IQ\_lsb}(t) = \cos(\alpha - \beta) = \cos[2\pi(f_{mod} - f_c)t] \quad (16\text{-}10)$$

And by the trigonometric identity

$$\cos(\alpha - \beta) = \cos(\beta - \alpha) \quad (16\text{-}11)$$

$$\text{Sum}_{IQ\_lsb}(t) = \cos[2\pi(f_c - f_{mod})t] = \text{lower-sideband signal} \quad (16\text{-}12)$$

If the combiner is a subtractor, then the new signal at the output of the combiner is

$$\text{Subtr}_{IQ\_usb}(t) = \cos(2\pi f_{mod} t) \cos(2\pi f_c t) - \sin(2\pi f_{mod} t) \sin(2\pi f_c t) \quad (16\text{-}13)$$

And by using the trigonometric identities (16-8) and (16-9), we get

$$\text{Subtr}_{IQ\_usb}(t) = \cos[2\pi(f_c + f_{mod})t] = \text{upper-sideband signal} \quad (16\text{-}14)$$

Note that to provide "perfect" single-sideband signals, not only the amplitudes of the input signals have to be exact, but also the phases of the carrier and input signals both must keep an exact 90-degree difference. In practice, the input signals may not be exactly the same amplitude-wise, and the mixers may not have identical conversion gain. However, adjustments can be made to the mixer's conversion gain and/or to the input's amplitude to match the overall amplitude for maximum cancellation of the undesirable sideband.

Now let's take a closer look at what happens if there is a deviation from a 90-degree signal of $\Delta\varphi$ in the carrier signal. However, this time we will work with a slight change in the product terms (Figure 16-9).

Figure 16-9 shows a single-sideband modulation system that provides a sine-wave output with an error of $\Delta\varphi$ in the Q carrier signal. Note that $\Delta\varphi \ll 1$, where $\Delta\varphi$ is measured in radians. Thus

$$\text{Sum}_{IQ\_usb}(t) = \cos(2\pi f_c t) \sin(2\pi f_{mod} t) + \sin(2\pi f_c t + \Delta\varphi) \cos(2\pi f_{mod} t)$$
$$= \tfrac{1}{2}\sin(2\pi f_c t + 2\pi f_{mod} t) + \tfrac{1}{2}\sin(2\pi f_c t + 2\pi f_{mod} t + \Delta\varphi)$$
$$+ \tfrac{1}{2}\sin(2\pi f_c t - 2\pi f_{mod} t) - \tfrac{1}{2}\sin(2\pi f_c t - 2\pi f_{mod} t + \Delta\varphi) \qquad (16\text{-}15)$$

The upper-sideband terms are

$$\tfrac{1}{2}\sin(2\pi f_c t + 2\pi f_{mod} t) + \tfrac{1}{2}\sin(2\pi f_c t + 2\pi f_{mod} t + \Delta\varphi)$$
$$\approx \sin(2\pi f_c t + 2\pi f_{mod} t) = \text{upper-sideband signal} \qquad (16\text{-}16)$$

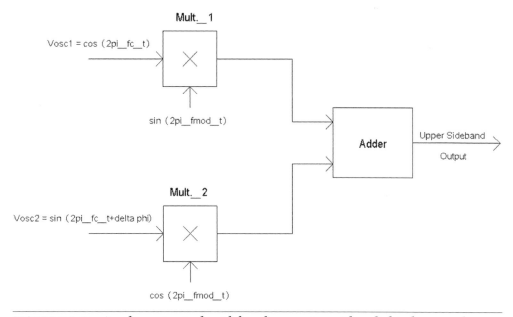

FIGURE 16-9   An alternate single-sideband generator with a slight phase-angle error in the Q carrier signal.

The lower-sideband signal would cancel out completely if $\Delta\varphi = 0$, but for small errors, we have

$$\tfrac{1}{2}[\sin(2\pi f_c t - 2\pi f_{mod}t) - \tfrac{1}{2}\sin(2\pi f_c t - 2\pi f_{mod}t + \Delta\varphi)] \qquad (16\text{-}17)$$

Both these sinusoidal waveforms represent "vectors" of the lower-sideband signal that have the same magnitude but nearly the same direction. In other words, both sinusoidal waveforms almost cancel out each other owing to the subtraction process.

For Equation (16-17), the trigonometric identity $\sin(\alpha) - \sin(\beta) = 2[\cos(\alpha + \beta)/2][\sin(\alpha - \beta)/2]$ will be useful in determining the magnitude (MAG) of the lower-sideband signal. Thus

$$\mathrm{MAG}[\sin(\alpha) - \sin(\beta)] = 2\left|\sin\left(\tfrac{\alpha-\beta}{2}\right)\right|$$

or equivalently,

$$\tfrac{1}{2}\mathrm{MAG}[\sin(\alpha) - \sin(\beta)] = \tfrac{1}{2}2\left|\sin\left(\tfrac{\alpha-\beta}{2}\right)\right| = \left|\sin\left(\tfrac{\alpha-\beta}{2}\right)\right|$$

If the phase angle of $\sin(2\pi f_c t - 2\pi f_{mod}t)$ is 0, then the residual magnitude of the lower-sideband signal is

$$\tfrac{1}{2}\mathrm{MAG}[\sin(2\pi f_c t - 2\pi f_{mod}t + 0) - \sin(2\pi f_c t - 2\pi f_{mod}t + \Delta\varphi)] = \tfrac{1}{2}\left|[2\sin\tfrac{\Delta\varphi}{2}]\right| = \left|\sin\tfrac{\Delta\varphi}{2}\right|$$

$$(16\text{-}18)$$

For small angles of $\Delta\varphi$, $\sin(\Delta\varphi) = \Delta\varphi$; thus

$$\sin\left(\tfrac{1}{2}\Delta\varphi\right) = \tfrac{1}{2}\Delta\varphi$$

therefore,

$$\tfrac{1}{2}\mathrm{MAG}[\sin(2\pi f_c t - 2\pi f_{mod}t + 0) - \sin(2\pi f_c t - 2\pi f_{mod}t + \Delta\varphi)] = \left|\tfrac{1}{2}\Delta\varphi\right| = \left|\left(\tfrac{1}{2}\right)\Delta\varphi\right|$$

$$(16\text{-}19)$$

Thus the residual lower-sideband signal has a magnitude of $(\tfrac{1}{2})\Delta\varphi$, where $\Delta\varphi$ is measured in radians. For example, if the Q phase error is 0.573 degree, or 0.01 radian, then the residual sideband will be $(\tfrac{1}{2})(0.01) = 0.005$, or 0.5 percent, of the upper-sideband signal's amplitude.

A graph of $100\{(\tfrac{1}{2})[\sin(2\pi f_c t - 2\pi f_{mod}t + 0) - \sin(2\pi f_c t - 2\pi f_{mod}t + \Delta\varphi)]\}$ is shown in Figure 16-10 to confirm Equation (16-18). From the preceding calculations, the graph should show a cosine function with a peak amplitude of 100×0.005, or 0.5.

Figure 16-10 shows a sinusoidal waveform with a peak amplitude of 0.5, which is the result of multiplying $(\tfrac{1}{2})[\sin(2\pi f_c t - 2\pi f_{mod}t + 0) - \sin(2\pi f_c t - 2\pi f_{mod}t + 0.01)]$ by 100. The Y axis is full scale at 2.5 from the origin.

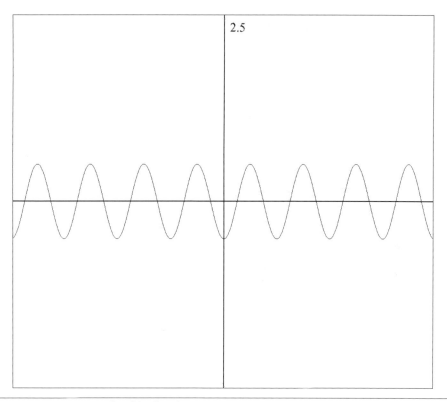

FIGURE 16-10   A graph of $(\frac{1}{2})[\sin(2\pi f_c t - 2\pi f_{mod} t + 0) - \sin(2\pi f_c t - 2\pi f_{mod} t + \Delta\varphi)]$ multiplied 100-fold for ease of viewing.

The subtraction of two sine functions of almost the same phase leads to a cosine function. Intuitively, one would have thought two sine functions that close in phase would lead to another sine function. However, this "unexpected" result also provides another way to generate a Q signal. Of course, if one looks at the trigonometric tables, then one will see that this result is indeed expected. That is, two sine functions of nearly the same phase when subtracted do produce a cosine function.

In practice, the mixers can be of switch-mode type that generates square waves, which may be difficult to subtract with a differential amplifier. So an equivalent function can be made by inverting an input signal to an input of a mixer instead and then combine by adding via resistors. If switch-mode mixers are used, then a low-pass or band-pass filter should be used to remove harmonics from the combining circuit.

Demodulating I and Q Signals

When a summed I and Q signal is sent to a demodulator, there are two channels of information coexisting with each other without interfering with the delivery of information from each channel. In standard AM broadcast signals, to send a 5-kHz audio channel with standard AM signals that contain upper- and lower-sideband signals plus the carrier signal, 10 kHz of bandwidth in the RF channel is needed. However, if I and Q signals are sent instead with each 5 kHz of information, each I and Q channel occupies 10 kHz within the same channel space. Thus, when demodulated, two channels of 5 kHz of information are delivered, which is twice the information received for the same RF bandwidth when compared with standard broadcast AM signals. Because the I and Q signals are orthogonal to each other, the two signals are almost in different but parallel universes.

When both I and Q signals are combined, they may look "messy," and at first glance, the I and Q signals seem inseparable. It is as if someone took a sheet of white paper and wrote an essay with blue ink from top to bottom and then wrote on the same page from top to bottom another essay in red ink. With the different essays written over each other, at first glance, one would find it difficult to read. But if a red filter is placed over the sheet, the writing in blue ink can be read, and likewise, if a blue filter is placed over the sheet, the writing in red ink can be read.

Demodulating I and Q signals is similar to the preceding description, and separating out each channel is done with multiplication of the I and Q signals with 0- and 90-degree signals.

Once the quadrature modulated signals are combined as described in Equation (16-6), a single wire or channel can be used to carry or transmit these signals to a receiver. To demodulate the two channels of information $V_1(t)$ and $V_2(t)$, an oscillator or signal source must have the same carrier frequency f_c during the modulation process. Normally, the signal source regenerates the frequency f_c. For example, in color TV signals, a phase-lock loop oscillator or regeneration oscillator along with a reference signal sent provides a demodulation signal with a frequency f_c at 0 and 90 degrees of phase.

Figure 16-11 shows a demodulation system consisting of two mixers and filters and two carrier signal sources. The output of the first mixer M1 then is

$$\text{Sum}_{IQ}(t) = V_1(t) \cos(2\pi f_c t) + V_2(t) \sin(2\pi f_c t)$$

To recover or demodulate $V_1(t)$ while "ignoring" $V_2(t)$, we multiply $\text{Sum}_{IQ}(t)$ by a signal source $= \cos(2\pi f_c t)$. Therefore,

$$\cos(2\pi f_c t) \, \text{Sum}_{IQ}(t) = \cos(2\pi f_c t)[V_1(t) \cos(2\pi f_c t) + V_2(t) \sin(2\pi f_c t)] = \text{M1}_{out}$$

$$\text{M1}_{out} = \cos(2\pi f_c t) \, \text{Sum}_{IQ}(t) = \cos(2\pi f_c t)V_1(t) \cos(2\pi f_c t) + \cos(2\pi f_c t) \, V_2(t) \sin(2\pi f_c t)$$
$$= \tfrac{1}{2}V_1(t)[\cos(2\pi 2f_c t) + \cos(0)] + \tfrac{1}{2}V_1(t) \, [\sin(2\pi 2f_c t) + \sin(0)]$$

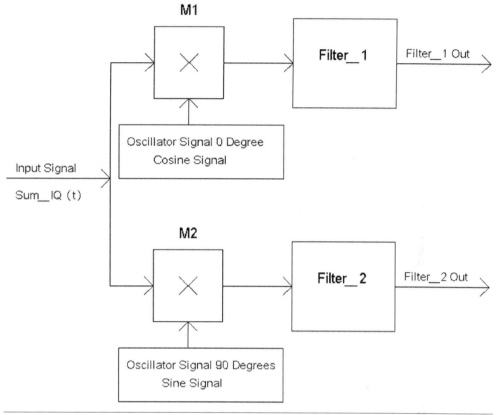

FIGURE 16-11 An IQ signal demodulation system.

The high-frequency terms relating to $2\pi 2f_c t$ are filtered out via low-pass or band-pass filtering and with $\cos(0) = 1$ and $\sin(0) = 0$. Also, whenever a sine wave and a cosine wave of the same frequency and phase are multiplied, the result is 0 after filtering out the high-frequency signals. The output of Filter_1 that is connected to M1 is then

$$\text{Filter1\_out} = \tfrac{1}{2}V_1(t)$$

Similarly to recover $V_2(t)$, we multiply $\text{Sum}_{IQ}(t)$ by signal $\sin(2\pi f_c t)$. Then

$$\text{M2\_out} = \sin(2\pi f_c t)\,\text{Sum}_{IQ}(t) = \sin(2\pi f_c t)V_1(t)\cos(2\pi f_c t) + \sin(2\pi f_c t)V_2(t)\sin(2\pi f_c t)$$
$$= \tfrac{1}{2}V_1(t)[\sin(2\pi 2f_c t) + \sin(0)] + \tfrac{1}{2}V_2(t)[-\cos(2\pi 2f_c t) + \cos(0)]$$

When the high-frequency terms from M2_out are filtered out, the output of Filter_2 results in

$$\text{Filter2\_out} = \tfrac{1}{2}V_2(t)\cos(0) = \tfrac{1}{2}V_2(t)$$

The demodulation of single-sideband signals, however, does not require an exact phase of the oscillator but does require an exact frequency. Most single-sideband signals are beyond the audio band of 10 kHz or 20 kHz. Thus the demodulation of these types of single-sideband signals is relatively easy by simple multiplication with the oscillator signal, whose frequency is the same as the oscillator frequency at the transmitting end.

For example, for demodulating an upper sideband signal

$$\text{Subtr}_{IQ\_usb}(t) = \cos[2\pi(f_c + f_{mod})t]$$

we need to multiply this signal, $[\text{Subtr}_{IQ\_usb}(t)]$, with $\cos(2\pi f_c t)$ and filter out the high-frequency signal. Thus we have

$$\cos(2\pi f_c t)\text{Subtr}_{IQ\_usb}(t) = \cos(2\pi f_c t)\cos[2\pi(f_c + f_{mod})t]$$
$$= \tfrac{1}{2}\{\cos[2\pi(f_c + f_{mod})t + \cos(2\pi f_{mod})t]\}$$

When the high-frequency signal $\cos[2\pi(f_c + f_{mod})t]$ is filtered out, the result is

$$\tfrac{1}{2}[\cos(2\pi f_{mod})t] = \text{a scaled version of the original modulating signal}$$

I and Q Signals Used in Software-Defined Radios (SDRs)

Another use for quadrature modulation is to transform a signal into two signals that are equal in amplitude and also have a relationship 0 and 90 degrees between each other. This transformation also changes the frequency from the original. For example, if a signal is at 10 MHz, quadrature modulation can provide two signals at 20 kHz that are 90 degrees apart from each other. Figure 16-12 shows a 0- and 90-degree phase shifter based on IQ or quadrature modulation.

The figure shows a single signal source $V_{RF}(t) = \cos(2\pi f_{RF}t)$ that is connected to the input of two mixers M1 and M2. A signal generator $\cos(2\pi f_{OSC}t)$ is connected to the remaining input terminal of mixer M1, and another signal generator $\sin(2\pi f_{OSC}t)$ is connected to the remaining input of mixer M2.

The outputs of M1 and M2 are connected to filters Filter_1 and Filter_2, respectively. These filters remove the high-frequency signals from the outputs of the mixers. Therefore, the output of mixer M1 is

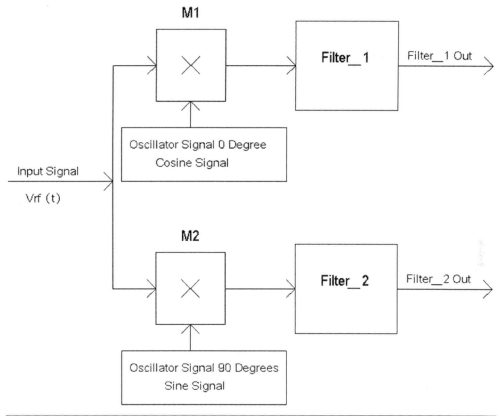

FIGURE 16-12 A 0- and 90-degree phasing circuit.

$$M1\_out = V_{RF}(t)\cos(2\pi f_{OSC}t) = \cos(2\pi f_{RF}t)\cos(2\pi f_{OSC}t)$$
$$= \tfrac{1}{2}[\cos(2\pi f_{RF}t + 2\pi f_{OSC}t) + \cos(2\pi f_{RF}t - 2\pi f_{OSC}t)]$$

Filter_1 removes the higher-frequency signal, which results in

$$Filter1\_out = \tfrac{1}{2}\cos(2\pi f_{RF}t - 2\pi f_{OSC}t) = \tfrac{1}{2}\cos(2\pi f_{OSC}t - 2\pi f_{RF}t) = I\ signal$$

For mixer M2, the output is

$$M1\_out = \cos(2\pi f_{RF}t)\sin(2\pi f_{OSC}t) = \sin(2\pi f_{OSC}t)\cos(2\pi f_{RF}t)$$
$$= \tfrac{1}{2}[\sin(2\pi f_{OSC}t + 2\pi f_{RF}t) + \sin(2\pi f_{OSC}t - 2\pi f_{RF}t)]$$

Filter_2 removes the higher-frequency signal, which results in

$$Filter2\_out = \tfrac{1}{2}\sin(2\pi f_{OSC}t - 2\pi f_{RF}t) = Q\ signal$$

Therefore, the signals from Filter1_out and Filter2_out are 90 degrees apart from each other no matter how low a frequency ($f_{OSC} - f_{RF}$) is, even down to near direct currect (DC). Now the fact that these two signals can maintain a 90-degree difference all the way down to nearly DC is truly amazing. The I and Q signals from the filters then allow a computer to digitize these signals and demodulate them.

References

1. U.S. Patent 1,855,576, Clyde R. Keith, "Frequency Translating System," filed on April 9, 1929.
2. U.S. Patent 2,086,601, Robert S. Caruthers, "Modulating System," filed on May 3, 1934.
3. U.S. Patent 2,025,158, Frank Augustus Cowan, "Modulating System," filed on June 7, 1934.
4. U.S. Patent 5,058,159, Ronald Quan, "Method and System for Scrambling and Descrambling Audio Information Signals," filed on June 15, 1989.
5. U.S. Patent 5,159,531, Ronald Quan and Ali R. Hakimi, "Audio Scrambling System Using In-Band Carrier," filed on April 26, 1990.
6. U.S. Patent 5,471,531, Ronald Quan, "Method and Apparatus for Low Cost Audio Scrambling and Descrambling," filed on December 14, 1993.
7. U.S. Patent 6,091,822, "Method and Apparatus for Recording Scrambled Video Audio Signals and Playing Back Said Video Signal, Descrambled, Within a Secure Environment," Andrew B. Mellows, John O. Ryan, William J. Wrobleski, Ronald Quan, and Gerow, D. Brill, filed on January 8, 1998.
8. Kenneth K. Clarke and Donald T. Hess, *Communication Circuits: Analysis and Design*. Reading: Addison-Wesley, 1971.
9. Thomas H. Lee, *The Design of CMOS Radio-Frequency Integrated Circuits*, 2nd ed. Cambridge: Cambridge University Press, 2003.
10. Donald O. Pederson and Kartikeya Mayaram, *Analog Integrated Circuits for Communication*. Boston: Kluwer Academic Publishers, 1991.
11. Allan R. Hambley, *Electrical Engineering Principles and Applications*, 2nd ed. Upper Saddle River: Prentice Hall, 2002.
12. Robert L. Shrader, *Electronic Communication*, 6th ed. New York: Glencoe McGraw-Hill, 1991.
13. Mischa Schwartz, *Information, Transmission, and Noise*. New York: McGraw-Hill, 1959.
14. Harold S. Black, *Modulation Theory*. Princeton: D. Van Nostrand Company, 1953.
15. B. P. Lathi, *Linear Systems and Signals*. Carmichael: Berkeley Cambridge Press, 2002.
16. Alan V. Oppenheim and Alan S. Wilsky, with S. Hamid Nawab, *Signals and Systems*, 2nd ed. Upper Saddle River: Prentice Hall, 1997.
17. Howard W. Coleman, *Color Television*. New York: Hastings House, Publishers, 1968.

18. Geoffrey Hutson, Peter Shepherd, and James Brice, *Colour Television System Principles*. 2nd ed. London: McGraw-Hill, 1990.

19. Mary P. Dolciani, Simon L. Bergman, and William Wooton, *Modern Algebra and Trigonometry*. New York: Houghton Mifflin, 1963.

Chapter 17

Intermediate-Frequency Circuits

The intermediate-frequency (IF) amplifier and filter determine selectivity and sensitivity for an amplitude-modulated (AM) superheterodyne radio. In commercially made radios, some of the first transistor radios needed two stages of IF amplification and three IF filters to match the sensitivity and selectivity of vacuum-tube receivers. Transistor radios made with only one IF amplifier generally performed worse in sensitivity (and selectivity) than the ones with two IF amplifiers. However, there are some ways to bring up the sensitivity of the radio by using longer antenna coils.

While IF amplifiers provide amplification, often an IF amplifier also serves to provide some type of automatic gain control. Thus another function of the IF amplifier is to reduce gain when the radio is tuned to a strong signal and increase the gain when a weak signal is received.

Therefore, the objectives of this chapter are

1. To investigate different types of IF amplifiers.
2. To introduce the early effect that results in a lossy resistance across the collector-emitter junction of a transistor.
3. To examine how gain control is achieved.
4. To look at how certain nonlinearities of an IF amplifier affect distortion.

IF Amplifiers

From the output current of a mixer (or converter) transistor comes a multitude of signals. Four signals from the output are the RF signal, the oscillator signal, the signal whose frequency is the sum of the RF and oscillator frequencies, and finally, the IF signal, whose frequency is the difference between the RF and oscillator frequencies. A simple inductor-capacitor (LC) IF filter is tuned to the IF frequency band but still passes some of the oscillator signal through. The reason is that the oscillator signal current is very strong in amplitude whether a simple mixer is used or a mixer oscillator converter circuit is used. Also, note that the IF signal is generally very small in amplitude compared with the oscillator signal.

In American and Japanese transistor radio designs from the 1950s to the 1980s, a single tuned IF transformer is coupled between the output of the mixer or converter and the first IF amplifier. However, some transistor radio designs from Russia use a double tuned IF stage following the converter or mixer.

At first look, one would think that the first IF amplifier transistor has a lot to handle with a high-amplitude oscillator signal and a small-amplitude IF signal at its input. Fortunately, there is sufficient attenuation of the oscillator signal, and the IF amplifier transistor can handle and amplify both signals in a mostly linear manner. A generalized IF system using a one-transistor IF amplifier is shown in Figure 17-1.

This figure shows an IF amplifier system in which the first IF filter is a single tuned circuit that couples into the base of Q1. A common emitter amplifier generally is used to allow for moderately high input resistance (e.g., 500 Ω to 2,000 Ω) so that the first IF filter is not excessively loaded down. A heavy load on the IF filter will cause low signal and reduced Q. Because the emitter is alternating-current (AC) grounded via capacitor CE, the small-signal transconductance of the amplifier is $g_m = I_{CQ1}/(0.026$ volt), where I_{CQ1} is the quiescent direct-current (DC) collector current.

The gain of the common emitter amplifier then is $-gm\ R_L$, where R_L is the equivalent load resistance at the IF (i.e., 455 kHz). Since the DC current is set up by the voltage VE across RE,

$$I_{CQ1} = \frac{VE}{RE}$$

Note that changing the DC current also changes the transconductance of Q1. Or put another way, changing the base voltage at Q1 results in a change in

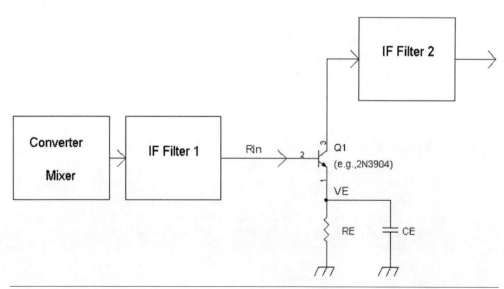

FIGURE 17-1 A block schematic diagram of a typical IF system for a radio.

transconductance. For example, increasing the DC voltage at the base increases the transconductance, whereas lowering the DC voltage at the base reduces the tranconductance.

In terms of input resistance of the amplifier, the input resistance looking into the base of Q1 is

$$\frac{\beta}{g_m} = \beta\frac{0.026V}{I_{CQ1}} = R_{in}$$

Note that while increasing the DC collector current increases the transconductance, it also lowers the input resistance.

For a collector current of 0.5 mA and a current gain β of 50,

$$R_{in} = 2.6 \text{ k}\Omega$$

If the collector current is increased to 5 mA,

$$R_{in} = 260 \text{ }\Omega$$

and if the collector current is decreased to 0.05 mA,

$$R_{in} = 26 \text{ k}\Omega$$

The collector of Q1 then normally is connected to another IF transformer, simple LC filter, or a ceramic filter. As mentioned previously, the ceramic filter requires some prefiltering or attenuation of the oscillator signal for proper band-pass filtering. Without the prefiltering from another LC circuit, the ceramic filter will have too much leak-through of the oscillator signal, which eventually will interfere with envelope detection of the AM signal.

Detection of the AM signal typically is taken from the output of the output filter. However, if more selectivity and sensitivity are needed, a second-stage IF amplifier is added after the output filter from Q1, and detection of the AM signal is taken from the output filter of the second IF amplifier.

Figure 17-2A is a schematic diagram of a typical IF amplifier such as from a commercially made transistor radio. The figure shows a one-transistor IF amplifier. IF transformer T1 couples a low-impedance signal source (e.g., 500 Ω or less) into the base of Q1.

The gain of the IF amplifier is $-g_m R_L$ where R_L is the equivalent load resistance at the IF (i.e., 455 kHz) from the collector. One should note that there is a capacitor across the collector base junction of Q1, Ccb. This capacitor is the internal capacitor in all transistors, and the capacitance depends on the supply voltage. Lowering the supply voltage results in higher capacitance for Ccb, and increasing the supply voltage results in lower capacitance for Ccb. In the earlier days of transistor radios, such as in the 1950s, capacitor Ccb had very high capacitance, which could cause two problems. These problems were a loss in gain and the possibility of causing oscillation. Therefore, to avoid losses in gain and to reduce the possibility of oscillation, the

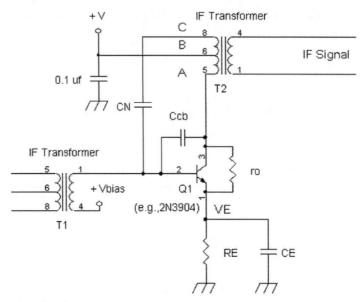

Note: Each IF Transformer has an internal capacitor.

FIGURE 17-2A   An IF amplifier circuit.

primary winding of the IF transformer had a tap for neutralizing or canceling out the Ccb capacitance.

In Figure 17-2A, the IF signal at the collector of Q1, node A, is 180 degrees out of phase from the signal at node C. Node B of IF transformer T2 is connected to the power supply. Therefore, a neutralizing capacitor CN is used to cancel out the capacitance from Ccb. For example, if the primary winding has equal turns from the center tap to the outside terminals, then the magnitude or amplitude of the signals at nodes A and C will be equal. The neutralizing capacitor will have the same capacitance as Ccb; thus CN = Ccb.

However, by the 1960s, transistors had improved, and the capacitance of Ccb was low enough to do without the neutralizing capacitance CN. But there are at least a couple of other reasons for the collector current output signal to feed into a tapped LC circuit. One reason is to avoid oscillation by setting up a lower-impedance point at the tap so that the overall gain of the amplifier is not too high. Recall that the impedance of a parallel LC circuit is very high and in practice can be in the hundreds of kiloohms at resonance.

Another reason is to avoid loading down the Q of the LC circuit. By using a tapped winding, any resistive losses from the collector to the emitter of the transistor will cause negligible problems. Since emitter capacitor CE is an AC short to ground, the collector to emitter resistance in the transistor will act as a resistor from the collector to ground.

This collector to emitter resistance is usually given as

$$r_0 = \frac{\Delta V_{CE}}{\Delta_{IC}} = \frac{\text{change in collector to emitter voltage}}{\text{change in collector current}} \tag{17-1}$$

$$r_0 = \frac{V_A}{I_{CQ}} \tag{17-2}$$

where I_{CQ} is the DC quiescent collector current, and V_A is the Early voltage.

Typically, the Early voltage V_A is 100 volts to 250 volts. For example, at a DC collector current of 1 mA for I_{CQ} and an Early voltage of 100 volts, the output resistance from collector to emitter is

$$r_0 = \frac{100V}{0.001\ A} = 100\ k\Omega$$

For a typical IF transformer, the lossy resistance across the whole primary winding from nodes A to C is about 200 kΩ. The turns ratio from nodes AB/AC is 1:3. This means that the loading by r_0 = 100 kΩ reflects back to nodes AC as 100 kΩ × 3^2 = 900 kΩ in parallel with the 200-kΩ lossy resistor. However, 900 kΩ in parallel with 200 kΩ = 163 kΩ. The 900-kΩ resistance lowers the Q of the coil (and the IF amplifier gain) by about 20 percent because 163 kΩ ≈ 80 percent of 200 kΩ.

If the collector of Q1 is connected to node A and not to a tap of the primary winding, the output resistance of the transistor will dramatically load down the Q (Figure 17-2B). In this figure, the collector is connected to node A, or the full winding of the coil. Therefore, r_0 = 100 kΩ is now in parallel with the 200-kΩ resistor across nodes A and C of the coil, which then results in an equivalent resistance of 67 kΩ across the full primary winding of T2, which is one-third of 200 kΩ. Because of this dramatic drop in resistance, the Q of the coil now has dropped by two-thirds, or 67 percent.

A typical 455-kHz IF transformer has an internal resonating capacitor of about 200 pF. Therefore, the Q of the coil is

$$Q = (2\pi 455,000\ Hz)RC = (2\pi 455,000\ Hz)R(200\ pF)$$

if R = 200 kΩ, then Q = 114. If R drops to 163 kΩ, then Q = 93, and if R drops to 67 kΩ, then Q = 38.

As can be seen, connecting the collector of the transistor to a tap on a tapped winding has an advantage of minimizing Q losses. Also recall that a loss in Q of the coil results in poorer selectivity and loss of gain.

The gain of the IF amplifier from the base to the collector is $-g_m R_L$, where R_L is the equivalent load resistance at the IF (i.e., 455 kHz) that is connected to the collector.

One other "side effect" of connecting the collector to the coil as shown in Figure 17-2B is that the transistor's Ccb can mistune the IF transformer when the power

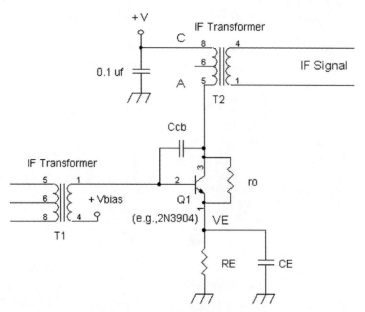

Note: Each IF Transformer has an internal capacitor.

FIGURE 17-2B An IF amplifier with the collector not connected to the tap of the primary winding.

supply drops in voltage. Since the base of the transistor is driven preferably by a low-impedance source, Ccb on the base side is almost grounded. Thus Ccb is almost like a capacitor from the collector to ground. Because the IF transformer's internal capacitor is typically 200 pF, a change of about 4 pF in Ccb when the supply voltage changes will cause a 1 percent shift in resonant frequency. Put another way, if the IF transformer is tuned with the Ccb of the transistor for 455 kHz, and if the supply voltage drops such that Ccb changes by 4 pF, the resonant frequency will shift downward to about 451 kHz. This shift in resonant frequency amounts to a slight mistuning of the IF transformer and can reduce the amplitude of the 455-kHz signal.

Thus the use of a tapped winding also reduces sensitivity to a shift in the resonant frequency when Ccb changes. For example, with a 3:1 turns ratio, a shift in capacitance of 4 pF with a tapped connection in the primary winding reduces the shift change in capacitance by 3^2, or ninefold (see below). That is,

$$\frac{4\text{pF}}{3^2} = 0.44 \text{ pF}$$

that is added to the 200 pF, which is much lower than 4 pF added to the 200-pF resonating capacitor.

Alternatives to the common-emitter amplifier are the common-base IF amplifiers seen in Figures 17-3A and 17-3B, the cascode amplifier seen in Figures 17-4A and 17-4B, and the differential-pair amplifier seen in Figures 17-5A and 17-5B.

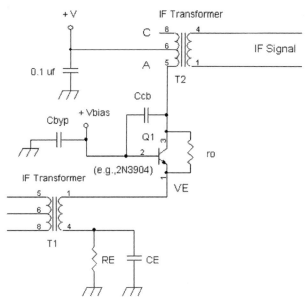

FIGURE 17-3A   Common-base IF amplifier.

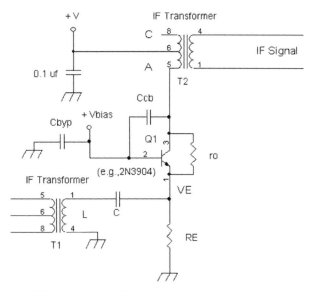

FIGURE 17-3B   Common-base amplifier that was used in General Electric transistor radios in the 1960s.

In the common-base amplifier, the gain from the emitter to the collector is $+g_m R_L$, where R_L is the equivalent load resistance at the IF (i.e., 455 kHz) from the collector. To avoid the problem of feedback capacitance from Ccb, a grounded-base amplifier is one solution. What is meant by grounded base is that the base of the transistor is grounded in terms of the AC signal. Therefore, bypass capacitor Cbyp is attached to the base and ground. Inspection of Figures 17-3A and 17-3B will show that Ccb is now grounded via the base lead of the transistor. Thus Ccb just adds capacitance to the tank circuit, but not much because of the tapped configuration of T2.

In a common-emitter amplifier, as seen in Figure 17-2B, the feedback capacitance Ccb along with inverting voltage gain $-A = -g_m R_L$ at the collector forms a Miller capacitance from the base of Q1 to ground. This Miller capacitance is equal to

$$Ccb(1 + A) = Cin$$

from the base to ground. Thus the Miller capacitance is a result of capacitance multiplication. Note that there will be further description of the Miller capacitance in Chapter 19. However, in a common-base amplifier, as shown in Figure 17-3B, there is no Miller capacitance because the base is grounded, and the input terminal is at the emitter. Note: The gain, A, is the magnitude of the inverting gain and A is positive.

In Figure 17-3A, the common-base amplifier is driven by the secondary winding of a parallel-tuned circuit. To preserve a reasonable amount of $Q > 30$, the step-down ratio of the IF transformer should be higher than those of the IF transformers used in coupling signal to a common-emitter IF amplifier. The input resistance in a common-base amplifier is $1/g_m$ instead of β/g_m, which is the input resistance to a common-emitter amplifier with the emitter AC grounded. Therefore, the input resistance of a common-base amplifier is lowered by a factor of β compared with the input resistance of a common-emitter amplifier.

For example, if the common-base amplifier's DC collector current is 0.2 mA, the input resistance is 130 Ω. In a tapped coil or a transformer, the impedance ratio is the square of the turns ratio. The turns ratio of the T1 should be on the order of at least 25:1 that yields an impedance ratio of 25^2:1, which means that the equivalent resistance across the tank circuit is $25^2 \times 130\ \Omega = 625 \times 130\ \Omega = 81\ k\Omega$ for $Q = 46$. Should the collector current increase, the turns ratio of the IF transformer must be increased as well to preserve sufficient Q for good selectivity.

In Figure 17-3B, the common-base amplifier is driven by a series resonant LC circuit. In a series resonant circuit, the desired goal is to load into a small input resistance. Thus the common-base amplifier fits the goal. However, series resonant IF transformers were not very common in AM transistor radios and were used mainly by the General Electric Company in the 1960s for some designs.

In terms of the collector to ground output resistance in a common-base amplifier, this output resistance $R_0 = r_0(1 + g_m R_s)$, where R_s is the source resistance of the signal source from the previous stage, and where $R_s \ll R_E$ and $g_m R_s \ll \beta$. For example, if $R_E = 1,000\ \Omega$ and $R_s = 100\ \Omega$, and if $g_m = 0.0384$ mho for 1-mA DC collector current, $g_m R_s = 3.8$. Let $r_0 = 100\ k\Omega$; then

$$R_0 = 100 \text{ k}\Omega(1 + 3.8) = 480 \text{ k}\Omega$$

Thus the output resistance is raised in a common-base amplifier compared with a common-emitter amplifier *with the emitter AC grounded.*

If the emitter is not AC grounded in a common-emitter amplifier, or if $C_E = 0$ (no capacitor in the circuit), then the output resistance from collector to ground in a common-emitter amplifier with emitter resistor RE is

$$R_0 = r_0 \left(1 + \frac{\beta RE}{r_\pi + RE}\right)$$

where $r_\pi = \beta/g_m$. For small values of RE, where $g_m RE \ll \beta$, then $R_0 = r_0(1 + g_m RE)$. Thus, if $R_E = 100 \ \Omega$ and $r_0 = 150 \text{ k}\Omega$ and the collector current is 1 mA, then

$$R_0 = 150 \text{ k}\Omega(1 + 3.8) = 150 \text{ k}\Omega(4.8) = 720 \text{ k}\Omega$$

Other characteristics of the common-emitter amplifier will be covered in Chapter 19, such as input resistance with the emitter not bypassed to ground.

Figure 17-4A shows a circuit to reduce Miller capacitance. The figure shows a combination of a common-emitter amplifier whose output collector from Q1 feeds

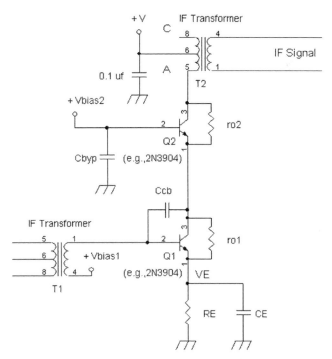

Note: Each IF Transformer has an internal capacitor.

FIGURE 17-4A   **A two-transistor cascode IF amplifier.**

into a common-base amplifier Q2. This configuration is called a *cascode amplifier.* The motivation for a cascode amplifier is to provide the input resistance of a common-emitter amplifier while minimizing Miller capacitance or minimizing feedback from the collector output of Q2 back to the base input of Q1. Note that for $\beta > 20$, for all practical purposes, the DC collector currents of Q1 and Q2 are the same. Thus the transconductances of both transistors Q1 and Q2 are the same; that is, $g_{mQ1} = g_{mQ2}$.

And the gain of the IF amplifier from the base of Q1 to the collector of Q2 is $-g_m R_L$, where R_L is the equivalent load resistance at the IF (i.e., 455 kHz) from the collector of Q2. In terms of DC bias voltages, Vbias2 is generally at least 1 volt DC above Vbias1. For example, if Vbias1 = 2 volts DC, Vbias2 is at least 3 volts DC.

The Miller capacitance at the input of Q1's base is twice the Ccb capacitance of Q1 because the gain from the base to collector of Q1 is –1, so A = 1. The voltage gain is $-g_{m\_Q1}/g_{m\_Q2}$. Thus the Miller capacitance = 2Ccb, where Ccb is the collector to base capacitance for Q1.

However, the Miller capacitance can be lowered further if Q2 is biased for more collector DC current than Q1 (Figure 17-4B). By connecting a resistor RE2 to ground,

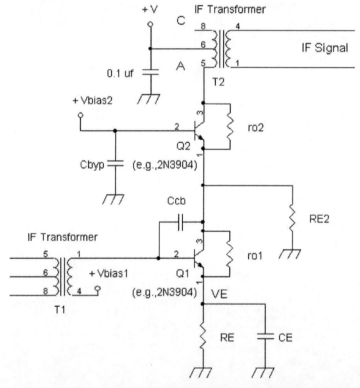

Note: Each IF Transformer has an internal capacitor.

FIGURE 17-4B A cascode amplifier with resistor RE to increase collector current of Q2.

the collector current of Q2 is increased; then $g_{m\_Q2} > g_{m\_Q1}$. For example, if Q1 is biased for 1 mA, and RE2 is chosen to drain 1 mA to ground, Q2 will have a total of 2 mA, 1 mA from Q1 and another 1 mA via resistor RE2. Thus $g_{m\_Q2} = 2g_{m\_Q1}$, so the gain is $-g_{m\_Q1}/g_{m\_Q2} = -g_{m\_Q1}/2g_{m\_Q1} = -(\frac{1}{2})$ or $A = (\frac{1}{2})$ which leads to the Miller capacitance $= 1.5Ccb$, where Ccb is the collector to base capacitance for Q1.

Obviously, there is a limit as to how much more you can increase current to Q2 because the lowest capacitance is 1 Ccb. In terms of output resistance from Q2's collector, with RE2 removed, Q2's emitter is "seeing" a resistor to ground from the collector to emitter resistance of Q1, r_{o1}, which is very high. Thus

$$R_0 = r_{o2}(1 + \beta)$$

is a very high resistance at Q2's collector, which makes the collector current from Q2 for all practical purposes an ideal current source.

Thus, to raise the output resistance of a transistor, field-effect transistor (FET), or metal-oxide semiconductor field-effect transistor (MOSFET), use a cascode circuit. In curve tracers, the output resistance causes the current curves to rise up at a positive slope. When the output resistance is increased, the current curves stay more flat in slope.

In practice, the benefit of the higher output resistance in a cascode amplifier with Q2's collector connected to a tap of the primary winding is negligible when compared with a common-emitter amplifier. However, the isolation from the output of Q2 to the input of Q1 reduces the possibility of oscillation.

Another way to reduce or avoid Miller capacitance is to use a differential amplifier, as seen in Figure 17-5A. Differential amplifiers in IF stages are used more commonly in communications receivers, stereo hi-fi receivers, and integrated circuits. They were rarely, if ever, used in transistor radios. This figure shows a differential amplifier with matched transistors Q1 and Q2. Q1 isolates and removes any Miller capacitance because its collector is tied to an AC ground, the power supply. The output at Q2 then feeds signal current into the IF transformer. The output resistance (referenced to ground) at Q2's collector is

$$R_0 = r_{o2}(1 + \frac{g_{m\_Q1}}{g_{m\_Q2}})$$

Since the collector currents are equal for Q1 and Q2, $g_{m\_Q1} = g_{m\_Q2}$, which leads to

$$R_0 = r_{o2}(1 + 1) = 2\,r_{o2}$$

$$g_{m\_Q1} = [\frac{V_{bias} - 0.7V}{2RE}]/0.026\ \text{volt}$$

And the gain is $(\frac{1}{2})g_{m\_Q1}R_L$, where R_L is the equivalent resistive load. The input resistance into the base of Q1 is $2\beta/g_{m\_Q1}$.

If matched transistors are not available, one can use discrete transistors, as shown in Figure 17-5B. This figure shows an alternative differential-pair amplifier that does

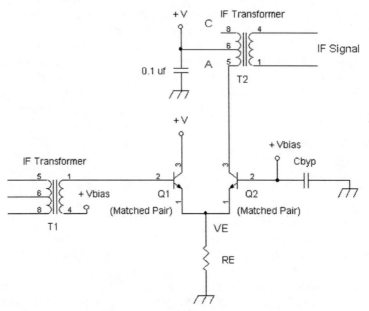

Note: Each IF Transformer has an internal capacitor.

FIGURES 17-5A A differential-pair amplifier for IF signal amplification.

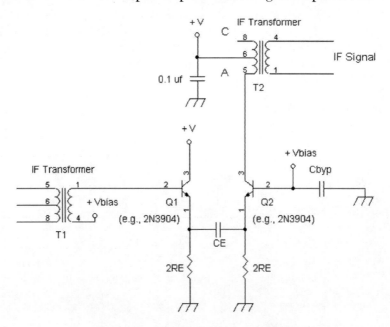

Note: Each IF Transformer has an internal capacitor.

FIGURES 17-5B A discrete transistor differential-pair amplifier using AC coupling between the emitters.

not require matched transistors for Q1 and Q2. Note that the emitter resistors are twice the value of RE in Figure 17-5A to result in the same collector current for a given Vbias. When the transistors are not matched, the base emitter voltages of Q1 and Q2 are different for the same collector current. Thus, if they are connected as shown in Figure 17-5A, there will be a great imbalance of DC collector current. By isolating the emitters with a capacitor CE in Figure 17-5B, each emitter of Q1 and Q2 is biased separately to essentially the same DC collector current via the emitter resistors. Capacitor CE is generally large. For example, at 455 kHz, a 0.01-µF capacitor has an impedance of about 50 Ω. For collector currents set to 1 mA, the emitter input and output resistances are about 26 Ω. Thus the capacitance of CE should be at least 10 times lower than the 26 Ω at the IF. For example, with collector currents at 1 mA, CE = 0.47 µF, which is about 1 Ω at 455 kHz.

To calculate R_L, the equivalent load resistor at the collector of the transistor, see Figure 17-6. Given a transformer with number of turns n1 and n2 at the primary and n3 in the secondary, and with the equivalent unloaded Q resistance Rpri across the whole winding of the primary, an equivalent resistance can be calculated across the n1 winding.

An equivalent secondary resistance Rsec in parallel to Rpri is

$$Rsec = Rin(\frac{n1 + n2}{n3})^2$$

where Rin is the input resistance that is loading the secondary winding. For example, Rin is the input resistance to a common emitter or common-base amplifier.

Rsec and Rpri are two resistors in parallel; that is,

$$Rsec||Rpri = \frac{RsecRpri}{Rsec + Rpri}$$

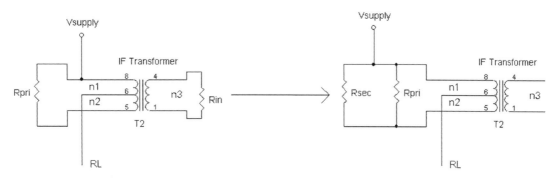

FIGURE 17-6   **An IF transformer with turns denoted at the primary and secondary windings.**

where

$$Rsec = Rin(\frac{n1 + n2}{n3})^2$$

Then the equivalent resistance across the winding n1 is

$$RL = [Rsec||Rpri](\frac{n1}{n1 + n2})^2$$

Gain-Controlled IF Amplifiers

Normally, it is desirable to control the gain of one or more IF amplifiers to equalize the loudness when tuned to a weak or strong radio-frequency (RF) signal. In AM receivers, the gain control is "automatic" via the detector stage. Often in communications radios the IF gain can be controlled manually or automatically via a detector stage.

Usually, the gain control should be set at the very first IF stage. In this way, any succeeding IF stage can avoid distortion of the IF signal that would result in distortion in the detected signal. For example, if a receiver has two IF amplifiers, if the second IF amplifier is adjusted for collector current as a way to control its gain, then a strongly received signal will cause the first IF stage to provide large signals into the second IF amplifier. However, the distortion of the IF amplifier is based on signal voltage into the input and not on collector current. Therefore, a high-level signal into the second IF stage can cause overload.

Normally, gain control on just the first IF amplifier should be adequate to equalize the loudness. The gain control is just a function of Vbias connected to the base via a winding of the IF transformer or via capacitive coupling. Therefore, the gain of each common emitter or cascode IF stage is

$$-[\frac{Vbias - 0.7V}{(RE)0.026V} R_L] = \text{gain as a function of Vbias for the common-emitter stage}$$

and

$$-[\frac{Vbias1 - 0.7V}{(RE)0.026V} R_L] = \text{gain as a function of Vbias1 for the cascode amplifier}$$

If a common base amplifier is driven by a very low-impedance input signal source, the preceding relationship for the common-emitter stage also holds.

For the differential pair amplifiers in Figures 17-5A or 17B,

$$\frac{1}{2} \frac{Vbias - 0.7V}{(2RE)0.026V} R_L = \text{gain as a function of Vbias}$$

Considerations of Distortion Effects on IF Amplifiers

As one who has made both tube and transistor AM radios, I find that the sound of tube radios has less distortion than that of transistorized receivers. That is, the sound is cleaner with tube radios. There could be a couple of factors responsible for this, such as a more linear IF amplifier or a class A audio amplifier in the tube radio versus a transistor radio with multiple stages of IF amplification and a class B audio amplifier. However, I am not talking about the "mysterious" tube versus transistor sound. What I would like to investigate instead is measurable distortion effects. Therefore, this section will look into the distortion characteristics of IF systems using tubes, FETs, and bipolar transistors.

In particular, what type of distortion in the IF system would most likely cause a demodulated AM signal to produce a distorted audio signal? That is, what role does second- and third-order distortion play in a demodulated signal?

Pentodes and FETs both have approximately square-law characteristics in terms of input voltage and output current. Therefore, both have very small amounts of third-order distortion. The following equation is for a generalized nonlinear amplifying device:

$$I_{out} = a_0 + (a_1)V_{sig} + (a_2)V_{sig}^2 + (a_3)V_{sig}^3 + \cdots + (a_N)V_{sig}^N \qquad (17\text{-}3)$$

The signal into the IF amplifier is characterized as an amplitude-modulated (AM) signal at the IF f_{IF}:

$$[1 + m(t)]\cos(2\pi f_{IF})t = V_{in\_IF} = V_{sig}$$

Note

m(t) denotes the modulating signal or audio signal that will be demodulated later. The spectrum that m(t) occupies is generally from about 20 Hz to about 5 kHz or 10 kHz.

For an approximate square-law device such as a pentode or FET,

$$I_{out\_SQ\_law} = a_0 + a_1[1 + m(t)]\cos(2\pi f_{IF})t + a_2\{[1 + m(t)]\ \cos(2\pi f_{IF})t\}^2 \quad (17\text{-}4)$$

In terms of the IF filter, the only signals that will pass through the IF bandpass filter are the ones that are centered around the frequency f_{IF}. All other signals, including DC terms or harmonics, will be filtered out by the IF transformer or filter. That is,

$$I_{out\_SQ\_law} = a_0 + a_1[1 + m(t)]\cos(2\pi f_{IF})t + a_2[1 + m(t)]^2\ [\cos(2\pi f_{IF})t]^2 \quad (17\text{-}5)$$

From a trigonometric identity, $[\cos(\beta)]^2 = (\frac{1}{2})[\cos(2\beta) + \cos(0)] = (\frac{1}{2})[\cos(2\beta) + 1]$, therefore,

$$I_{out\_SQ\_law} = a_0 + a_1[1 + m(t)] \cos(2\pi f_{IF})t + a_2[1 + m(t)]^2 [\frac{1}{2}][\cos(4\pi f_{IF})t + 1]$$

And by expanding out the terms related to a_2, we get

$$\begin{aligned} I_{out\_SQ\_law} = a_0 &+ a_1[1 + m(t)] \cos(2\pi f_{IF})t + a_2[1 + m(t)]^2 [\frac{1}{2}] \\ &+ a_2[1 + m(t)]^2 [\frac{1}{2}]\cos(4\pi f_{IF})t \end{aligned} \qquad (17\text{-}6)$$

Note $\cos(4\pi f_{IF})t = \cos(2\pi 2f_{IF})t$ is a second harmonic of the IF signal.

Now let's examine each term that makes up $I_{out\_SQ\_law}$:

1. a_0 is a DC term that is filtered out by the IF filter.
2. $a_1[1 + m(t)] \cos(2\pi f_{IF})t$ is the linear "amplified amplitude-modulated IF signal" centered around the IF that passes through the IF filter.
3. $a_2[1 + m(t)]^2[\frac{1}{2}]$ is a baseband audio signal and DC term that is filtered out by the IF filter. For example, this baseband audio signal has frequencies below 20 kHz, and a 455-kHz IF filter will remove these audio and DC signals.
4. $a_2[1 + m(t)]^2[\frac{1}{2}]\cos(4\pi f_{IF})t]$ is an AM signal that is centered around twice the IF. Thus the IF filter will remove signals related to the second harmonic of the IF.

Therefore, after IF filtering, all that is remaining is

$$I_{out\_SQ\_filtered} = a_1[1 + m(t)] \cos(2\pi f_{IF})t \qquad (17\text{-}7)$$

This result is quite remarkable. What it says is that a device can have as much second-order distortion in an IF amplifier, but all the second-order distortion is filtered out by the IF filter, leaving only "distortion-less" amplification. Therefore, a perfect square-law device contributes no distortion to an IF amplifier.

Now let's take a look at third-order distortion in amplifiers that include a cubic term such as a bipolar transistor or a differential-pair amplifier. Since the terms relating to a_0, a_1, and a_2 have been analyzed, let's examine the term relating to a_3:

$$a_3 (V_{sig}^3) = a_3 ([1 + m(t)]\cos(2\pi f_{IF})t)^3 \qquad (17\text{-}8)$$

Equation (17-8) then equals:

$$\begin{aligned} &a_3 ([1 + m(t)]\cos(2\pi f_{IF})t) ([1 + m(t)]\cos(2\pi f_{IF})t)^2 = \\ &a_3 ([1 + m(t)]\cos(2\pi f_{IF})t)[1 + m(t)]^2 [\frac{1}{2}][\cos(4\pi f_{IF})t + 1] = a_3 (V_{sig}^3) \qquad (17\text{-}9) \end{aligned}$$

Equation (17-9) then equals:

$$a_3 \cos(2\pi f_{IF})t)[1 + m(t)]^2 \, [^1/_2][\cos(4\pi f_{IF})t + 1] +$$

$$a_3[m(t)]\cos(2\pi f_{IF})t)[1 + m(t)]^2 \, [^1/_2][\cos 4\pi f_{IF})t + 1] = a_3 \, (V_{sig}^3) =$$

$$a_3 \, [\cos(2\pi f_{IF})t[1 + m(t)]^2 \, [^1/_2][\cos(4\pi f_{IF})t] +$$

$$a_3 \, \cos(2\pi f_{IF})t)[1 + m(t)]^2 \, [^1/_2] +$$

$$a_3[m(t)][\cos(2\pi f_{IF})t][1 + m(t)]^2 \, [^1/_2][\cos(4\pi f_{IF})t] +$$

$$a_3[m(t)]\cos(2\pi f_{IF})t)[1 + m(t)]^2 \, [^1/_2] = a_3 \, (V_{sig}^3) \qquad (17\text{-}10)$$

Note $\cos(4\pi f_{IF})t = \cos(2\pi 2f_{IF})t$ is a second harmonic of the IF signal.

Now let's take a look at each of the four terms in Equation (17-10) related to third-order distortion:

1. $a_3 \, [\cos(2\pi f_{IF})t][1 + m(t)]^2 \, [^1/_2][\cos(4\pi f_{IF})t]$ represents sum and difference signals, one at f_{IF} and another at the third harmonic $3f_{IF}$. After IF filtering, the signal related to f_{IF} is: $a_3 \, [\cos(2\pi f_{IF})t][1 + m(t)]^2 \, [^1/_4] = a_3 \, [\cos(2\pi f_{IF})t)[1 + 2m(t) + [m(t)]^2][^1/_4]$. However the modulating signal is squared, which means that there will be second harmonics of the modulating signal generated, thus causing distortion when the signal is demodulated.

2. $a_3 \, [\cos(2\pi f_{IF})t][1 + m(t)]^2 \, [^1/_2] = a_3 \, [\cos(2\pi f_{IF})t)[1 + 2m(t) + [m(t)]^2][^1/_2]$ represents an amplitude modulated signal at a frequency of f_{IF}, which is a signal that passes through the IF filter. However, since there is a squared modulating signal term, there will be second-order audio distortion when the signal is demodulated.

3. $a_3 \, [m(t)][\cos(2\pi f_{IF})t][1 + m(t)]^2 \, [^1/_2][\cos(4\pi f_{IF})t]$ represents sum and difference signals, one signal at f_{IF} and another signal at $3f_{IF}$. After using the IF filter we have: $a_3 \, [m(t)][\cos(2\pi f_{IF})t][1 + m(t)]^2 \, [^1/_4] = a_3 \, \cos(2\pi f_{IF})t[m(t) + 2[m(t)]^2 + [m(t)]^3][^1/_4]$. Note that the modulating signal now has been squared and cubed, which will lead to second- and third-order distortion on $m(t)$ when the signal is demodulated.

4. $a_3 \, [m(t)][\cos(2\pi f_{IF})t][1 + m(t)]^2 \, [^1/_2] = a_3 \, \cos(2\pi f_{IF})t[m(t) + 2[m(t)]^2 + [m(t)]^3][^1/_2]$ represents an amplitude modulated signal at frequency but contains second- and third-order distortions on the modulating signal $m(t)$.

Thus third-order distortion in an IF amplifier results in both second- and third-order distortion on the modulating signal, which results in a distorted demodulated audio signal. Thus typically a single-ended or differential-pair transistor amplifier in the IF section will cause second- and third-order distortion at the output of the detector stage. It should be noted that intuitively it makes sense as to why third order distortion in the IF stage causes a distorted demodulated signal. An AM signal has a carrier, upper-sideband, and lower-sideband signals that are close together in frequencies. If two signals of close frequencies, F1 and F2, are connected to amplifier with third-order distortion, the resulting third-order intermodulation distortion products will have frequencies of (2F1 − F2) and (2F2 − F1) that fall within the IF bandwidth (band-pass), which will cause distortion in the demodulated signal. However, an amplifier with second-order distortion will have intermodulation products whose frequencies are (F1 − F2) and (F1 + F2) that fall outside the IF bandwidth (band-pass), and thus will not cause distortion in the demodulated signal.

Recall that second-order distortion in an IF amplifier does not add any distortion at all to the modulating signal, so when the IF signal is demodulated, the recovered audio signal is free of distortion. Therefore, square-law devices such as pentodes and FETs provide "distortion-free" demodulated signals.

Question: Is it possible to make a single-ended amplifier free of third-order distortion? The answer is yes.

One of the most useful facts I remember from my days taking classes in nonlinear integrated circuits design from Professor R. G. Meyer is that if an emitter degeneration resistor RE is set to $(\frac{1}{2})(1/g_m)$ in a common-emitter amplifier (*without* bypassing the emitter to ground), the third-order distortion disappears! For example, if the DC collector current is at 100 µA, set the emitter resistor RE to 130 Ω, and "magically" the third-order distortion products go to zero. Actually, the "magic" was in his analysis on how a small amount of local feedback in a common-emitter amplifier would cancel out the third-order term. In practice, the transistor already has some lossy internal resistances, so one may have to adjust the emitter resistor for minimum third-order distortion. Because the DC voltage drop across a resistor of $(\frac{1}{2})(1/g_m)$ is too low to bias the collector current reliably, see Figure 17-7 for an example of how to implement a third-order distortion-canceling amplifier.

In this figure, Vbias is fixed and is not varied for gain control. Typically, Vbias is set for at least 1 volt across RE. Capacitor CE is large such that its impedance is typically 1 Ω or less at frequency f_{IF}. Series resistor RE3 is selected such that the parallel combination of resistors RE and RE3 is $(\frac{1}{2})(1/g_m)$. For example, if Vbias is set such that there is 1 volt across RE, or VE = 1 volt DC and RE = 2,000 Ω, then the collector current is 0.5 mA, and $(\frac{1}{2})(1/g_m)$ is 26 Ω. Because RE >> $(\frac{1}{2})(1/g_m)$, we can set RE3 to about 27 Ω ±10 percent for minimum third-order distortion. In practice, there is a source resistance that drives the base of the transistor, and the transistor has an internal series base resistor. Typically, in an IF transformer, the secondary winding has an equivalent source resistance driving the base with about a few hundred ohms. Also, the transistor may have an internal base series resistance of 50 Ω to 1,000 Ω. With the transistor's current gain β large (e.g., β > 50), the source resistance Rs and the base series resistance Rbb can be reflected back to the emitter as an extra series emitter resistance of (Rs + Rbb)/β.

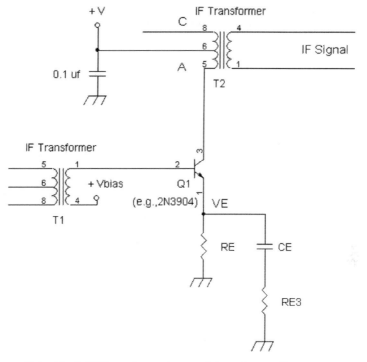

Note: Each IF Transformer has an internal capacitor.

FIGURE 17-7   An example of how to implement third-order distortion cancellation in a common-emitter amplifier.

Suppose that Rbb + Rs = 700 Ω, β = 100, and the collector current is 0.5 mA; then the equivalent "internal emitter resistor" is $(700/100)\Omega = 7\ \Omega$. Then, to minimize third-order distortion,

$$\text{RE3} = \frac{1}{2}\frac{1}{g_m} - \frac{\text{Rs} + \text{Rbb}}{\beta}$$

$$= 26\ \Omega - 7\ \Omega = 19\ \Omega = \text{RE3}$$

References

1. Class Notes EE140, Robert G. Meyer, UC Berkeley, Fall 1975.
2. Class Notes EE240, Robert G. Meyer, UC Berkeley, Spring 1976.
3. Paul R. Gray and Robert G. Meyer, *Analysis and Design of Analog Integrated Circuits*, 3rd ed. New York, John Wiley & Sons, 1993.

4. Kenneth K. Clarke and Donald T. Hess, *Communication Circuits: Analysis and Design*. Reading: Addison-Wesley, 1971.

5. Robert L. Shrader, *Electronic Communication*, 6th ed. New York: Glencoe/McGraw-Hill, 1991.

6. General Electric Company, *Essential Characteristics of Receiving Tubes*. 13th ed. Owensboro: General Electric Company, 1969.

7. Mary P. Dolciani, Simon L. Bergman, and William Wooton, *Modern Algebra and Trigonometry*. New York: Houghton Mifflin, 1963.

8. William E. Boyce and Richard C. DiPrima, *Elementary Differential Equations and Boundary Valued Problems*. New York: John Wiley & Sons, 1977.

Chapter 18

Detector/Automatic Volume Control Circuits

Although the title of this chapter may seem to cover simple circuits, the analysis of some detector circuits and automatic volume control (AVC) systems can be quite detailed and extremely tedious. In particular, a peak-envelope amplitude-modulation (AM) detector such as that shown in Figure 18-1 with D3, C13, R8, and R9 is actually very hard to analyze in terms of calculating for distortion, bandwidth, and reduction of Q of the tank circuit. Depending on the amount of modulation on the envelope, the resistor-capacitor discharge current varies.

An approximation for a peak detector is to analyze it as a sample-and-hold circuit. Each peak cycle of the carrier is rectified to a positive pulse that charges a capacitor that acts as if each peak were sampled and held. But still the analysis of the peak envelope detectors is tricky and really long.

Here is a quote from Harold S. Black from his 1953 book, *Modulation Theory*, regarding peak envelope detectors: "Detailed analysis of the current wave that flows, or the spectrum of the voltage across the load resistance, tends to get tedious and not to produce results in an interesting form."

Therefore, this chapter will take a look at average envelope detection for AM signals, which does not include a charging capacitor such as C13 in Figure 18-1. However, for an excellent and detailed analysis of AM detectors and in particular peak envelope detectors, please read Chapter 10 of *Communications Circuits Analysis and Design*, by Clarke and Hess.

In terms of AVC circuits, the analysis of this feedback system is also very involved. Unlike many standard feedback systems, where a reference signal forces the output to a predetermined amplitude level, AVC systems are unique in that there is no reference signal. The output of an amplifier is converted to a direct-current (DC) signal and fed back to a gain-control terminal of the radio-frequency (RF), intermediate-frequency (IF), and or mixer circuits.

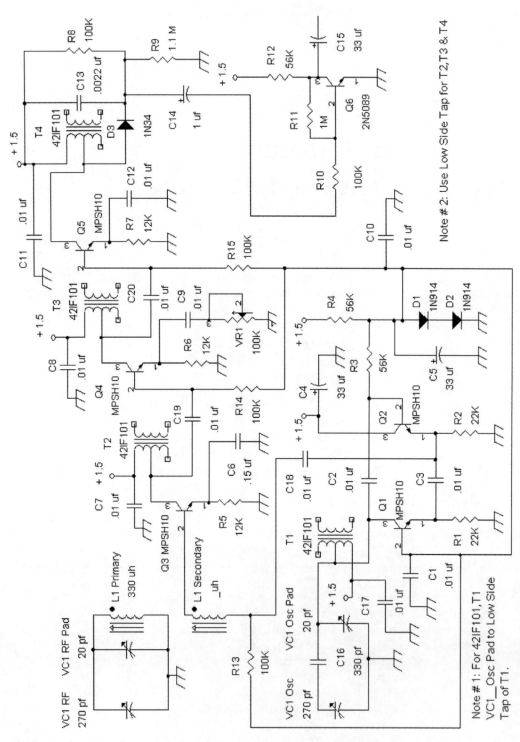

FIGURE 18-1 A superheterodyne radio with a peak envelope detector D3.

Therefore, this chapter will investigate certain aspects of AM detectors. The objectives are

1. To examine averaging envelope detectors and power detectors.
2. To look into synchronous detector circuits for single- and double-sideband signals.
3. To explain how AM radio I and Q signals are demodulated without envelope or synchronous detection.
4. To measure the average carrier signal for an AVC system.

Average Envelope Detectors

Figure 18-2 shows an envelope detector consisting of a diode D1 and resistor R1. For analysis of this circuit, the diode is "perfect" in that there is no turn voltage such as 0.1 volt for germanium or 0.5 volt for silicon diodes. Figure 18-3 shows a biasing scheme consisting of resistor R2 and diode D2 to overcome the turn on voltage of germanium or silicon diodes. (*Note:* R2D2 can be bought at your local electronics parts vendor, so you do not have to travel to a galaxy far, far away to get them—just kidding!) Diodes D1 and D2 are of the same semiconductor material.

For analyzing a rectifying demodulator for a standard AM signal such as the one shown in Figure 18-4, we assume the following:

1. The diode has no voltage drop.

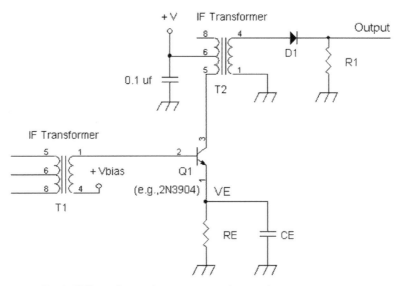

Note: Each IF Transformer has an internal capacitor.

FIGURE 18-2   A typical average envelope detector for standard AM signals.

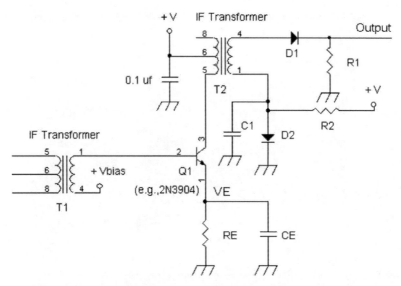

FIGURE 18-3   A biasing circuit to overcome the turn on voltage of a diode or transistor.

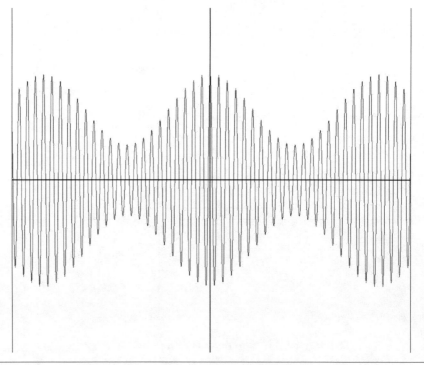

FIGURE 18-4   An AM signal that is fed into the diode detector circuit.

2. The diode detector is driven by a sufficiently low resistance that the load resistor R does not load down the IF signal.

The waveforms shown, unless noted, have the X axis for time and the Y axis for amplitude.

Figure 18-4 is characterized by equation

$$[1 + m(t)] \cos(2pf_{IF}) = V_{in\_IF} = V_{sig}$$

If the modulating signal $m(t) = 0$, then all we have is the carrier and an unmodulated AM signal.

For an unmodulated signal, the output of the diode rectifier provided a half-wave-rectified signal. This rectified signal should not be confused with any exponential characteristic of the typical semiconductor diode such as

$$I_D = I_s e^{VD/(KT/q)} = \text{diode current as a function of the forward bias voltage}$$

In fact, the diode is basically switching or gating only one polarity of the AM signal into the load resistor. Therefore, any power-series analysis of the exponential function is not applicable because switching or gating effects result in discontinuities in the half-wave-rectified waveform that are not associated with the exponential function.

The output of the rectifier for an unmodulated AM signal is seen in Figure 18-5. The half-wave-rectified signal (HWRS) of peak amplitude V_p, as seen in Figure 18-5, can be described by the following Fourier series:

$$V_p[\tfrac{1}{\pi} + \tfrac{1}{2}\cos(2\pi f_{IF})t + \tfrac{2}{3\pi}\cos(2\pi f_{IF})t - \tfrac{2}{15\pi}\cos(4\pi f_{IF})t \ldots] = \text{rectified signal}$$

$$(18\text{-}1)$$

Figure 18-6 shows a rectified modulated signal. The figure shows an AM signal that is being demodulated by half-wave rectification. To relate to this waveform, it is equivalent to multiplying the half-wave unmodulated waveform from Figure 18-5 with the same modulating signal $[1 + m(t)]$.

Therefore, the formula that describes a demodulated AM signal for average envelope detection is

$$[1 + m(t)]V_p[\tfrac{1}{\pi} + \tfrac{1}{2}\cos(2\pi f_{IF})t + \tfrac{2}{3\pi}\cos(2\pi f_{IF})t - \tfrac{2}{15\pi}\cos(4\pi f_{IF})t \ldots]$$

$$(18\text{-}2)$$

To recover the modulated signal and its average carrier level, the high-frequency signals related to the frequency f_{IF} and harmonics of f_{IF} are removed by filtering. As a result of filtering, the output of the demodulator is

$$[1 + m(t)]V_p\tfrac{1}{\pi}$$

$$(18\text{-}3)$$

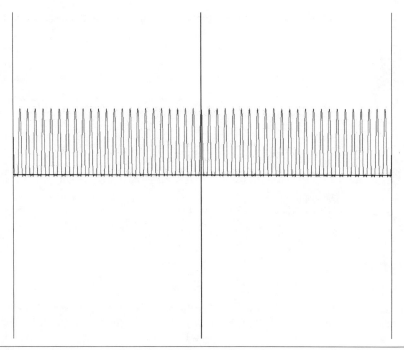

FIGURE 18-5 Output signal of the diode rectifier of an unmodulated AM signal.

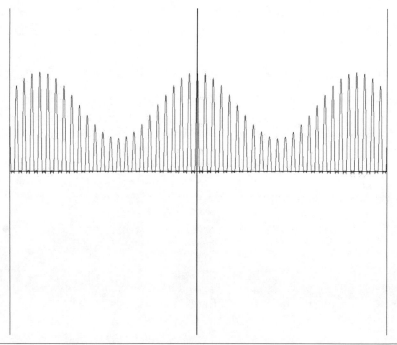

FIGURE 18-6 A modulated AM signal is half-wave rectified via a diode circuit.

Therefore, the output of the demodulator for an average envelope detector is

$$m(t)V_p\frac{1}{\pi}$$

which means that the audio output signal is scaled by $(1/\pi) = 0.318$. So as compared with a peak detector, the average envelope detector provides only about 32 percent of the audio level of a peak detector. For example, if $m(t) = (\frac{1}{2})\cos(2\pi f_{mod})t$, the output of the demodulator is

$$[1 + \frac{1}{2}\cos(2\pi f_{mod})t]V_p\frac{1}{\pi} \tag{18-4}$$

Figure 18-7 shows the half-wave-rectified AM signal with the filtered modulating signal. Had a full-wave rectifier been used instead of the half-wave circuit, the output of the average envelope detector would be twice the audio level, or a scaling factor of $(2/\pi)$, which then would approach the demodulated output of a peak envelope detector.

However, full rectification of the IF signal for detection is rarely used because matched diodes and a center-tapped secondary winding of the IF transformer are

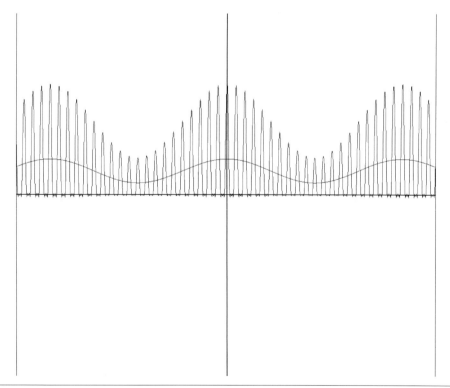

FIGURE 18-7   Demodulated AM signal with audio signal at $(1/\pi)$ of the carrier level.

needed, which adds to the cost. Most AM demodulators just use a peak envelope detector.

Power Detectors

Power detectors for AM broadcast signals have been around since the days of vacuum-tube regenerative radios. The basic premise of a power detector is to take advantage of the "inherent" nonlinearity in active devices such as tubes, field-effect transistors (FETs), or transistors. Power detectors also have an advantage over diode demodulators. They provide a higher load resistance to the IF tank circuit, which maintains a higher Q or selectivity in the last IF stage.

In Chapter 17 we looked at the consequences of second- and third-order distortions of IF signals. For a square-law device, the output current for an AM signal is

$$I_{out\_SQ\_law} = a_0 + a_1[1 + m(t)] \cos(2\pi f_{IF})t + a_2[1 + m(t)]^2[\cos(2\pi f_{IF})t]^2$$

Note that the input signal into the square-law device is

$$V_{sig} = [1 + m(t)] \cos(2\pi f_{IF})t$$

V_{sig} is represented in Figure 18-4. Also note that since second-order distortion rises as the square of the input signal, increasing this input signal results in more distortion and more output in terms of the demodulated signal. However, for small amplitudes of the input signal $[1 + m(t)] \cos(2\pi f_{IF})t$, there should be less "conversion gain" for audio signals.

In particular, the second-order distortion produced signals such as baseband audio signals and second harmonic IF signals that can be filtered out by the IF transformer.

We now turn our attention to the baseband signal. In the second-ordered term, there is a squared cosine term $[\cos(2\pi f_{IF})t]^2 = (\frac{1}{2})[1 + [\cos(2\pi 2f_{IF})t]$. When a low-pass filter is applied, the high-frequency signal $\cos(2\pi 2f_{IF})t$ is removed, leaving only the $(\frac{1}{2})(1) = (\frac{1}{2})$ term [see Equation (18-5)].

For a device that generates second-order harmonic distortion with the high-frequency IF signal and its harmonics filtered out, the output current is given by

$$I_{out\_SQ\_lowpassfiltered} = a_2[1 + m(t)]^2[\frac{1}{2}] = a_2[1 + 2m(t) + m(t)^2][\frac{1}{2}] \quad (18\text{-}5)$$

As can be seen, after filtering, the modulating signal $m(t)$ is recovered. However, there is a distortion term $m(t)^2$ that adds second-harmonic distortion to the recovered audio signal. At first glance, the second-harmonic distortion terms looks really bad because given 100 percent modulation for a sine-wave-modulating signal, the second-harmonic distortion can be 25 percent of the amplitude of the recovered audio signal. For example, if $m(t) = \cos(2\pi f_{mod})t$ for 100 percent modulation, then

$$[m(t)]^2 = [\cos(2\pi f_{mod})t]^2 = (\tfrac{1}{2})[\cos(2\pi 2f_{mod})t + 1]$$

However, the relative level of the recovered audio signal is $2m(t) = \cos(2\pi f_{mod}t)$, which leads to the fact that there is a 4:1 ratio between the recovered audio signal and its second harmonic.

However, when one listens subjectively to a complex waveform such as speech or music, the power detector provides very tolerable and intelligible audio reproduction. Figures 18-8 and 18-9 show typical bipolar transistor power detector circuits. The power detector shown in Figure 18-8 is basically a common-emitter amplifier that has its emitter alternating-current (AC) grounded via the 12-µF capacitor. Grounding the emitter via capacitor CE is essential to take advantage of the inherent nonlinearity of the transistor. Figure 18-9 shows a slightly simplified version of the power detector circuit in Figure 18-8. Figure 18-10 shows the output current to a bipolar transistor power detector with 50 percent modulation.

Although bipolar transistors have third-, fourth-, and higher-harmonic distortion, in practice, when they are used as a power detector, the bipolar transistor's exponential curve approximates square-law power detection very closely. That is, when the audio signal is measured at the output of the collector of Figures 18-8 or 18-9, the third harmonic is typically 10 times lower than the second harmonic.

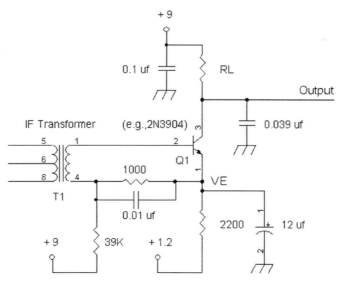

Note: IF Transformer has an internal capacitor.

FIGURE 18-8   **A power detector circuit from the Motorola 56T transistor radio from 1956.**

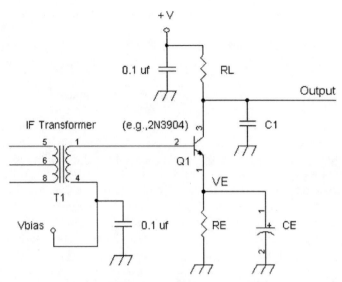

Note: IF Transformer has an internal capacitor.

FIGURE 18-9 A more simplified power detector circuit.

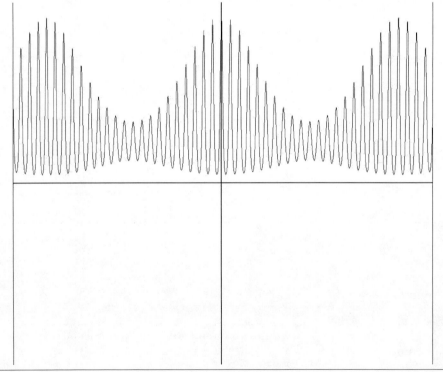

FIGURE 18-10 Collector current with an AM signal at the base.

Measurements were done on the circuit in Figure 18-9 and produced the following parameters:

- C1 = 0.022 µF
- CE = 4.7 µF (*Note:* The capacitance of CE can be larger, such as 10 µF to 33 µF.)
- RE = R_L = 1,000 Ω
- VE = 0.72 V
- I_{CQ} = 0.72 mA
- g_m = 0.0277 mho or S
- f_{mod} = 500 Hz
- f_{IF} = 455 kHz
- 50 percent modulation was used.
- V_{sig} = 100 mV {[1 + 0.5 cos($2\pi f_{mod}$)t] cos($2\pi f_{IF}$)t}
- Power detector output = 380 mV[0.5 cos($2\pi f_{mod}$)t]
- The second-harmonic distortion was about 7 percent, whereas the third-harmonic distortion was about 0.7 percent.
- A lower-level IF signal was coupled into the power detector.
- V_{sig} = 40 mV [1 + 0.5 cos($2\pi f_{mod}$)t] cos($2\pi f_{IF}$)t
- Power detector output = 110 mV [0.5 cos($2\pi f_{mod}$)t]
- The second-harmonic distortion was about 12 percent, whereas the third-harmonic distortion was about 1.2 percent.

Note that when the input signal dropped from 100 mV to 40 mV, a factor of 2.5:1, the detector ouput dropped from 380 mV to 110 mV, a factor of 3.45:1. This shows that reducing the IF signal level also reduces the "conversion gain" on recovering the audio signal. Of course, this makes sense because when the IF signal is very small, the common-emitter amplifier does not create the distortion needed for AM detection and thus provides very little or no demodulated signals.

One may notice that the power detector provides demodulated audio signal and amplification of the demodulated signal. For the input signal

$$V_{sig} = 100 \text{ mV } [1 + 0.5 \cos(2\pi f_{mod})t] \cos(2\pi f_{IF})t$$

the power detector provided a demodulated output signal of 380 mV peak. In contrast, a peak envelope detector will provide only 50 mV peak (100 mV × 0.5).

And more gain from the power detector can be provided by any combination of increasing the DC collector current or the collector load resistor. For example, increasing the load resistor from 1,000 Ω to 2,000 Ω will double the amplitude of the demodulated signal to provide 760 mV in peak amplitude of the audio signal.

In practical terms, a superheterodyne receiver that uses a power detector must provide adequate IF signal levels to the detector to deliberately drive the transistor into distortion. Thus better performance will be achieved by radios with two IF stages than just a single IF amplifier.

Synchronous Detectors

Synchronous detectors are often used in three situations. One is for the detection of Morse code transmission. Another is for the demodulation of suppressed carrier signals such as single- or double-sideband signals. In Chapter 17, synchronous detection of double- and single-sideband AM signals was discussed. The third common use for a synchronous detector is to provide much better demodulation fidelity of standard AM signals. When envelope detectors are used, generally fairly good distortion performance can be achieved. However, synchronous detectors can be used to further lower distortion when demodulating an AM signal. For analog-transmitted television signals, amplitude modulation is used. And in the television receiver, generally a diode envelope detector is used. However, with integrated circuits (ICs), including many complex circuits as part of a television signal system, synchronous detectors are used instead of envelope detectors.

At the IC level, adding more transistors and resistors is not much of a problem, and the cost is negligible. But on a printed-circuit-board implementation, adding extra functions with more parts takes up board space and adds cost. Therefore, when more circuits were literally integrated into the chips, complex demodulation systems were designed routinely.

The basic synchronous detector requires two circuits at a minimum. One circuit is the carrier regeneration circuit that provides a continuous-wave (CW) signal that represents the carrier signal, and the other circuit is a multiplier to provide the demodulation.

Figure 18-11 shows a typical synchronous detector used for demodulating AM signals. A phase-lock loop circuit consisting of the phase detector, loop filter

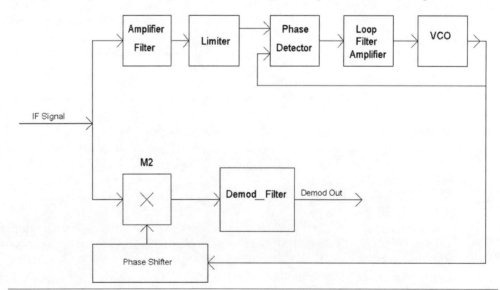

FIGURE 18-11 A synchronous detector with a phase-lock loop oscillator and mixer for detecting AM signals.

amplifier, and voltage-controlled oscillator (VCO) regenerates the IF carrier signal. Demodulation of a standard AM signal is achieved by multiplying a phase-shifted version of the VCO output with the IF signal.

The AM IF signal is fed to an amplifier and a band-pass filter for increasing the level of the IF signal and filtering out some of the noise that may be present. With a sufficient level of the IF signal, a limiter further amplifies and detects the zero crossings of the IF signal. These zero crossings provide the phase information needed for regeneration of the carrier signal. In this particular example, the output of the limiter circuit provides a square-wave signal, which then is connected to an input of a phase comparator. The output of the phase comparator, such as an Exclusive OR gate, will provide an average voltage based on the phase difference between the IF signal and the VCO. Normally, if an Exclusive OR gate is used, the phase that is desired for the VCO to lock is 90 degrees away from the incoming IF signal, which then outputs an average voltage from the Exclusive OR gate of one-half the supply voltage. For example, if the supply voltage is 5 volts, when the VCO is typically locked, the average output voltage from the Exclusive OR phase detector is 2.5 volts DC.

The output of the phase detector then is coupled to an amplifier and low-pass filter (loop filter amplifier), which then feeds an error-correction signal into the VCO. The phase-lock loop circuit provides a CW signal. This CW signal then has the average phase of the IF signal carrier. However, care must be taken such that the VCO itself does not generate too much phase noise. Thus, normally, the VCO consists of a high-stability oscillator such as an LC oscillator. The VCO provides a signal whose frequency depends on the voltage into the VCO. The output of the VCO then is coupled to the remaining input of the phase detector.

Therefore, the output of the VCO provides a regenerated carrier signal of constant amplitude. Because the phase detector locks the VCO to the IF signal's carrier with a 90-degree offset, a phase-shifter circuit is provided to cancel the 90-degree offset and match the IF signal carrier's phase. Thus the output of the phase-shifter circuit is connected to one input of a multiplying circuit, and the other input of the multiplier circuit is connected to the IF signal. The output of the multiplier then is low-pass-filtered to recover the audio signal.

Of course, there are many variations on how carrier regeneration can be done. Figure 18-11 is just one example. For readers who are interested in a "jungle chip" that demodulates AM TV signals, the Mitsubishi M51366SP integrated circuit is an example.

IQ Detectors for AM Broadcast Signals

For software-defined radios (SDRs), demodulating a standard AM signal, a double-sideband signal with carrier, is implemented entirely differently. There is no envelope detector, and although synchronous detection can be used, detection of the standard AM signal is done via a Pythagorean process. That is, a key to understanding this type of demodulation process will include the trigonometric identity $[\cos(\theta)]^2 + [\sin(\theta)]^2 = 1$ (Figure 18-12).

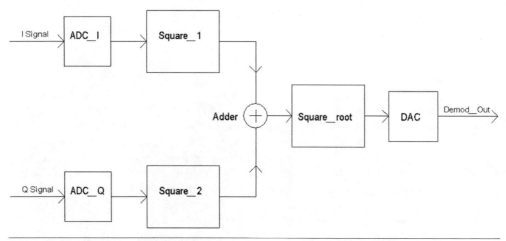

FIGURE 18-12   Demodulating AM signals via a Pythagorean method.

Figure 18-12 shows two signals—the I Signal V_{sig_I} and the Q Signal V_{sig_Q}—that are individually squared and added. For simplicity, the analysis will be done in the continuous time domain. The analog-to-digital converters (ADC_I and ADC_Q) and digital-to-analog converter (DAC) can be omitted for the analysis below. The square root of the output of the adder then provides the demodulated signal, such as an audio signal. For a general situation, the two signals are defined as follows:

$$V_{sig\_I} = [1 + m(t)] \cos(2\pi f_{IF})t \text{ and } V_{sig\_Q} = [1 + m(t)] \cos(2\pi f_{IF})t$$

The output of squaring function SQ_1 is

$$\{[1 + m(t)] \cos(2\pi f_{IF})t\}^2$$

and the output of squaring function SQ_2 is

$$\{[1 + m(t)] \sin(2\pi f_{IF})t\}^2$$

And the output of the adder is

$$\{[1 + m(t)] \cos(2\pi f_{IF})t\}^2 + \{[1 + m(t)] \sin(2\pi f_{IF})t\}^2 = [1 + m(t)]^2[\cos(2\pi f_{IF})t]^2$$
$$+ [1 + m(t)]^2[\sin(2\pi f_{IF})t]^2 = [1 + m(t)]^2\{[\cos(2\pi f_{IF})t]^2 + [\sin(2\pi f_{IF})t]^2\}$$

However, we know that $[\cos(\theta)]^2 + [\sin(\theta)]^2 = 1$, and we let $\theta = (2\pi f_{IF})t$. Therefore, $[\cos(2\pi f_{IF})t]^2 + [\sin(2\pi f_{IF})t]^2 = 1$, and the summed or added output is

$$[1 + m(t)]^2\{[\cos(2\pi f_{IF})t]^2 + [\sin(2\pi f_{IF})t]^2\} = [1 + m(t)]^2(1)$$
$$= [1 + m(t)]^2 = \text{added output}$$

The output of the adder is connected to the input of a square-root function, so the positive output of the square-root function is

$$\sqrt{[1 + m(t)]^2} = [1 + m(t)] = \text{demodulated output signal } m(t)$$

and the output of the square-root function provides a DC offset signal (1) that represents the carrier level at 0 modulation.

Figure 18-12 also includes the analog-to-digital converters and a digital-to-analog converter to denote that normally the I and Q signals are digitized, and the squaring, summing, and square-root functions are implemented in a computational manner.

At this point in this chapter, it would seem that the explanation on how I and Q signals are demodulated for AM signals is done. Well, not so fast. There is one equation that may need further explanation. In high school trigonometry classes, the trigonometric identity $[\cos(\theta)]^2 + [\sin(\theta)]^2 = 1$ is "taken for granted." After all, the angle addition and subtraction laws prove this identity. But what does it really represent in terms of time-varying signals? The idea of taking the sum of the squares of the cosine and sine functions of the same angle and having this sum equal to 1 seems a bit abstract. Let's see what is really going on by graphing out the cosine and sine functions.

Figure 18-13 presents graphs of $\cos(2\pi f_{IF})t$ and $[\cos(2\pi f_{IF})t]^2$. The squared cosine signal has a twice-frequency signal that is half the amplitude of the cosine signal, and

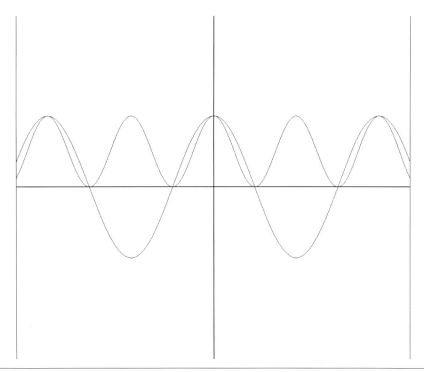

FIGURE 18-13 Graphs of a cosine signal and a squared cosine signal.

the squared cosine signal includes a DC signal at one-half the peak amplitude of the cosine signal. Since the peak amplitude of the cosine signal is 1, the DC signal has a value of 0.5.

Figure 18-14 shows graphs of a sine signal and the square of the sine signal. Because the sine signal is one-quarter of a cycle or 90 degrees shifted from the cosine signal, the squared sine signal will produce a twice-frequency signal that is one-half cycle or 180 degrees in relationship to the twice-frequency cosine signal.

Very much like the squared cosine signal, the squared sine signal has twice the frequency and half the amplitude of the sine signal. Also, the squared signal has a DC signal of 0.5.

Now let's take a look at two squared signals. Figure 18-15 displays the two squared signals. When the squared cosine signal and the squared sine signal are added or summed, the twice-frequency signals are 180 degrees from each other and will cancel, whereas the DC signals that are each 0.5 in amplitude will add to form a DC signal of 1.0. Thus the sum of the squares of the cosine and sine signals of the same frequency is a constant 1 (Figure 18-16).

As seen in Figure 18-16, the sum of the two twice-frequency signals with 0.5 DC offsets result in a DC value of 1.0. Thus, in terms of time functions, the sum of the

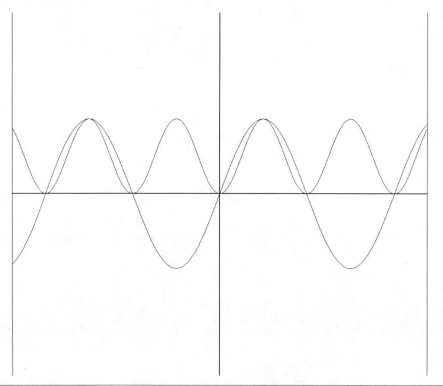

FIGURE 18-14   Graphs of a sine signal and a squared sine signal.

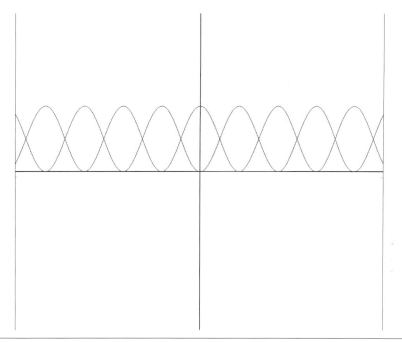

FIGURE 18-15   The squared cosine signal and the squared sine signal shown together.

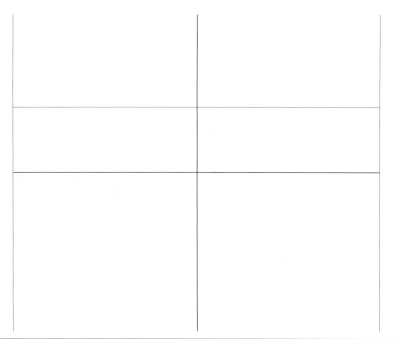

FIGURE 18-16   The sum of the waveforms shown in Figure 18-15.

squares of 0-degree cosine and 90-degree sine signals produces a frequency-doubling effect that results in twice the phase shift of 180 degrees. The twice-frequency AC signals cancel out to zero, whereas the DC components of 0.5 add to form 1, which is the result for $[\cos(\theta)]^2 + [\sin(\theta)]^2 = 1$ for any θ.

Measuring an Average Carrier or Providing Automatic Volume Control

In designing an AVC system for an AM radio, there is one very important factor to consider, and that is defining the maximum carrier level from the output of the IF transformer. By defining the maximum carrier level, the maximum audio level also is defined. For example an average carrier level of 1.0 volt peak provides 1.0 volt peak of audio at 100 percent modulation (Figure 18-17). From Equation (16-1) in Chapter 16, $\{1 + m[\cos(2\pi f_{mod})t]\} \cos(2\pi f_{carrier})t =$ AM signal.

An AVC system is shown in Figure 18-18. The figure shows an IF system and detector of a superheterodyne radio with the level-shifter resistor and capacitor values for R2, C3, R3, and R4 similar to those in the Sony Model 3F-66W radio.

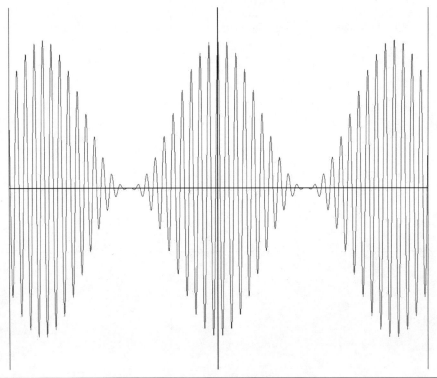

FIGURE 18-17 An AM signal with 1.0-V peak average carrier level with an audio signal of 1.0 volt.

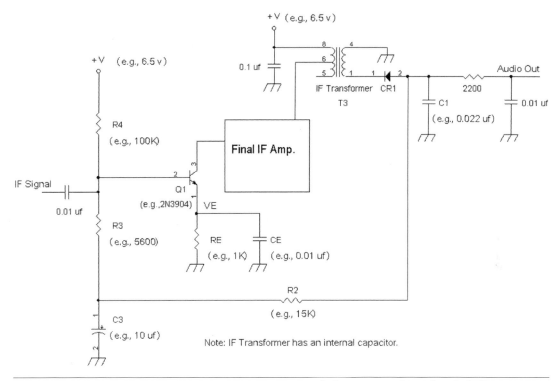

FIGURE 18-18   An AVC system using a negative peak detector and DC level shifter.

The AM signal is demodulated by a negative peak detector CR1 and C1 that follows the audio signal. The output of the negative peak detector at the anode of CR1 provides the negative going peaks of the AM signal that is level-shifted by about +0.6 volt owing to the turn-on voltage of diode CR1.When the AM signal is very weak or near zero, the output of CR1 is about +0.6 volt. When the AM signal is strong, the output of CR1 is less than +0.6 volt because of the negative peak detector CR1 and hold capacitor C1.

Resistor R2 and capacitor C3 remove the audio signal and provide a DC voltage that is indicative of the negative peak voltage of an unmodulated AM signal.

To design the AVC system as shown in Figure 18-18 for a particular audio level V_{pk_audio}, and given a silicon diode and transistor with about a 0.6-volt forward voltage drop, then all that is needed is a level-shifting circuit that provides V_{peak_audio} of DC bias to the emitter voltage V_E. For example, if 400 mV peak audio is required, then 0.4 volt DC is needed for V_E under a weak- or no-signal condition. The level-shifting circuit, consisting of CR1, R2, R3, and R4, along with the power supply, then will provide a Q1 base voltage of about 0.6 volt + 0.4 volt = 1.0 volt.

In Figure 18-18, assume that the diode and transistor turn-on voltages are the same. And thus the level-shifting voltage at a no-signal condition is the same voltage as the emitter voltage V_E. If at 100 percent modulation, 500 mV peak audio is needed, and the voltage drop across R3 and R2 is about 500 mV DC given that R4 >> R3 + R2. Normally, the resistance of R4 is at least four times the series resistance of R2 and R3. When a weak station is received, the average DC voltage at C1 is about 0.6 volt. With the 0.5-volt level-shifting voltage from R2 and R3, the base voltage is about 1.1 volts for an emitter voltage of about 0.5 volt to set up a DC collector current of $(+0.5 \text{ volt})/RE = I_{CQ1}$. However, when a very strong station is received, the negative peak value of the IF signal will be negative 0.5 volt. This negative 0.5 volt then is added to the +0.6-volt DC level-shifting voltage of CR1, resulting in a net average voltage of +0.1 volt DC at C1. With the +0.5-volt level-shifting voltage across R2 and R3, the base of Q1 then has a voltage of +0.6 volt. However, the base-to-emitter voltage of Q1 is about +0.6 volt, which results in the emitter voltage approaching 0 volt across RE that results in nearly 0 collector current to reduce the gain of Q1.

Figure 18-18 also includes in parenthesis values taken from a schematic of the Sony 3F-66W transistor radio from 1969. In this radio, Q1 is a silicon transistor, but the detector diode CR1 is germanium with a turn-on voltage of about 100 mV. With about 6.5 V powering R4, there is about 53 µA of current through the resistors R4, R3, and R2. The total resistance of R2 and R3 is 15 kΩ + 5.6 kΩ = 20.6 kΩ. The voltage drop across R2 and R3 under no-signal conditions is 53 µA × 20.6 kΩ = 1.09 volts and is also the level-shifting voltage that will be added to the CR1 diode drop of 0.1 volt. Thus the total voltage at the base of Q1 is 1.09 volt + 0.1 volt = 1.19 volt. The voltage at VE given a +0.6-volt base-emitter voltage drop for Q1 then is 1.19 volts – 0.6 volt = +0.59 volt = V_E. Thus the maximum expected audio amplitude is about 0.59 volt peak when a strong station is tuned.

References

1. Class Notes EE140, Robert G. Meyer, UC Berkeley, Fall 1975.
2. Class Notes EE240, Robert G. Meyer, UC Berkeley, Spring 1976.
3. Paul R. Gray and Robert G. Meyer, *Analysis and Design of Analog Integrated Circuits*, 3rd ed. New York, John Wiley & Sons, 1993.
4. Kenneth K. Clarke and Donald T. Hess, *Communication Circuits: Analysis and Design*. Reading: Addison-Wesley, 1971.
5. Robert L. Shrader, *Electronic Communication*, 6th ed. New York: Glencoe/McGraw-Hill, 1991.
6. Howard W. Sams & Co., *Sams Photofact Transistor Radio Series*, TSM-2. Indianapolis: November 1958.
7. Howard W. Sams & Co., *Sams Photofact Transistor Radio Series*, TSM-100. Indianapolis: May 1969.
8. Mary P. Dolciani, Simon L. Bergman, and William Wooton, *Modern Algebra and Trigonometry*. New York: Houghton Mifflin 1963.
9. William E. Boyce and Richard C. DiPrima, *Elementary Differential Equations and Boundary Valued Problems*. New York: John Wiley & Sons, 1977.

Chapter 19

Amplifier Circuits

My thought on writing a chapter on amplifiers is that only certain topics about amplification will be explained. There are literally hundreds of books on amplifier design, including operational amplifiers, radio-frequency (RF) amplifiers, low-noise preamplifiers, audio power amplifiers, servo amplifiers, instrumentation amplifiers, and so on. Some of these books cover detailed material concerning input stages, voltage-gain stages, and output stages. And when video amplifiers are covered, a new set of specifications and signals is introduced, such as differential gain and phase, 2T pulse response, color bar, and multiburst signals.

As I am writing this chapter, I know that some readers would like more material on amplifiers. Given the constraints of this book, though, which relates to radio circuits that include audio amplifiers, the following subjects will be covered.

1. Operational amplifiers
2. Basic characteristics of amplifiers, such as gain and frequency response
3. Connecting multiple amplifiers and biasing concerns
4. Practical considerations when using amplifiers

Introduction to Operational Amplifiers

An operational amplifier (or op amp) generally has two input terminals and an output terminal. Of course, there are usually two power-supply terminals as well (Figure 19-1).

For controlled amplification, the op amp must have some type of feedback element from the output terminal to one of the input terminals. The form of feedback must be negative feedback. In some instances, the output terminal is connected to an inverting device such as a common-emitter amplifier or inverting amplifier, which means that the output of the inverting device must be connected to the positive (+) input of the op amp.

However, more commonly, the output of the op amp is connected through a resistor and or capacitor to the negative (−) input of the op amp (Figure 19-2).

315

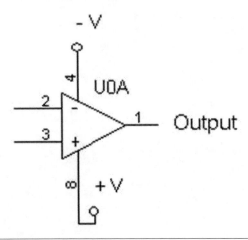

FIGURE 19-1 An operational amplifier.

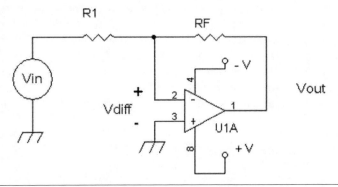

FIGURE 19-2 An op amp configured for inverting gain.

The general rules for analyzing an op amp circuit are as follows:

1. The op amp must have a net negative feedback. That is, if there is both positive and negative feedback from the output to the (–) and (+) inputs, there has to be more negative feedback than positive feedback. In most cases, there are only feedback elements from the output to the (–) input.
2. The raw or open-loop gain of the op amp is very large (>10,000).
3. Once negative feedback is achieved, two conditions hold. One is that the current into the (–) and (+) inputs is zero, and the other is that the potential difference between the two inputs (–) and (+) is zero. That is, the (–) input and the (+) input have the same voltage.

Note If there is net positive feedback in an op amp circuit, usually there is some combination of oscillation or the output terminal being stuck near one of the power-supply voltages.

In Figure 19-2, the (+) input is grounded, and given feedback resistor RF, there is negative feedback. The voltage at the (+) input is the same voltage at the (–) terminal. This means that the (–) terminal looks like a ground potential. Therefore, to find the gain, or we sum the currents into the (–) terminal to equal 0. That is,

$$\frac{Vin}{R1} + \frac{Vout}{RF} = 0 \tag{19-1}$$

$$\frac{Vin}{R1} = -\frac{Vout}{RF} \tag{19-2}$$

$$\frac{VinRF}{R1} = -\frac{Vout}{1}$$

$$\frac{-RFVin}{R1} = \frac{Vout}{1}$$

$$\frac{-RF}{R1} = \frac{Vout}{Vin} = \text{gain} \tag{19-3}$$

Thus gain is just set by the two resistors RF and R1. Also, because the (–) input is a virtual ground, the input resistance in the view of Vin is just R1 because the (–) input connected to R1 is virtually grounded since the (+) input is grounded.

What the virtual ground at the (–) input also implies is that the input resistance at the (–) terminal is nearly 0 Ω to ground when there is a negative-feedback resistor connected from the output to the (–) terminal of the op amp and when the open-loop gain of the op amp is very large.

Now let's take a look at two characteristics of this op amp circuit. One is to establish the input resistance at the (–) terminal with a given negative-feedback resistor RF and a given open-loop gain of the amplifier. And the other is to examine why the voltages are the same for the (–) and (+) inputs when negative feedback is applied.

For a negative-feedback amplifier with a feedback resistor RF, Figure 19-3 shows an amplifier with gain –A to mimic the configuration as shown in Figure 19-2. To calculate the input resistance, we need to know

$$\frac{Vintest}{Iintest} = \text{Rin\_test}$$

However, we can equivalently show Figure 19-3 as Figure 19-4. If Vintest = 1 volt DC and –A = –2 and RF = 1000 Ω, the voltage at the resistor, RF, is +1 volt DC on one end and –2 volts DC on the other end of RF, thereby forming 3 volts across

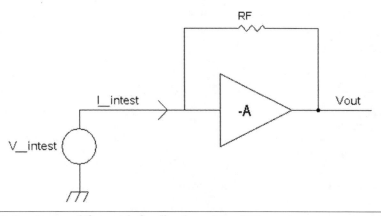

FIGURE 19-3 An amplifier with feedback resistor RF.

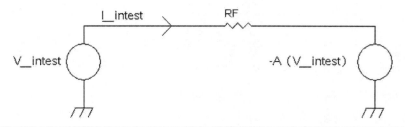

FIGURE 19-4 An equivalent circuit to Figure 19-3 for determining V_{in_test}/I_{in_test}.

RF, or 3 milliamps of drain when 1 volt DC is applied at the input. At first glance, this seems counter intuitive since most of the time we are used to believing the worse case concerning current drain for a resistor is to have one end grounded, which would result in 1 volt across the resistor to yield 1 milliamp of resistor current.

So what happens if by "magic" we make the amplifier's gain $-A = -1,000,000$? With Vintest = 1 volt DC, this would mean that there is 1,000,001 volts across the resistor RF = 1,000 Ω, so Iintest = 1,000.001 A. And this excessive current would be equivalent to Vintest loading into a resistor of about 0.001 Ω, or for all practical purposes, Vintest is loading into a short circuit.

Thus the general equation for finding the input current drain with feedback resistor RF in Figure 19-4 is

$$\text{Iintest} = \frac{\text{Vintest} - -A\text{Vintest}}{\text{RF}} = \frac{\text{Vintest} + A\text{Vintest}}{\text{RF}} = \frac{\text{Vintest}\,(1 + A)}{\text{RF}} \quad (19\text{-}4)$$

$$\frac{\text{Vintest}}{\text{Iintest}} = \frac{\text{RF}}{(1 + A)} \quad (19\text{-}5)$$

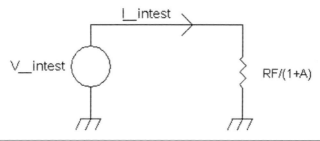

FIGURE 19-5   The equivalent input resistance caused by a feedback resistor and inverting amplifier.

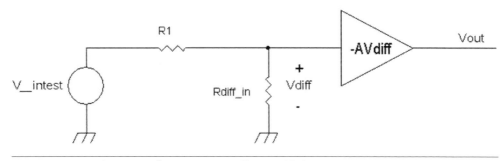

FIGURE 19-6   An equivalent circuit to Figure 19-2.

Thus, if the amplifier's gain is very large, the input resistance at the (–) input terminal of the op amp approaches a short circuit. Figure 19-5 shows an equivalent input resistance at the input terminal. Another way to model the inverting op amp circuit in Figure 19-2 is via Figure 19-6. In this figure, the input resistor R1 is voltage divided via Rdiff_in = RF/(1 + A) to form the input voltage Vdiff.

$$\text{Vout} = -A\ \text{Vdiff} \tag{19-6}$$

$$\text{Vdiff} = \text{Vin}[\tfrac{RF}{(1+A)}] / [\tfrac{RF}{(1+A)} + R1] \tag{19-7}$$

$$\text{Vout} = -A\ \text{Vin}[\tfrac{RF}{(1+A)}] / [\tfrac{RF}{(1+A)} + R1] \tag{19-8}$$

$$\tfrac{\text{Vout}}{\text{Vin}} = -A[\tfrac{RF}{(1+A)}] / [\tfrac{RF}{(1+A)} + R1] \tag{19-9}$$

$$\tfrac{\text{Vout}}{\text{Vin}} = [\tfrac{-A(RF)}{(1+A)}] / [\tfrac{RF}{(1+A)} + R1] \tag{19-10}$$

For large values of A, that is, the op amp has a very large open-loop gain $-A/(1 + A) = -1$ and $RF/(1 + A) = 0$, then

$$\frac{Vout}{Vin} = -RF/R1$$

This is the same answer as calculated before. Also note that

$$\frac{Vout}{-A} = Vdiff$$

Again, for large values of $-A$,

$$\frac{Vout}{-A} = 0 = Vdiff$$

This means for a large open-loop gain, the voltage across the (–) and (+) input terminals, which is Vdiff, goes to zero.

Figure 19-3 showed that when a resistor RF is connected in a negative-feedback configuration of an amplifier whose gain is –A, the equivalent resistance looking into the input of the amplifier is $RF/(1 + A)$. The negative feedback thus causes a "divided" or "reduced" resistance effect by pulling more current at the input of the amplifier than expected.

Now suppose that the resistor RF is replaced with a capacitor CF whose impedance is $Z_{CF} = 1/j2\pi fCF$ and the voltage source Vin is replaced with an alternating-current (AC) signal source Vac (Figure 19-7).

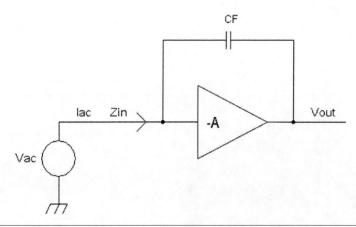

FIGURE 19-7 A Miller capacitance multiplier circuit that uses a feedback capacitor.

Note $j = \sqrt{-1}$, an imaginary number. And impedance, which is measured in ohms, is any combination of resistance and/or reactance, where reactance includes the term $j2\pi f$. Resistance is limited to the real numbers, and impedance includes the complex numbers (real and/or imaginary numbers).

We should expect that the negative feedback will cause more current drain at the input, and the AC input impedance is

$$\frac{Z_{CF}}{1 + A} = \frac{1/j2\pi CF}{1 + A} = \frac{1}{(1 + A)j2\pi fCF} = \frac{1}{j2\pi f[CF(1 + A)]} \tag{19-11}$$

Thus the impedance Z_{in} at the input terminal of the amplifier with gain of $-A$ looks like the impedance of a capacitor that has a capacitance of $[CF(1 + A)]$. Or equivalently, the input to the amplifier with a feedback capacitor CF is capacitive with a capacitance of CF multiplied by $(1 + A)$.

Multiplying capacitance lowers the impedance, which is in accordance with Figures 19-3 and 19-5, which show that a resistive feedback element results in a reduced or lowered resistance at the input of the amplifier. Stated another way, Figure 19-8 shows the equivalent AC voltage across capacitor CF and its AC current I_{ac}.

The AC current is

$$Iac = \frac{Vac - -A(Vac)}{Z_{CF}} = \frac{Vac + A(Vac)}{Z_{CF}} = \frac{Vac(1 + A)}{Z_{CF}}$$

After rearranging the equation and noting that $Z_{CF} = 1/j2pf\,CF$, the impedance at the input of the amplifier is

$$\frac{Vac}{Iac} = \frac{Z_{CF}}{(1 + A)} = \frac{1/j2\pi fCF}{1 + A} = \frac{1}{(1 + A)j2\pi fCF} = \frac{1}{j2\pi f[CF(1 + A)]} \tag{19-12}$$

Thus the Miller capacitance is $CF(1 + A)]$, which is the capacitance at the input of the amplifier to ground (see Figure 19-9).

In common-emitter amplifiers, the Miller capacitance exists and multiplies the collector-to-base capacitance by $(1 + A)$ where A is the inverting gain from the base to

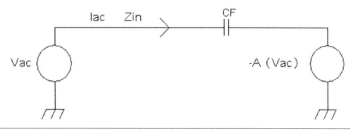

FIGURE 19-8 An equivalent circuit showing the voltage sources across capacitor CF.

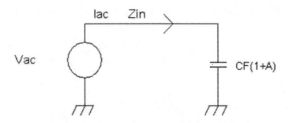

FIGURE 19-9 An equivalent capacitance at the input owing to a feedback capacitor and an inverting amplifier.

the collector. This Miller capacitance usually robs the amplifer's high-frequency gain and reduces its input impedance and can be considered a "bad" effect. As shown in Chapter 17, the Miller capacitance can be reduced or eliminated by using a cascode amplifier or a differential amplifier. Another way to reduce the effects of Miller capacitance in a common-emitter amplifier is to connect an emitter follower before the common-emitter amplifier. The emitter follower will be designed to drive the increased capacitance at the input and thus maintain the desired high-frequency response.

In op amps, often there are two voltage-gain stages, and in order to maintain unity-gain stability, the Miller capacitance is often used. The second stage usually includes extra capacitance across the base and collector of a common-emitter amplifier for a deliberate high-frequency roll-off to the overall open-loop gain of the op amp such that when negative feedback is applied, there are no oscillations.

Now a noninverting op amp circuit will be examined (Figure 19-10). In analyzing the inverting amplifier of Figure 19-2, when the open-loop gain A is large, the voltage

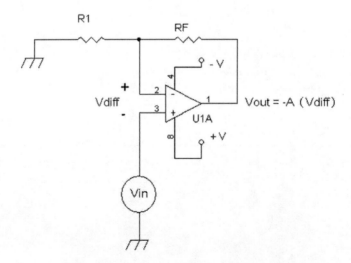

Note: DC open loop gain of the op amp = A

FIGURE 19-10 A noninverting amplifier with feedback resistors RF and R1.

at the (–) and (+) terminals of the amplifier is the same. Therefore, the voltage at the (–) terminal in Figure 19-10 is Vout[R1/(R1 + RF)], and the voltage at the (+) input terminal is just Vin. Therefore,

$$\text{Vout}\frac{R1}{R1 + RF} = \text{Vin} \tag{19-13}$$

or

$$\frac{\text{Vout}}{\text{Vin}}\frac{R1}{R1 + RF} = 1$$

which leads to

$$\frac{\text{Vout}}{\text{Vin}} = \frac{R1 + RF}{R1} = \frac{RF + R1}{R1} = \text{noninverting gain} \tag{19-14}$$

However, if the open-loop gain A is just some gain value, then Vout/Vin can be determined as a function of A. See Figure 19-10.

$$\text{Vout} = \text{Vdiff}\,(-A)$$

$$[\text{Vout}\frac{R1}{(R1 + RF)} - \text{Vin}] = \text{Vdiff} \tag{19-15}$$

$$\text{Vout} = [\text{Vout}\frac{R1}{(R1 + RF)} - \text{Vin}][-A] \tag{19-16}$$

$$\text{Vout} = [-A][\,\text{Vout}\frac{R1}{(R1 + RF)}] + A[\text{Vin}] \tag{19-17}$$

$$\text{Vout} - [-A][\,\text{Vout}\frac{R1}{(R1 + RF)}] = [\text{Vin}][A] \tag{19-18}$$

$$\text{Vout}[1 + \frac{AR1}{(R1 + RF)}] = [\text{Vin}]A \tag{19-19}$$

$$\frac{\text{Vout}}{\text{Vin}}[1 + \frac{AR1}{(R1 + RF)}] = A \tag{19-20}$$

$$\frac{\text{Vout}}{\text{Vin}} = \frac{A}{1 + [AR1/(R1 + RF)]} = \frac{A(R1 + RF)}{R1 + RF + AR1} = \frac{A(R1 + RF)}{RF + (A + 1)R1} \tag{19-21}$$

$$\frac{\text{Vout}}{\text{Vin}} = \frac{A(R1 + RF)}{RF + (A + 1)R1} \tag{19-22}$$

Now let's check the equation. If R1 = 0, then we should have just the open-loop gain

$$\frac{A(RF)}{RF} = A = \frac{Vout}{Vin}$$

In reality, if R1 is shorted to ground, the output of the op amp will head toward one of the power-supply rails. And if A is very large such that (A + 1)R1 >> RF, we will get

$$\frac{R1 + RF}{R1} = \frac{Vout}{Vin}$$

Amplifier Characteristics

In this section, only two topics will be covered. One is the input resistance to a common-emitter amplifier with an emitter resistor when the emitter is not AC grounded. The other is the gain of the common-emitter amplifier with this emitter resistor.

While there are plenty of other topics that can be covered, there are many excellent books on op amp applications and transistor amplifiers. Two of them that come to mind are

1. *Intuitive IC Op Amps from Basics to Useful Applications*, by Thomas M. Frederiksen (National's Semiconductor Technology Series, Santa Clara: National Semiconductor Corporation, 1984).
2. *Bipolar and MOS Analog Circuit Design*, by Alan B. Grebene, New York: John Wiley & Sons (1984).

The reader can consult these books for much more detail on amplifiers in general.

Figure 19-11 shows a simple common-emitter amplifier with an emitter resistor RE. Note that there is no capacitor from the emitter of Q1 to ground.

Analyzing the common-emitter amplifier in Figure 19-11 will be done a little bit differently from most college courses that use hybrid π transistor models. Instead, some simple equations will be used.

Recall the following if the transistor's current gain β is large: The output resistance looking into the emitter with the base connected to a voltage source is $1/g_m$, which is equal to the output resistance of an emitter follower when the base is driven by a low-source-resistance signal voltage.

$$\frac{\beta}{g_m} = \frac{\Delta VBE}{\Delta IB}$$

$$g_m = \frac{\Delta IC}{\Delta VBE} = \frac{ICQ}{0.026V}$$

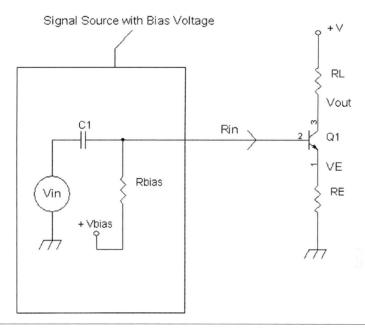

FIGURE 19-11 Common-emitter amplifier with a local negative-feedback resistor RE.

where ICQ = quiescent collector DC current.

Now referring to Figure 19-11, one will see that Vin = VBE + VE. Therefore the AC signals can be characterized as:

$$\Delta Vin = \Delta VBE + \Delta VE$$

To find the input resistance, we need to find

$$\frac{\Delta Vin}{\Delta IB} = Rin$$

Therefore,

$$\frac{\Delta Vin}{\Delta IB} = \frac{\Delta VBE}{\Delta IB} + \frac{\Delta VE}{\Delta IB}$$

But we know that the voltage at the emitter is VE = (IE)RE. Therefore, the change in emitter voltage VE over the change in emitter current is RE.

$$\frac{\Delta VE}{\Delta IE} = RE$$

and that IE = (β + 1)IB. Therefore, ΔIE = (β + 1)ΔIB, and

$$\frac{\Delta VE}{(\beta + 1)\Delta IB} = RE$$

which then leads to

$$\frac{\Delta VE}{\Delta IB} = (\beta + 1)RE$$

by substitutions,

$$Rin = \frac{\Delta Vin}{\Delta IB} = \frac{\beta}{gm} + (\beta + 1)RE$$

The emitter resistor can have beneficial effects in terms of raising the input resistance. For example, if ICQ = 26 mA, β = 100, and R_E = 0, β/g_m = 100 Ω, and Rin = (100 + 0) Ω, a low input resistance. However, if RE is 10 ohms, then (β + 1)RE = (101)10 Ω = 1,010 Ω, and Rin = (100 + 1,010) Ω = 1,110 Ω, a medium input resistance. Of course, having RE at 10 Ω also reduces the overall transconductance of the transistor, as we will see in the following analysis.

Transconductance of the transistor is given by

$$g_m = \frac{\Delta IC}{\Delta VBE}$$

When there is an emitter resistor as shown in Figure 19-11, we define a new transconductance, g'_m, as:

$$g'_m = \frac{\Delta IC}{\Delta Vin}$$

For large β, $\Delta I_E = \Delta I_C$.

The AC emitter current is

$$\frac{\Delta VE}{RE}$$

However, the AC voltage at the emitter has to take into account the output resistance 1/g_m of the emitter follower. The emitter then provides the input AC signal ΔV_{in} with a series resistance of 1/g_m to drive into resistor RE. As a result, a voltage divider is formed, and the AC voltage at RE is ΔV_E:

$$\left[\frac{RE}{RE + \frac{1}{gm}}\right]\Delta Vin = \Delta VE$$

and

$$\frac{\Delta VE}{RE} = \Delta IE = \Delta IC \qquad \text{for large } \beta$$

$$\Delta VE = RE\ \Delta IC$$

By substitution,

$$\left[\frac{RE}{RE + \frac{1}{gm}}\right]\Delta Vin = RE\ \Delta IC$$

and dividing RE from both sides of the equation gives

$$\left[\frac{1}{RE + \frac{1}{gm}}\right]\Delta Vin = \Delta IC$$

which leads to

$$\left[\frac{1}{RE + \frac{1}{gm}}\right] = \frac{\Delta IC}{\Delta Vin}$$

and multiplying by g_m at the numerator and denominator,

$$\frac{\Delta IC}{\Delta Vin} = \frac{g_m}{g_m RE + 1} = \frac{g_m}{1 + g_m RE} = g'_m$$

or alternatively and probably more useful to do quick calculations is:

$$\left[\frac{1}{RE + \frac{1}{gm}}\right] = g'_m$$

Thus the emitter resistor reduces or scales the original transistor's transconductance by a factor of

$$\frac{1}{1 + g_m RE}$$

For example, with 26 mA of collector current, $g_m = 1$ mho, and with RE = 10 Ω, then

$$\frac{1}{1 + 10}\ \text{mho} = 0.0909\ \text{mho} = g'_m$$

Therefore, in this particular example, the "penalty" of having a 10-Ω emitter feedback resistor is a reduction in the transistor's transconductance by 11-fold. This reduction in transconductance means that the gain of the amplifier will be reduced 11-fold. But the "bright" side is that the input resistance in this example had increased about 11-fold. Also, the linearity of the amplifier with the 10-Ω emitter resistor increased dramatically, and thus having the emitter resistor also reduces distortion for the same output voltage.

Another advantage is the output resistance from collector to ground for this example is increased by about 11-fold given the 10-Ω emitter resistor. For example, given an Early voltage of 100 volts at 26 mA, the collector-to-emitter resistance is about 3.84 kΩ. If the emitter is grounded via a large emitter capacitor to ground, the circuit has an internal 3.84-kΩ resistor in parallel with the collector load resistor. However, if RE = 10 Ω, the equivalent output resistance from collector to ground is now 11 times larger, or 42.3 kΩ. And thus the collector load resistor will be in parallel with a 42.3-kΩ resistor instead of a 3.84-kΩ resistor.

Before we leave this section, here is an intuitive explanation of why an emitter resistor raises the output resistance of the transistor. See Figure 19-12 with internal resistor r_0. When the emitter of Q1 is grounded, by inspection, the lossy or extra resistance is r_0 from the collector to ground. However, with an emitter resistor RE, let's examine what happens when the supply voltage is raised. The collector-to-emitter resistor r_0 actually adds current to the emitter of the transistor to partially reduce the base-to-emitter voltage. So, as the voltage is increased at the collector, the transistor's emitter receives a current via r_0 to slightly lessen the base emitter voltage and thus cause the collector current to reduce. The total collector current, including the current flow into r_0, then is stabilized and does not increase as much compared with the emitter being grounded.

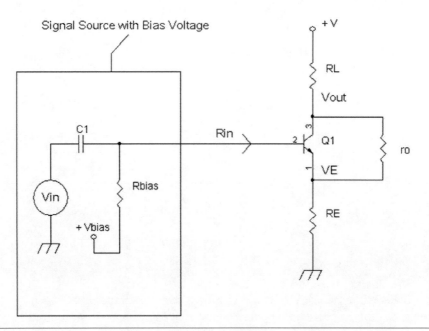

FIGURE 19-12 Common emitter amplifier with internal resistor.

Connecting Multiple Amplifiers for an Amplifier System

Often a single-stage amplifier or single-transistor amplifier may not provide enough gain, or high enough input resistance, or power output into a low-resistance load. Therefore, multiple transistor stages are required. In other cases, although the amplifier's gain may meet the requirements, the harmonic distortion may be too high or the high-frequency response is inadequate.

For example, in terms of a distortion specification, suppose that a high-gain amplifier of 1,000 is needed for low levels of audio signals that do not exceed 2 millivolts peak with less than 1 percent distortion second harmonic distortion for a sine-wave input signal. Assume a 12-volt power supply with resistive loads and an input resistance greater than 10,000 Ω, output resistance of 1,000 Ω, minimum output swing of 2 volts peak to peak into an open circuit, and the current gain $\beta = 100$ for each transistor. Also assume that the transistors (e.g., MPSH10) have very low capacitance such that the audio-frequency response is not a problem and that there is no Early effect. Also, Vbias = 1.6 volts, and the VBE turn on voltage = 0.6 volt.

A first try is a common-emitter amplifier, as shown in Figure 19-13. The second-harmonic distortion for a common-emitter amplifier where the emitter is AC grounded via capacitor CE is about 1%/1 mV peak sine-wave input.

Since, the specification is less than 1 percent second harmonic distortion (second harmonic distortion = HD_2), this means that the AC-grounded emitter amplifier

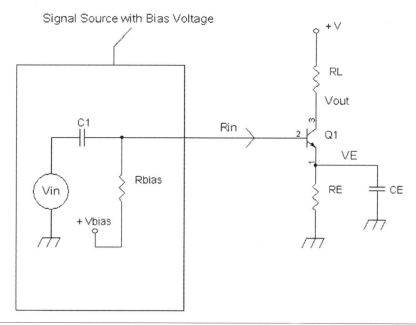

FIGURE 19-13   Common emitter amplifier with an AC-grounded emitter.

cannot have an input level of less than 1 mV peak. However, the specification calls for a 2-mV input, which then would result in 2 percent second-harmonic distortion.

If there is an emitter resistor in the common-emitter amplifier and the emitter is not AC grounded, the second-harmonic distortion as a function of input voltage is approximately

$$\frac{1\%}{1 \text{ mV peak sine-wave input}} \times \frac{1}{(1 + g_m RE)^2}$$

See Figure 19-11. For example, if $g_m RE = 2$, then $(1 + g_m RE)^2 = (1 + 2)^2 = 9$, and then the second harmonic distortion is

$$\frac{1\%}{9 \text{ mV peak sine-wave input}}$$

This would mean with a 2-millivolt peak sine-wave signal at the input would result in less than 1 percent second-harmonic distortion. Note that second-harmonic distortion is in proportion to input voltage, so the expected second-harmonic distortion at the output of the common-emitter amplifier is

$$\frac{2}{9}\% = 0.22\%$$

The input resistance into the base amplifier is greater than 10 kΩ. One can precisely calculate the collector current and the associated value of R_E. However, in practice, it is easier to just make an estimate. If the collector current is 1 mA, the input resistance is

$$\text{Rin} = \frac{\Delta \text{Vin}}{\Delta \text{IB}} = \frac{\beta}{g_m} + (\beta + 1)RE = 2.6 \text{ k}\Omega + 101\,RE$$

But $RE = 2/g_m = 52 \ \Omega$, so

$$\text{Rin} = \frac{\Delta \text{Vin}}{\Delta \text{IB}} = \frac{\beta}{g_m} + (\beta + 1)RE = 2.6 \text{ k}\Omega + (101)52 = 7.8 \text{ k}\Omega < 10,000 \ \Omega$$

Let the collector current now equal 0.7 mA; then $RE = 74 \ \Omega$ and

$$\text{Rin} = 3.7 \text{ k}\Omega + (101)74 = 11 \text{ k}\Omega > 10,000 \ \Omega$$

The load resistor RL gain of the common-emitter amplifier thus is

$$\left[\frac{-RL}{RE + \frac{1}{g_m}}\right] = -g'_m RL = \left[\frac{-RL}{74 + 37}\right] = \left[\frac{-RL}{111}\right] = \text{gain}$$

If we want a gain of 1,000, RL = 111 kΩ. However, the DC collector current is 0.7 mA, which would develop a voltage of 0.7 mA × 111 kΩ = 77 volts, which exceeds the 12-volt supply. Therefore, we have to design this amplifier with more than one stage.

At this point, let's just make the first stage have a gain of 100 instead and then follow up with a second voltage-gain stage of 10 to provide a total gain of 1,000. Thus R_L = 11.1 kΩ for a gain of –100, and the voltage drop across the 11.1-kΩ resistor is 7.7 volts.

Figure 19-14 shows a practical implementation of setting RE to 74 Ω. Note in practice the use of a 75-Ω resistor. When Vbias is set to 1.6 volts, the base voltage of Q1 is 1.6 volts DC, and the emitter voltage is about 1.0 volt. Thus REQ1 = 1,400 Ω will provide a collector current of about 0.7 mA. With a RL = 11.1 kΩ (note that the nearest 1 percent value for an 11.1-kΩ resistor is 11.0 kΩ), this means that the output resistance of the common-emitter amplifier is also 11.1 kΩ. Therefore, the gain from the base of Q1 to the collector of Q1 is about –100. Note that capacitor CE in Figure 19-14 has sufficiently large capacitance to provide a short-circuit impedance at audio frequencies.

To minimize further loading effects on RL, an emitter follower buffers this common-emitter amplifier, and the output of the common-emitter amplifier is fed to a second common-emitter amplifier with a gain of about –10 (see Figure 19-15).

When multiple-stage amplifiers are coupled in the manner shown in Figure 19-15, usually the later stages will generate the most distortion. So it is usually

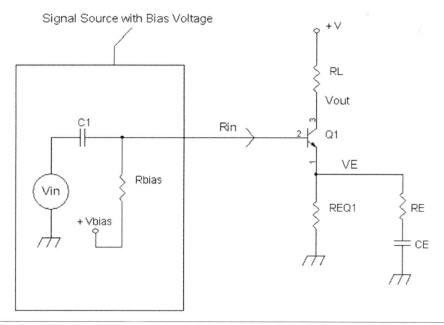

FIGURE 19-14   An input amplifier stage.

preferred to ensure that the first stage's distortion characteristic is well below the specification (e.g., 0.22 % << 1 %).

The base of emitter follower Q2 is connected to the output of Q1 via RLQ1. The output of the emitter follower then is connected to the base of Q3. Because there is a 7.7-volt drop across the load resistor RL, with a 12-volt supply the DC voltage at the collector of Q1 is 4.3 volts, and the emitter of Q2 is 3.7 volts, which leads to Q3's emitter voltage of 3.1 volts. Since the voltage swing at the output has to be 2 volts peak, the Q3 collector current will be set for 4 milliamps so that the maximum output swing is about 4 volts peak. However, at 4 volts peak, the distortion is likely to exceed 1 percent. Therefore, half the maximum voltage swing at the collector of Q3 gives a good chance that the distortion specification will be met.

Therefore, REQ3 = 775 Ω such that 3.1 volts/775 Ω = 4 mA. For the 1,000-Ω output resistance specification, RLQ3 = 1,000 Ω. For a gain of –10, the parallel combination of REQ3 and RE3 must equal about 96 Ω. Thus RE3 = 110 Ω.

The emitter resistor of Q2 is 5.1 kΩ, and the input resistance to Q3 is about $(\beta + 1)96\ \Omega$ = 9.69 kΩ. Thus the total load resistance at the emitter of Q2 is 5.1 kΩ parallel to 9.69 kΩ = 3.3 kΩ. For the current gain β = 100, the approximate input resistance into the base of Q2 is about $(\beta + 1)3.3$ kΩ = 333 kΩ. Because the output resistance of first stage is about 11.1 kΩ, the 333-kΩ load at the base of Q2 provides a negligible effect on the signal amplitude from the collector of Q1.

So now let's calculate the distortion at Q3. The gain from Q1 and unity-gain amplifier Q2 is 100, which means that for a 2-mV peak input signal, there is a 200-mV peak signal at the emitter of Q2 and base of Q3. The equivalent emitter resistor is RE_equiv = 96 Ω, the parallel combination of the two emitter resistors REQ3 and RE3 connected to Q3. At 4 mA, g_{m_Q3} = 0.1536 mho.

$$g_{mQ3}RE\_equiv = 0.1536(96) = 14.746$$

$$1 + g_{mQ3}RE\_equiv = 1 + 14.746 = 15.746$$

and thus the distortion from Q3's output is

$$\frac{1\%}{1\ mV\ peak\ sine\text{-}wave\ input} \times \frac{1}{(1 + g_{mQ3}RE\_equiv)^2} =$$

$$\frac{1}{1\ mV\ peak\ sine\text{-}wave\ input} \times \frac{1}{(15.746)^2} = \frac{1\%}{247\ mV\ peak\ sine\text{-}wve\ input}$$

Since 200 millivolts peak is less than 247 millivolts, the output of Q3 will have less than 1 percent second order harmonic distortion. And since the second harmonic distortion is proportional to the input level, the expected second harmonic distortion is

$$\frac{200}{247}\% = 0.807\%$$

Actually, there may or may not be a bit more because there is about 0.22 percent distortion from Q1, but the total distortion rarely adds in a linear fashion from all the amplifier stages. The amplifier was built and with a 2 mV peak sinusoidal waveform at its input, the second harmonic distortion was measured at about 0.65 percent. It appears that the second harmonic distortions of the first and third stages partially cancel instead of adding. Intuitively, the partial cancellation makes sense. As the first stage is increasing in Q1's collector current, which increases the transconductance of the first stage, the output voltage from the first stage is lowered that causes the collector current of the third stage to lower. This in turn causes the transconductance of the third stage to lower as well. Therefore, the combination of increasing and decreasing transconductances of the first and third stages result in a more constant transconductance, which then increases linearity of the combined amplifying stages.

The distortion can be lowered further just by increasing the collector current of Q3 from 4 mA to 5 mA, which will reduce the distortion of the Q3 amplifier by about 35 percent further to about 0.52 percent second harmonic distortion.

Thus, Figure 19-15 is just one example of designing an amplifier or amplifier system consisting of multiple stages of amplification. This example was chosen to show the interaction of input resistance, output resistance, gain, and distortion considerations in designing an amplifier. Obviously, there are other parameters one must take into account, such as power consumption, sensitivity to power-supply noise, small- and large-frequency response, cost, and board space, to name a few.

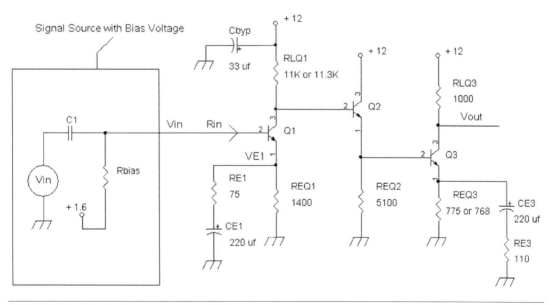

FIGURE 19-15   A three-stage amplifier.

Practical Considerations for Using Amplifiers

In Chapter 18 concerning automatic volume control (AVC) amplifiers, it was shown that for a common-emitter amplifier, a minimum base voltage is required for turning on the transistor for amplification. In op amps, one must consider the minimum voltage needed to bias on the transistors for amplification. This section will cover biasing conditions, output swing, and frequency response of op amps.

Figure 19-16 shows basically two types of input stages. One important consideration for using op amps is determining the voltage range at the input terminals that allows proper biasing within the op amp. This voltage range is named the *common-mode input voltage*, which is based on a given supply voltage. For now, assume that the –V supply voltage is grounded.

In Figure 19-16, the NPN differential amplifier input stage consisting of Q1 and Q2 has an emitter current source Q3. This type of input stage can be found in the NE5534 op amp. Generally, for each transistor allow 0.7 volt minimum for VBE base-to-emitter turn-on voltage, and allow the collector-to-base voltage to be generally 0 volt or higher. Therefore when the negative power supply volt is grounded such that –V = 0 volt, the minimum input voltage is at least or equal to 0.7 volt from the VBE base-to-emitter turn-on voltage of f Q1 or Q2 plus another 0.7 volt from the collector Q3. So the minimum voltage at the input terminal is at least 1.4 volts. However, to be on the safe side, the minimum input voltage should be about 2 volts, which is about 1 VBE drop extra to guarantee operation over low temperature.

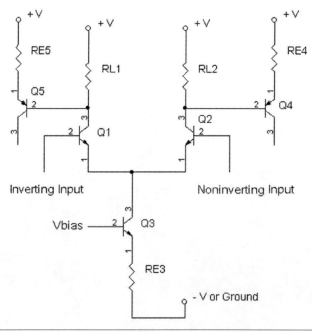

FIGURE 19-16 An op amp input stage using NPN transistors.

For the maximum voltage at the input, the current source provides a voltage drop across the load resistors RL1 and RL2 for Q1 and Q2. If the input voltage is raised too high, the base collector junctions of Q1 or Q2 will forward bias, which will cause Q1 or Q2 to saturate. Therefore, in this example, the input voltage has to be smaller than +V, the supply voltage. The NE5534 specification sheet shows that there must be an input voltage of at least 2 volts below the positive supply voltage or +V –2 volts.

Figure 19-17 shows an LM358 op amp input stage using PNP devices Q1 to Q4. A current mirror active-load circuit is provided by NPN transistors Q5 and Q6. The PNP devices allow the input range to ground or even a little bit below ground, such as –0.5 volt. Q1 and Q2 are PNP emitter follower circuits that level shift the input signal up by about +0.6 volt into the differential voltage gain amplifier Q3 and Q4. The emitter voltage of Q3 and Q4 is about 1.2 volts, which is about 0.6 volt higher than the emitter voltages of Q1 and Q2. This means that the collectors of Q3 and Q4 can operate safely at about +0.6 volt. Note that Q5 with its diode connection provides 0.6 volt at the collector of Q3. The collector of Q4 is allowed to swing from 0 volt to almost 1.2 volts, but normally operates at about 0.6 volt. Thus, the common input voltage for this example works down to 0 volt or to about –0.5 volt.

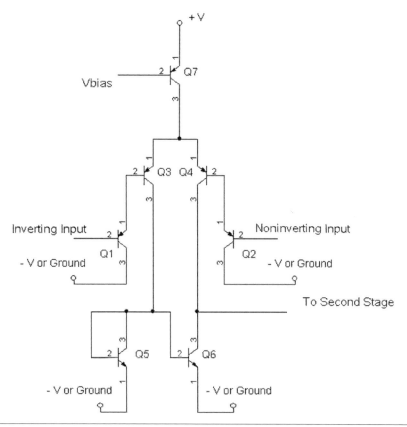

FIGURE 19-17   An op amp input stage using PNP transistors.

In terms of the input voltage near the +V supply, the emitters of Q3 and Q4 are connected to a PNP current source Q7. So the input voltage is level-shifted by 1.2 volts up to the collector of the PNP current source Q7. If the current source allows its collector to still operate correctly at about 0.2 volt below +V, then the positive or high input range about +V – (VBEQ1 + VBEQ3 + 0.2 volt) = +V – 1.4 volts. The LM358 specification sheet states the high input range if +V – 1.5 volts.

Input range becomes important especially when working with a single supply. If an NE5532 is used with a single supply, then the inputs should be biased to about one-half the supply voltage, and the input range has to be 2 volts below the supply voltage and 2 volts above ground. For an LM358, the inputs can be biased at half or slightly lower than half the supply voltage, and the input range will work at 1.5 volts below the (+) power-supply voltage to 0 volt.

It should be noted that the voltage at the (+) input terminal is the one that determines the voltage at the (–) input terminal of the op amp. In terms of op amp output stages, Figure 19-18 shows a typical push-pull output stage using NPN and PNP output transistors. The figure shows an output stage with push-pull emitter followers Q10 and Q11. The maximum output signal then is on the order of a base-emitter turn on voltage plus the saturation voltage of the transistors driving the bases of Q10

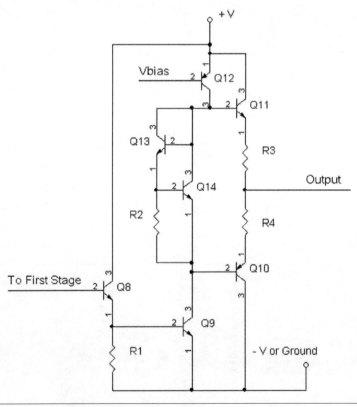

FIGURE 19-18 An RC4558 op amp output stage.

and Q11, which is about 0.2 volt. The output-transitor base emitter turn-on voltage is about 0.7 volt. Thus the "loss" is about 0.7 volt plus the loss of 0.2 volt for positive and negative swings of transistors Q12 and Q9, respectively. This results in a maximum high output voltage of about +V –0.9 volt and the low-side voltage is about 0.9 volt from ground. The specification sheet states that the output swing loses about 1 volt for the high-side output and the low-side output.

Fortunately, today there are many op amps that will have rail-to-rail output capability, such as the ISL28208 and ISL28218. Also, op amps such as the THS4281 and LM6152 have both rail-to-rail input and output capability.

Now we shall turn to frequency-response considerations in op amps. Op amps are generally unity-gain stable. That is, the op amp can be configured for a unity-gain follower for a gain of +1 without oscillations. Normally, the voltage follower is configured as shown in Figure 19-19.

Most op amps also are specified with a gain-bandwidth product (GBWP). Alternatively, the unity-gain bandwidth is the gain-bandwidth product. For example, if the gain-bandwidth product is 50 MHz, at a gain of +1, the frequency response would be flat and exhibit a –3 dB loss or a drop in gain to 0.707 at 50 MHz. If the gain of the op amp is set for 100, the response will be flat, and at 500 kHz, there will be a drop in gain to 70.7 (see Figure 19-20).

Figure 19-21 illustrates setting a gain of 100 for a single supply. This figure shows a single-supply implementation for an AC gain of +100. R1, which normally goes to ground, is coupled to ground via a capacitor C2. The (+) input of the op amp is biased to one-half the supply voltage, and the AC input signal is coupled to the (+) input via input capacitor C1. The output of the op amp can be coupled to another op amp stage, as in Figure 19-21, because the DC voltage at the output is one-half the supply voltage. Otherwise, the output of the op amp might require an output capacitor C3.

Strictly speaking, the frequency response and gain are not always tied to the gain-bandwidth product. Rather, the frequency response is related to how the negative-feedback network is applied.

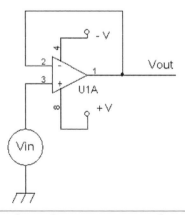

FIGURE 19-19 Op amp configured as a unity-gain voltage follower amplifier.

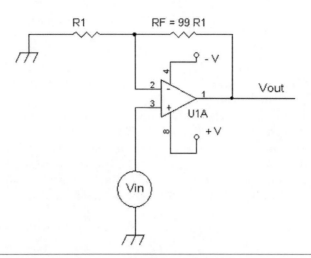

FIGURE 19-20 A noninverting-gain op amp set for 100.

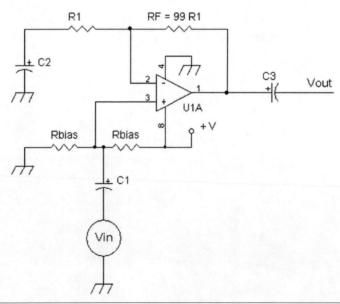

FIGURE 19-21 A single-supply implementation for a gain of +100.

$$GBWP = (OLBW)(OLGain)$$

where OLBW is the open-loop bandwidth, and OLGain is the open-loop gain at DC.

When a resistive negative-feedback network is connected between the output and the (–) input of the op amp, the bandwidth or –3-dB frequency response in a feedback system is

$$(GBWP)(FBNet) + OLBW = bandwidth$$

FBNet is the resistive attenuation factor from the output to the (–) input.

Figure 19-22 shows a resistive network for an inverting amplifier. The attenuation factor then is R1/(RF + R1), and thus the bandwidth of the inverting amplifier is

$$GBWP\frac{R1}{RF + R1} + OLBW = bandwidth$$

However, usually the OLBW $\ll$ GBWP[R1/(RF + R1)]. So, commonly, the bandwidth is expressed as GBWP[R1/(RF + R1)] = bandwidth. For example, for a unity inverting gain amplifier, RF = R1, and

$$GBWP\frac{R1}{R1 + R1} = GBWP(\tfrac{1}{2}) = bandwidth$$

Note that even though the gain is –1, the bandwidth is half the voltage follower.

Now let's look at an inverting summing amplifier, where all the resistors are equal RF = R1 = R2 = · · · = RN (see Figure 19-23). The attenuation factor from the output to the input is

$$\frac{R1/N}{R1 + R1/N} = \frac{1}{N + 1}$$

because when all the input sources are 0 or grounded, all the resistors from R1 to RN are paralleled, which is the resistance (R1/N).

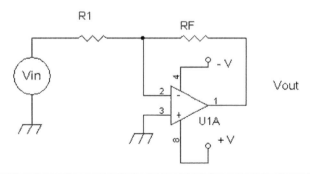

FIGURE 19-22   An inverting amplifier with feedback resistor RF and R1.

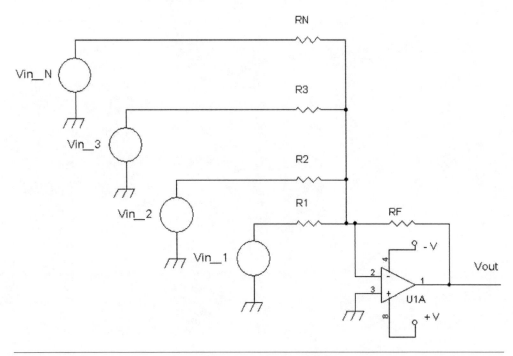

FIGURE 19-23 An *N*-input inverting summing amplifier.

Therefore,

$$\text{GBWP}\frac{1}{N+1} = \text{bandwidth}$$

So, if $N = 100$, then there are 100 inputs to the inverting summing amplifier, and the bandwidth is $1/101$ of the gain-bandwidth product. For example, if the op amp's GBWP is 1 MHz and there are 100 inputs with all resistors equal, then the bandwidth is 9.9 kHz, and the gain is –1 for any input.

One can look at the preceding problem another way: if we ground 99 of the inputs and feed R1 with the input signal. The input resistors R2 to R100 are parallel and go to ground. Thus the inverting summer is redrawn in Figure 19-24. For simplicity of analysis, let all resistors have a resistance R. There is a voltage divider from the input side of R and $R/99$. This resistive divider can be simplified with a Thevenin equivalent circuit, which states that the Thevenin resistance is just the resistance across the resistor going ground, $R/99$, in parallel with the driving resistance R (see Figure 19-25).

Therefore, with $N = 100$, the Thevenin resistance is $R/100$. And the Thevevin voltage is just the divided voltage times the input voltage. So the Thevenin voltage is

$$\frac{V_{in}}{100}$$

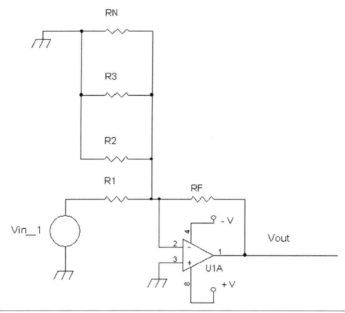

FIGURE 19-24   An inverting summing amplifier redrawn with other inputs grounded.

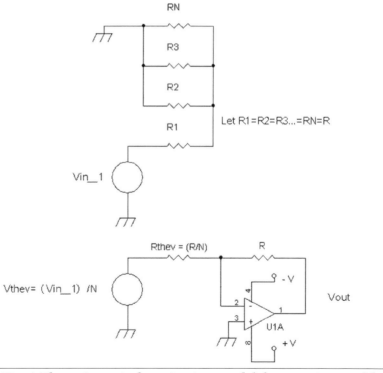

FIGURE 19-25   A Thevenin equivalent circuit to model the summing amplifier with N input resistors.

To calculate the gain Vout/Vin,

$$\text{Vout} = -\frac{\text{Vin}}{100} \times \frac{R}{R(1/100)} = -\frac{\text{Vin}}{100} \times \frac{100}{1} = -\text{Vin}$$

$$\frac{\text{Vout}}{\text{Vin}} = -1$$

which is expected. However, the attenuation factor is

$$\frac{R\frac{1}{100}}{R + R(\frac{1}{100})} = \frac{1}{101} = \text{FBNet}$$

and thus

$$\text{GBWP}\frac{1}{100} = \text{bandwidth}$$

which agrees with the previous analysis. Also note that using the Thevenin equivalent voltage shows that the Thevenin voltage source Vin/100 is being amplified by 100 because the Thevenin resistor, which is the input resistor, is 1/100 of the feedback resistor. By using the Thevenin circuit, the original input signal is attenuated by 100, but that attenuated input signal is being amplified by 100 via the feedback resistor and the Thevenin resistor. So the amplifier is really providing an inverted gain of –100, which accounts for the loss in bandwidth.

To reiterate, though, with a resistive negative-feedback network,

$$\text{Bandwidth} = \text{GBWP}(\frac{R1}{RF + R1}) + \text{OLBW}$$

This makes sense because when R1 = 0, the (–) input is shorted or tied to ground, and there is no feedback; thus the bandwidth is just the open-loop bandwidth OLBW.

In op amps, a DC offset voltage at the output is caused by the input bias currents flowing through the resistors connected to the (+) and (–) inputs. More details of input offset voltages and input current bias effects can be found in Section 3.5 of Thomas M. Frederiksen's book, *Intuitive IC Op Amps* (Santa Clara: National Semiconductor Corporation, 1984). However, using an op amp with a field-effect transistor (FET) input stage will eliminate concerns of input bias currents causing offset voltages.

References

1. Class Notes EE141, Paul R. Gray and Robert G. Meyer, UC Berkeley, Spring 1975.
2. Class Notes EE241, Paul R. Gray and Robert G. Meyer, UC Berkeley, Fall 1975.

3. Class Notes EE140, Robert G. Meyer, UC Berkeley, Fall 1975.
4. Class Notes EE240, Robert G. Meyer, UC Berkeley, Spring 1976.
5. Paul R. Gray and Robert G. Meyer, *Analysis and Design of Analog Integrated Circuits*, 3rd ed. New York: John Wiley & Sons, 1993.
6. Kenneth K. Clarke and Donald T. Hess, *Communication Circuits: Analysis and Design*. Reading: Addison-Wesley, 1971.
7. Robert L. Shrader, *Electronic Communication*, 6th ed. New York: Glencoe/McGraw-Hill, 1991.
8. William G. Oldham and Steven E. Schwarz, *An Introduction to Electronics*. New York: Holt Reinhart Winston, 1972.
9. Allan R. Hambley, *Electrical Engineering Principles and Applications*, 2nd ed. Upper Saddle River: Prentice Hall, 2002.
10. Gene F. Franklin, J. David Powell, and Abbas Emami-Naeini, *Feedback Control of Dynamic Systems*, 3rd ed. Reading: Addison-Wesley, 1995.
11. Texas Instruments, *Linear Circuits Amplifiers, Comparators, and Special Functions*, Data Book 1. Dallas: Texas Instruments Incorporated, 1989.
12. Motorola Semiconductor Products, Inc., *Linear Integrated Circuits*, Vol. 6, Series B., 1976.
13. Thomas M. Frederiksen, *Intuitive IC Op Amps From Basics to Useful Applications* (National Semiconductor Technology Series). Santa Clara: National Semiconductor Corporation, 1984.
14. Alan B. Grebene, *Bipolar and MOS Analog Circuit Design*. New York: John Wiley & Sons, 1984.
15. Geoffrey Hutson, Peter Shepherd, and James Brice, *Color Television System Principles*. London: McGraw Hill, 1990.
16. Margaret Craig, *Television Measurements PAL Systems*. Beaverton: Tektronix, Inc., 1991.
17. Margaret Craig, *Television Measurements NTSC Systems*. Beaverton: Tektronix, Inc., 1994.

Chapter 20

Resonant Circuits

Resonant circuits, whether they are constructed from capacitors, inductors, ceramic, or quartz materials, play an important role in passing signals in a particular frequency range and attenuating those signals outside the particular frequency range. Generally, resonant circuits are used in oscillators as well as band-pass filters.

Less commonly known, resonant circuits can be used in all-pass networks to provide group delay equalization or phase equalization to band-pass, high-pass, and low-pass filters. The group delay (GD) can be expressed as GD = $\Delta\varphi/\Delta\omega$, where $\Delta\varphi$ is the change in phase (in radians), $\Delta\omega$ is the change in frequency (in radians per second), and $\omega = 2\pi f$ for a plot of the filter in terms of frequency and phase.

For example, prior to digital finite impulse response (FIR), low-pass filters that provided a square-wave signal with symmetric pre-ringing and post-ringing effects, analog brick-wall filters used in professional (broadcast) video recorders in the 1960s or earlier routinely included a phase equalizer to provide symmetric ringing to pulse waveforms. An all-pass network generally has constant amplitude over a range of frequencies while providing various phase shifts as a function of frequency.

In this book, the radio projects used mainly inductor-capacitor (LC) oscillators and band-pass filters. However, types of resonant circuits such as crystal or ceramic resonator oscillators were shown in Chapters 4 and 12. Ceramic resonators and crystals can be modeled as a series LC circuit with a very large inductor and a very small capacitor, along with a parasitic body capacitance across the two terminal leads of the crystal.

For this chapter, then, the objectives are

1. To examine simple parallel and series LC circuits
2. To understand how resonant circuits play a role in oscillators
3. To show some examples of band-pass and band-reject filters

Simple Parallel and Series Resonant Circuits

Before we start analyzing parallel and series resonant circuits, we need to know the basic characteristics of an inductor and a capacitor. An inductor has the following voltage characteristic with inductance L: Voltage across the inductor L = $V_{inductor}$ = L($\Delta i/\Delta t$), where $\Delta i/\Delta t$ is the change in inductor current divided by the change in time, and L is the inductance in henries. The current flowing into a capacitor is $I_{capacitor}$ = C($\Delta v/\Delta t$), where $\Delta v/\Delta t$ is the change in voltage across the capacitor divided by the change in time, and C is the capacitance in farads.

Or, more generally, the voltages and currents of the inductor and capacitor vary with time and can be expressed as functions of time:

$$V_{inductor}(t) = L\frac{\Delta i}{\Delta t} \tag{20-1}$$

$$I_{capacitor}(t) = C\frac{\Delta v}{\Delta t} \tag{20-2}$$

These basic equations, when applied to differential equations and Laplace/Fourier transforms, result in the complex impedance of the inductor and capacitor *when using sinusoidal signals* as

$$Z_L = \text{impedance of the inductor} = j(2\pi f)L \tag{20-3}$$

$$Z_C = \text{impedance of the capacitor} = \frac{1}{j(2\pi f)C} = -j\frac{1}{(2\pi f)C} \tag{20-4}$$

where $j = \sqrt{-1}$.

The mysterious j or j *operator* is defined as the square root of a negative number. Actually, j represents a 90-degree phase shift on a sinusoidal signal. When j is squared, $j^2 = -1$, or think of a sinusoidal signal that has been phase-shifted 90 degrees and another 90 degrees for a total phase shift of 180 degrees. However, 180 degrees of shift on any sinusoidal signal is just an inverted version of the original sinusoid. Hence this is how the –1 term comes in.

Thus $j^N = N$ quarter-turns, or $N \times$ 90-degree phase shift of a sinusoidal signal. Also, whenever j is used in the context of complex impedances, only sinusoidal or circular functions/signals are used. For complicated signals such as pulses, one can use the j operator with the complicated signals that are equivalently characterized as the summation of sinusoidal signals via Fourier analysis. Then each of the sinusoidal signals can be applied with the j operator.

Figure 20-1 shows a parallel LC circuit that is voltage driven with a resistor. If a direct-current (DC) voltage is applied to an inductor via a resistor, initially the inductor current will be zero. After a period of time, the inductor pulls current and acts like a wire. Thus, in an inductor, applying a voltage across the inductor results

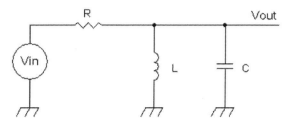

FIGURE 20-1   **An LC band-pass filter.**

in the inductor current having a delay in pulling current. Thus the current in an inductor lags the voltage across the inductor.

Conversely, if a DC voltage is applied to a capacitor via a resistor, the capacitor will initially act like a short circuit or wire and pull current immediately. After a period of time, though, the capacitor's voltage starts to match the voltage of the DC voltage source and thereby reduce capacitor current. Thus the current in a capacitor leads the voltage across the capacitor.

When a sinusoidal voltage is applied to an inductor, the inductor current lags by 90 degrees in reference to the inductor voltage. Conversely, a sinusoidal voltage applied to a capacitor results in the capacitor current leading the capacitor voltage by 90 degrees. And when a sinusoidal voltage is applied to an LC parallel circuit where the frequency is chosen such that the magnitude of the impedances of the inductor and capacitor is the same, the inductor current and the capacitor current have the same magnitude, but the currents are a net 180 degrees from each other owing to the inductor's current having a –90-degree phase shift of the sinusoid voltage and the capacitor's current having a +90-degree phase shift of the sinusoidal voltage. Because of this 180-degree difference between the inductor and capacitor currents, there is no net current drain looking into the parallel LC circuit at resonance, where

$$\text{Mag}(Z_L) = \text{Mag}(Z_C)$$

where Mag = magnitude. The resonant frequency is

$$\frac{1}{2\pi\sqrt{LC}} = f_{res}$$

$$Z_L = \text{impedance of the inductor} = j(2\pi f)L \qquad (20\text{-}3)$$

$$Z_C = \text{impedance of the capacitor} = \frac{1}{j(2\pi f)C} = \frac{j}{j}\frac{1}{j(2\pi f)C} = \frac{j}{-1(2\pi f)C} = -j\frac{1}{(2\pi f)C}$$

$$Z_C = \text{impedance of the capacitor} = -j\frac{1}{(2\pi f)C} \qquad (20\text{-}4)$$

Also note that the impedances of inductors and capacitors have opposite signs.

Thus, with an ideal inductor and capacitor, at resonance, the parallel LC circuit drains no alternating current (AC) and thus has infinite resistance or impedance. For signals that are above the resonant frequency, the impedance drops and looks capacitive because at high frequencies the inductor's impedance rises, whereas the capacitor's impedance trends toward a short circuit.

At frequencies below the resonant frequency, the parallel LC circuit starts to look more inductive because the inductor's impedance drops while the capacitor's impedance increases. For example, if the signal is DC, the inductor looks like a wire, and the capacitor looks like an open circuit.

In terms of calculating the transfer function of the parallel LC band-pass filter, we will treat the circuit as a voltage divider with the paralleled LC impedance going to ground with the driving resistance R. Thus

$$\frac{Vout}{Vin} = (Z_L || Z_C)/(R + Z_L || Z_C)$$

where $Z_L || Z_C = \frac{Z_L Z_C}{Z_L + Z_C}$, and

$$Z_L = \text{impedance of the inductor} = j(2\pi f)L \qquad (20\text{-}3)$$

$$Z_C = \text{impedance of the capacitor} = \frac{1}{j(2\pi f)C} = -j\frac{1}{(2\pi f)C} \qquad (20\text{-}4)$$

After some algebraic manipulations, for the transfer function of Figure 20-1,

$$\frac{Vout}{Vin} = \frac{1}{1 + jRC2\pi f - j\frac{R}{(2\pi f)L}} \qquad (20\text{-}5)$$

Thus the maximum gain of the transfer function in Equation (20-5) is 1, but in practice, it is less than 1. The reason is that the inductor has resistive and core losses that equivalently connect an extra resistor in parallel with the inductor and capacitor (Figure 20-2).

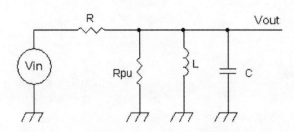

FIGURE 20-2   A parallel LC circuit including the equivalent lossy resistor Rpu.

The resistor R_pu represents equivalent parallel resistance based on the unloaded Q of the coil Q_u:

$$Q_u = 2\pi f_{res} RpuC$$

In general, the Q of the parallel tank circuit is

$$Q_p = 2\pi f_{res}(Rpu \parallel R)C \tag{20-6}$$

And the bandwidth measured from 0.707 of the maximum gain is

$$BW = \frac{f_{res}}{Q} \tag{20-7}$$

For a current-driven parallel LRC circuit, see Figure 20-3.

Examples of current sources driving an LRC circuit are common-emitter, common-base, and differential-pair amplifiers. The impedance of the LRC circuit is

$$Zin = Z_{c\_parallel}$$

$$Z_{c\_parallel} = \frac{R}{1 + jRC2\pi f - j\dfrac{R}{(2\pi f)L}} = \frac{1}{\dfrac{1}{R} + jC2\pi f - j\dfrac{1}{(2\pi f)L}} \tag{20-8}$$

Equation (20-8) states that the maximum impedance when driven with a perfect current source is infinity if the inductor and capacitor have no losses. However, the early effect from the transistor current source and the internal equivalent resistance Rpu will be the limiting resistances across the parallel LC circuit. In terms of the phase relationship with the input signal, at the resonant frequency, the phase shift is 0 degree, and for signals below the resonant frequency, the phase shift is leading or positive up to +90 degrees, whereas for signals above the resonant frequency, the phase shift is lagging and may be as much as –90 degrees. At the frequencies when the signal is 0.707 of the maximum output or –3 dB of bandwidth, the phase shift is +45 degrees at below the resonant frequency and –45 degrees at above the resonant frequency.

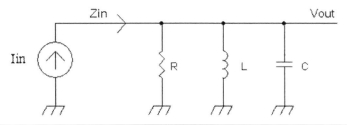

FIGURE 20-3 **A current-source-driven LRC circuit.**

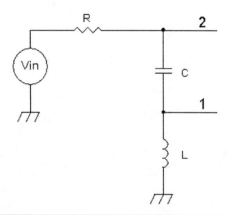

FIGURE 20-4   A series resonant circuit at node 1 as a high-pass filter and at node 2 as band-eject filter.

Figure 20-4 shows a series resonant circuit that can provide different types of filtering. Unlike the parallel resonant circuit, which can be configured for band-pass and band-reject characteristics, a series resonant circuit can be designed for four types of filters: high-pass, low-pass, band-reject, and band-pass filters. For example, the output from node 1 of Figure 20-1 provides a high-pass-filter response. Depending on the driving resistance R, the high-pass filter also can provide a peaked response at the resonant frequency.

Series resonant circuits are also common in ceramic resonators and crystal oscillators, where these types of devices often operate as series resonators. In the parallel resonant circuit, with a voltage source applied across the inductor and capacitor, the currents flowing into the inductor and capacitor are a net 180 degrees from each other. However, in a series resonant circuit with current flowing through the resistor, capacitor, and inductor, the voltage across the inductor and the voltage across the capacitor are a net 180 degrees from each other, and therefore, the net voltage across the capacitor and inductor as seen in node 2 of Figure 20-4 is zero at the resonant frequency.

This zero voltage at the resonant frequency happens because even though there are AC voltages individually across the capacitor and inductor, the "polarity" is out of phase. It is like having two flashlight cells in series connected at either both positive terminals or both negative terminals, which results in the net voltage across the two batteries being zero.

In Figure 20-5, with a voltage source driving the RCL circuit, the current flow is the same through all three devices. The voltage across an inductor leads the current by +90 degrees at resonance, whereas the voltage across a capacitor lags the current by –90 degrees at resonance. Therefore, the net phase difference voltage-wise across the inductor and capacitor is 180 degrees. If the current were in phase, such as a voltage source driving two series resistors, the voltages across each resistor would add up. However, when the phase is opposite, as in the case of the series RLC circuit at the

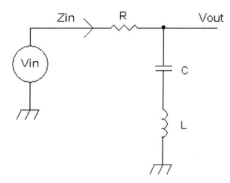

FIGURE 20-5 Series resonant circuit where at resonance the voltage drops to zero across the series capacitor-inductor circuit.

resonant frequency, equal and opposite AC signals are occurring, and a cancellation effect results when the voltage is taken at Vout where the inductor and capacitor are in series.

The impedance of the series RLC circuit is $Zin = Z_{LC\_series}$, which is

$$Z_{LC\_series} = R + Z_L + Z_C = R + j(2\pi f)L + \frac{1}{j(2\pi f)C} \qquad (20\text{-}9)$$

The Q of a series RLC circuit is

$$Q_{S\_LL} = \frac{(2\pi f)L}{R} \qquad (20\text{-}10)$$

where Q_{S_LL} is the Q for a lossless inductor L.

However, as stated previously, an inductor includes losses, and in a series resonant circuit, the coil has an internal resistance based on coil resistance and core losses as Rser. For Equations (20-9) and (20-10), Rser = 0.

Generally, for low- or high-pass filters in series resonant circuits, the Q requirement is very low, <1 (e.g., 0.6 to 1). Therefore, Rser is not as important in the overall frequency response of high- passor low-pass filters compared with band-pass or notch filters, where the Q typically is greater than 10.

For a band-pass series RLC circuit, the voltage source is driven as shown in Figure 20-6 with Rser. With the resistor as the output terminal, the transfer characteristic for the series resonant band-pass filter is

$$\frac{Vout}{Vin} = \frac{R}{R + Rser + j(2\pi f)L - j\frac{1}{(2\pi f)C}} \qquad (20\text{-}11)$$

and

$$Q_S = \frac{(2\pi f)C}{R + Rser} \qquad (20\text{-}12)$$

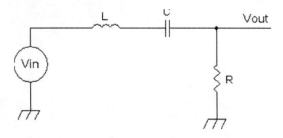

FIGURE 20-6 A series LC band-pass filter.

R_ser is the equivalent internal series lossy resistor of the inductor.

The frequency response of either parallel or series resonant band-pass filters can be characterized as a function of Q_p or Q_s. The –3-dB bandwidth is

$$BW_{-3\,dB} = \frac{f_{res}}{Q}$$

where Q is either Q_p or Q_s depending on whether a parallel or series resonant circuit is being used.

To find the relationship between the amplitude at the center frequency, which is the resonant frequency f_{res}, and the amplitude of signals whose frequencies are below or above f_{res} by Δf, see Equation (20-13):

$$\text{Attenuation factor from the center frequency} = \frac{1}{\sqrt{1 + [\Delta f/(0.5BW_{-3\,dB})]^2}}$$

(20-13)

For example, suppose that the center frequency f_{res} = 1,000 kHz and the Q is 50, then

$$BW_{-3\,dB} = \frac{1,000}{50} = 20 \text{ kHz and } 0.5\,BW_{-3\,dB} = 10 \text{ kHz}$$

Then

$$\text{Attenuation factor from the center frequency} = \frac{1}{\sqrt{1 + (\Delta f/10 \text{ kHz})^2}}$$

If Δf = 10 kHz, then the attenuation factor is 0.707 at 1,010 kHz and 990 kHz. If Δf = 100 kHz, then the attenuation factor is 0.10 at 1,100 kHz and 900 kHz. When

$$\frac{\Delta f}{0.5BW_{-3\,dB}} > 2$$

the attenuation factor from the center frequency is characterized as $\approx$

$$\frac{1}{\frac{\Delta f}{0.5BW_{-3\ dB}}} = \frac{0.5BW_{-3\ db}}{\Delta f} \tag{20-14}$$

In a series resonant circuit, at resonance, the impedances of the inductor and capacitor cancel, and thus the series impedance of the capacitor and inductor is zero. This means that if the driving resistance is very low, very large amounts of AC current can flow through the capacitor and inductor. What is also important to observe is that at resonance, the AC voltage across the inductor or capacitor can exceed the input signal. In a sense, there is a step-up voltage effect when using the series resonant circuit in a particular manner.

For example, at node 1, the AC voltage across the inductor is determined by the AC current flowing through it. At resonance, this AC current is just

$$I_L = \frac{Vin}{R}$$

However, the magnitude of the AC voltage across the inductor L is the inductor current multiplied by the magnitude of the impedance of the inductor, as seen in Equation (20-15):

$$I_L (2\pi f)L = \frac{Vin}{R}(2\pi f)L = Vin\ Q_s \tag{20-15}$$

For example, if $Q_s = 10$, by setting R to a low resistance value, the AC voltage across the inductor will be 10 times the AC voltage of the input signal generator. In a sense, the series RLC circuit can be thought of as a step-up transformer. One should note, however, that although the voltage across the inductor is larger than the input signal, the phase of the inductor voltage leads the phase of the input by +90 degrees at resonance, where $f_{res} = 1/(2\pi\sqrt{LC})$. This positive phase shift at the inductor makes sense because the voltage across the inductor forms a high-pass filter, which provides a positive or leading-phase relationship when compared with the input signal.

It should be noted that the impedance of the series RLC circuit is more capacitive below the resonant frequency (because at DC the inductor is like a wire), resistive at the resonant frequency with resistance R, and inductive (because at high frequencies the capacitor is like a short circuit) at above the resonant frequency.

The high-pass filtering characteristic of the series RLC circuit with voltage "gain" or step-up capability along with the positive phase shift is serendipitous for providing oscillations with an amplifier whose voltage gain is less than 1. If the series RLC circuit is configured as a low-pass filter, as seen in Figure 20-7, a voltage gain or step-up capability can be achieved along with a negative phase shift.

At resonance, the phase shift is –90 degrees. Also, depending on the driving resistance R, the "gain" can be greater than 1 at the resonant frequency, similar to how

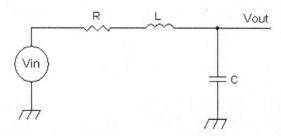

FIGURE 20-7 Series resonant circuit configured for a low-pass filter effect.

Q affects the "gain" in the high-pass filter circuit. To reiterate, see Equation (20-12) on the Q of a series resonant circuit:

$$Q_s = \frac{(2\pi f)L}{R + R_{ser}} \tag{20-12}$$

Oscillators that use an inverter gate and capacitors at the output and input along with an inductor or crystal or ceramic resonator between the output and input use a series resonant low-pass filter for oscillation.

Resonant Circuits in Oscillators

This section will examine three types of oscillators that use series resonant circuits. Figure 20-8 shows a series resonant oscillator with a unity-gain amplifier. In the figure, although the amplifier has a voltage gain of 1 or less than 1, it has the capability to provide current, which will be used to step up AC signal voltages. The output of the unity-gain amplifier is low-pass-filtered with an RC filter, which provides a phase lag

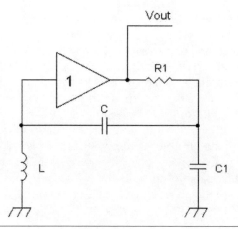

FIGURE 20-8 A series resonant oscillator.

or negative phase shift. A series high-pass filter circuit consisting of capacitor C and inductor L is connected to the output of the low-pass filter at C1. The inductor of the high-pass filter is connected to an AC ground, which when driven by the low-pass filter output at C1 provides an increase in signal voltage and positive phase shift to cancel the negative phase shift of the low-pass filter. The cancellation in phase shifts provides a zero net phase shift of signals between the input and output of the unity-gain amplifier. And the unity-gain amplifier has current gain and is able to drive the high-pass filter C and L sufficiently to step up the signal voltage such that there is an equivalent voltage gain at the input of the unity-gain amplifier. Thus oscillation occurs at or near the resonant frequency $f_{res} = 1/(2\pi\sqrt{LC})$.

Figure 20-9 shows a practical implementation of Figure 20-8 using a transistor. Emitter follower Q1 is biased to a base voltage of Vbias by connection of the inductor L. With a 0.6 volt-base emitter voltage drop, the collector current of Q1 is

$$\frac{Vbias - 0.6V}{RE} = I_{CQ1}$$

Because the base is driven via an AC voltage across an inductor and not a pure voltage source, the actual output resistance is higher than the approximation of the output resistance of Q1 at the emitter of

$$\frac{1}{g_m} = R_{out}$$

where $g_m = \dfrac{I_{CQ1}}{0.026V}$

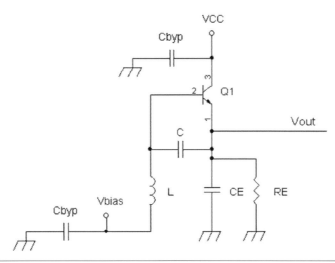

FIGURE 20-9   An emitter follower oscillator using a series resonant circuit.

The capacitor CE forms a low-pass filter with R_{out} at the emitter with a cutoff frequency of about

$$\frac{1}{2\pi\frac{CE}{g_m}} = f_{cutoff}$$

For example, let RE = 12 kΩ, V_{bias} = 1.8 volts DC, L = 220 μH, C = 220 pF, and CE = 1,800 pF. Then

$$\frac{Vbias - 0.6V}{RE} = \frac{1.8V - 0.6V}{12 \text{ k}\Omega} = 100 \text{ μA} = I_{CQ1}$$

$$g_m = \frac{100 \text{ μA}}{0.026V} = 0.00384 \text{ mho}$$

$$\frac{1}{g_m} = \frac{1}{0.00384 \text{ mho}} = 260 \text{ }\Omega = R_{out}$$

For a starting point, capacitor CE is chosen to a value such that f_{cutoff} is about the resonant frequency or a little below the resonant frequency:

$$\frac{1}{2\pi(CE/g_m)} = \frac{1}{2\pi(1,800 \text{ pF}/0.00384 \text{ mho})} = \frac{1}{2\pi(130\Omega)(1,800 \text{ pf})} = 680 \text{ kHz}$$

$$f_{res} = \frac{1}{2\pi\sqrt{LC}} = \frac{1}{2\pi\sqrt{(220 \text{ μH})(220 \text{ pF})}} = 723 \text{ kHz}$$

which is the approximate frequency of oscillation.

The measured frequency of oscillation is 769 kHz, but remember that although CE >> C, capacitor CE can contribute to the oscillation frequency because it is in series with capacitor C. Let's calculate the series capacitance of CE and C, and use that value to recalculate the resonant frequency. The series capacitance of CE and C is

$$\frac{(1,800 \text{ pF})(220 \text{ pF})}{1,800 \text{ pF} + 220 \text{ pF}} = 196 \text{ pF}$$

$$f_{res} = \frac{1}{2\pi\sqrt{(220 \text{ μH})(196 \text{ pF})}} = 766 \text{ kHz}$$

which is closer to the measured frequency of 769 kHz.

One question is, could the value of CE be increased further? The answer is yes and no. Yes, as long as there is a reliable oscillation. No, because there will be a point where CE is too large in capacitance, and the gain loss from the roll-off in response will stop the oscillation. For example, if CE were changed from 1,800 pF to 18,000 pF, there would be about a 10-fold loss in gain, and if the Q of the LC circuit is not high enough, there will be insufficient gain to sustain an oscillation. Recall that a gain of at least 1 is needed with a net phase shift of 0 degree.

The oscillator circuit shown in Figure 20-10 also can explain why at times some common-emitter amplifiers or some emitter follower circuits self-oscillate. Common-emitter amplifiers without the emitter AC grounded or emitter follower circuits generally load into some small capacitance C_load at the emitter to cause a phase lag at high frequencies. All transistors have an internal base emitter capacitance C_π = Cpi, which, although small, still serves as a capacitor from the emitter back to the base. The base lead or wires connected to the base include a parasitic or stray inductance, which then resonates with C_π or Cpi. Usually, the self-oscillation frequency is at 100 MHz or higher, and placing a series resistor of 47 Ω to 470 Ω very near to the base reduces the Q of the parasitic/stray inductance by adding resistance to the stray inductor. Therefore, the added base resistor reduces the voltage-gain effect of a series resonant circuit that stops the oscillation. Recall that the Q for a series resonant
circuit is

$$Q_s = \frac{(2\pi f)L}{R}$$

(20-10)

Figure 20-10 shows a base oscillation-stopping resistor, R_base stop, for an emitter follower circuit.

Another oscillator circuit that uses series resonant circuits is a crystal oscillator, as shown in Figure 20-11. Figure 20-12 shows an equivalent LC oscillator circuit to that in Figure 20-11. In Figure 20-12, the crystal is modeled as a series resonant circuit with an inductor Lcrystal and a capacitor Ccrystal. Typically, the value of Ccrystal is in femtofarads or on the order of less than 0.1 pF, and inductor Lcrystal has the equivalent inductance in millihenrys.

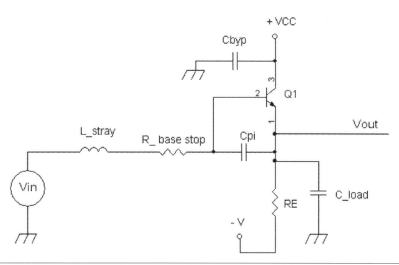

FIGURE 20-10   Emitter follower circuit with a base-stopping resistor to avoid oscillation.

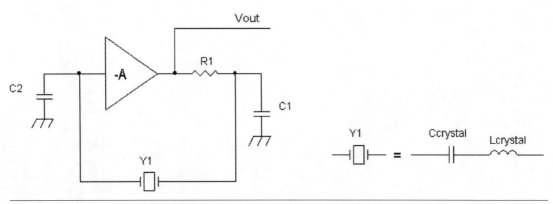

FIGURE 20-11 A crystal oscillator using an inverting amplifier.

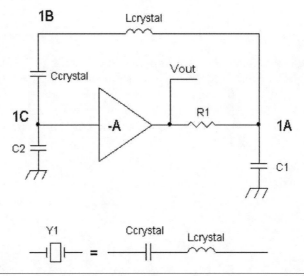

FIGURE 20-12 An equivalent circuit to Figure 20-11 with an inductor and capacitor to model the crystal.

Figure 20-13 shows series and parallel modes for crystals. In Figure 20-13, basically, the difference between series and parallel modes is just that the parallel-mode crystal is cut such that when the crystal is in series with a loading capacitor C_load, with a value such as 18 pF, the crystal will oscillate at the given frequency. A series-mode crystal will operate at its given frequency without a series capacitor. Either parallel or series mode can oscillate with other values of series capacitors to "pull" the crystal frequency on order of about 0.15 percent.

A 28.636-MHz crystal rated at 18 pF loading provided a frequency of 28.636 MHz, with C_load being an 18-pF loading capacitor. However, with C_load equal to 100 pF,

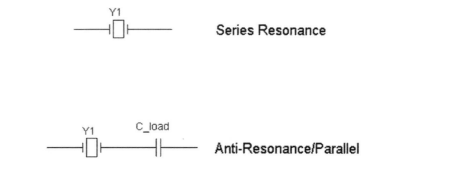

FIGURE 20-13   Series crystal and parallel/antiresonant crystal modes.

the oscillator provided a frequency of 28.600 MHz. From these two measurements, the internal capacitance of Ccrystal was calculated to be about 0.063 pF.

In Figure 20-12, the capacitors Ccrystal and C2 provide a capacitive voltage divider from node 1B. A capacitive voltage divider keeps the same phase of the signal from nodes 1B and 1C. However, the signal at node 1C is an attenuated version of the signal at node 1B.

The inverting amplifier's output is connected to a low-pass filter R1 and C1 to provide a lagging phase shift because the phase shift from R1 and C1, plus the phase shift from Lcrystal and Ccrystal must provide 180 degrees of phase shift. Typically, the phase shift from R1 and C1 lags about 60 degrees to 80 degrees, and the phase shift from Lcrystal and Ccrystal can be in the range of 100 degrees to 135 degrees lagging. Thus the combination of R1 and C1 and the crystal provides the required phase shift for oscillation.

Although the attenuation from the capacitive voltage divider is around 100 or more, the gain of the inverting amplifier and the extremely high Q of the crystal (e.g., Q > 1,000) provide sufficient overall gain to overcome the attenuation from the capacitive voltage divider (Ccrystal and C2) to provide a reliable oscillation.

Figure 20-14 presents examples of crystal and ceramic resonator oscillators based on Figures 20-11 and 20-12. Although the Figure 20-14 shows using CMOS gates for the inverting amplifiers, other implementations include using a single-transistor or field-effect-transistor (FET) common-emitter or common-source amplifier instead. For example, the FET amplifier will have a drain load resistor and a capacitor (e.g., 18 pF to 47 pF) from the drain to ground. The crystal is connected between the drain and the gate of the FET. At the gate, there is a high-resistance biasing resistor (e.g., >500 kΩ) and another capacitor (e.g., 18 pF to 47 pF) from the gate to ground. This type of oscillator is often referred to as a Pierce oscillator.

We now turn our attention to another type of series resonant crystal oscillator, the two-gate oscillator (Figure 20-15). With a two-gate oscillator as shown in Figure 20-15, at series resonance, the impedance of the crystal drops sufficiently to ensure positive feedback. The advantage of this type of crystal oscillator circuit is that the frequency of the parallel mode (otherwise known as an *antiresonant mode*) allows dropping the

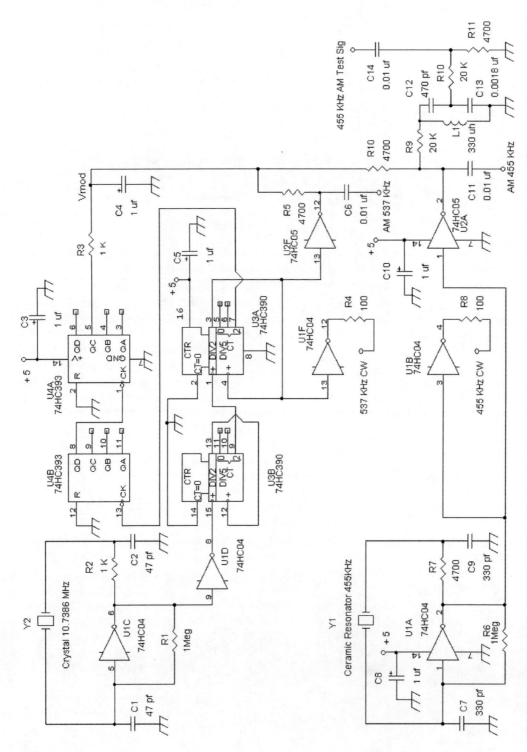

FIGURE 20-14 Crystal and ceramic resonator oscillators U1C and U1A.

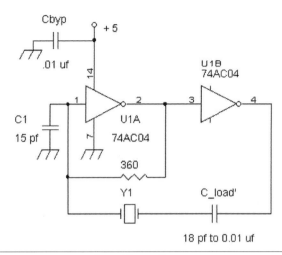

FIGURE 20-15   A two-gate crystal oscillator for series resonant mode.

crystal frequency by at least 0.15 percent. For example, if the series capacitor C_load' is set to the loading capacitance of about 18 pF, the oscillator will provide a frequency as shown on the parallel-mode crystal. However, if the frequency needs to be lowered, then a larger series capacitance via C_load', such as 56 pF to 1,000 pF, can be used to lower the oscillator's frequency.

Likewise, if a series crystal is used, normally the series capacitor C_load' is an AC short circuit, such as a 0.01-μF capacitor. However, if the frequency needs to be raised by as much as 0.15 percent, a series capacitor with a value as low as 18 pF can be used.

In practical terms, as mentioned previously, a standard parallel-mode 28.636-MHz crystal with 18-pF loading can be used for the 40-meter amateur radio band simply by installing it in the oscillator circuit of Figure 20-6 with the series capacitor C_s with a value of about 100 pF instead of the 18-pF capacitor. With the 100-pF series capacitor, the oscillation frequency will be pulled down to 28.600 MHz (a 36-kHz downward shift). Afterwards, just divide the frequency of the 28.6000-MHz signal by 4 to provide 7.150 MHz.

Examples of Band-Pass and Band-Reject Filters

At times, a simple band-pass filter may not provide sufficient selectivity in a radio. A way to increase selectivity is by cascading or linking band-pass filters. Figure 20-16 shows two parallel LC circuits linked together with a capacitor.

The parallel circuit L1 and C1 is driven with resistor R via a voltage source Vin. From the output of the first section, a coupling capacitor C3 links the signal from L1

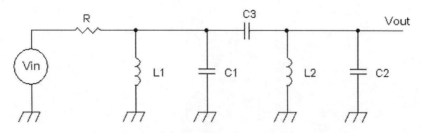

FIGURE 20-16   A two-stage band-pass filter.

and C1 to another parallel tank circuit C2 and L2. In many instances, L1 = L2, C1 = C2, and C3 = 0.5 to 3 percent of C1. For example, if C1 = 500 pF, C3 = 15 pF (3 percent of 500 pF). Driving resistor R can be chosen for a Q of 15 to 20 for a starting point depending on the bandwidth desired. There are many variations of this circuit, such as a variable capacitor for C1 or C2, a variable inductor for L1 or L2, or a loading resistor across L2 to shape the bandwidth. The output at L2/C2 should be coupled to a high-resistance amplifier to maintain the Q of the L2/C2 tank circuit and make up for signal losses.

Figure 20-17 shows examples of band-reject filters, one with a parallel LC circuit and another with a series LC circuit. Notch or band-reject filters pertaining to standard amplitude-modulation (AM) radios are not used commonly. In frequency-modulated (FM) radios, a notch filter such as a 10.7-MHz series resonant filter is placed at the input of the mixer or the output of the radio-frequency (RF) amplifier to ensure that the mixer output has a 10.7-MHz intermediate-frequency (IF) signal from the result of mixing RF signals from 88 MHz to 108 MHz with the FM radio's local oscillator's signal.

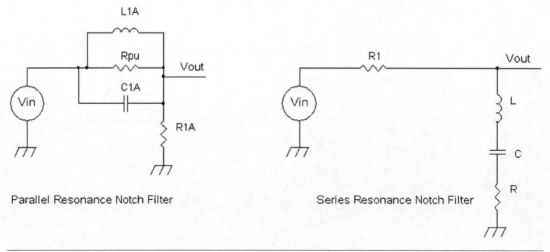

Parallel Resonance Notch Filter   Series Resonance Notch Filter

FIGURE 20-17   A parallel resonant notch filter and a series resonant notch or band-reject filter.

Recall that the impedance of a parallel LC network is

$$Z_{LC\_parallel} = \frac{Rpu}{1 + jRpuC2\pi f - j\frac{Rpu}{(2\pi f)L}} = \frac{1}{\frac{1}{Rpu} + jC2\pi f - j\frac{1}{(2\pi f)L}}$$

where Rpu in this case is the equivalent lossy parallel resistance owing to coil resistance and core losses.

For a parallel resonant notch filter,

$$\frac{Vout}{Vin} = \frac{R1A}{R1A + \frac{1}{\frac{1}{Rpu} + jC2\pi f - j\frac{1}{(2\pi f)L}}} \qquad (20\text{-}16)$$

At resonance, the parallel LCR circuit has the L and the C impedances canceling out, thus leaving only the resistance R_{pu}. So

$$\frac{Vout}{Vin} = \frac{R1A}{R1A + Rpu}$$

Therefore, it is advantageous to have Rpu as high a resistance as possible for maximum attenuation at the resonant frequency:

$$f_{res} = \frac{1}{2\pi\sqrt{LC}}$$

If there are no losses in the inductor L, Rpu = infinity = ∞, and

$$\frac{Vout}{Vin} = \frac{R1A}{R1A + \frac{1}{jC2\pi f - j\frac{1}{(2\pi f)L}}}$$

which means for any given R1A, at resonance, the notch or drop in signal voltage at the output is zero. However, this is not the case in practice. Typical starting values for R1A can be in the range of a few hundred ohms to 1,000 Ω.

For the series LC notch circuit in Figure 20-17,

$$\frac{Vout}{Vin} = \frac{R + \frac{1}{j(2\pi f)C} + j(2\pi f)L}{R1 + R + \frac{1}{j(2\pi f)C} + j(2\pi f)L} \qquad (20\text{-}17)$$

At resonance, the impedance of the series combination of L and C is zero, leaving

$$\frac{Vout}{Vin} = \frac{R}{R1 + R}$$

where R is the equivalent series lossy resistance owing to coil resistance and core losses. Thus it will be advantageous to obtain a coil or inductor with the lowest equivalent R.

And, of course, for a perfect inductor where R = 0, at resonance, and R1 > 0,

$$\frac{\text{Vout}}{\text{Vin}} = \frac{0}{\text{R1} + \text{R}} = 0$$

For choosing R1, the driving resistance for a series resonant notch filter, a resistor in the range of 1,000 Ω to 10,000 Ω is a good starting point. One should keep in mind that the series resonant capacitor C also forms a one-pole low-pass filter with R1 if the inductor L has a low inductance. So, once the resonant frequency f_{res} is known and the value of R1 is chosen, generally try to choose C in such a way that the low-pass filter cutoff frequency is at or higher than the resonating frequency. That is,

$$\frac{1}{2\pi f(\text{R1})\text{C}} = f_{low-pass} \geq f_{res} = \frac{1}{2\pi\sqrt{\text{LC}}}$$

Once the value of C is chosen, then the inductance of L can be calculated. This recommendation is to avoid having the roll-off of the notch filter occur prematurely, before the "notch" effect comes in. It is better to have a relatively flat frequency response until very close to the notch frequency and then let the series resonant effect take over and cause a steep dip in amplitude.

Figure 20-18 shows combination band-pass and band-eject (notch) filter. During the earlier days when there were very few radio stations in an area, superheterodyne receiver designs had a very low IF, on the order of 100 kHz or less. The image-frequency problem, where another station at twice the IF above the tuned desired station could mix back into the IF band, was not common as long as the stations' frequencies were separated far enough from each other. However, later on, more and more radio stations were crowding the AM band, and the image-frequency problem became very real. Before the 455-kHz IF was used as the standard IF, the earlier radios used a much lower IF band. And in order to address the image-frequency problem during those early days of radio, a front-end radio circuit consisted of two types of variable-frequency filters. One was a band-pass filter to tune to the desired frequency, and another was a notch or band-reject filter that resonates at a frequency above the desired frequency with a separation of twice the IF.

For example, suppose that the IF is 100 kHz. If a desired station is tuned to 600 kHz for the band-pass filter, the notch filter is tuned to 2 × 100 kHz above 600 kHz, or at 800 kHz, for a band-reject (notch) filter. In this way, the image-frequency signal at 800 kHz would be attenuated or removed.

Figure 20-18 shows a band-pass filter consisting of the secondary winding of an RF transformer T1 that resonates with the variable capacitor VC1A that tunes to the desired frequency. The output signal from VC1A then is coupled with a very small capacitance C_o (e.g., <6 pF) to inductor L, which is an inductor that has less inductance than the secondary winding of T1. VC1A and VC1B form a twin-gang variable capacitor. Variable capacitor VC1B then resonates with inductor L to notch-out or band-reject frequencies related to the image-frequency signal. The output then is coupled to another RF amplifier stage or to a mixer.

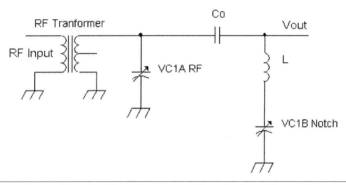

FIGURE 20-18   A combination band-pass and band-eject filter using capacitive coupling.

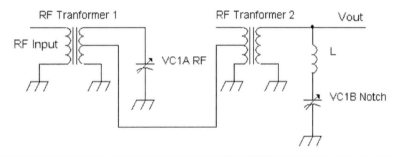

FIGURE 20-19   A combination band-pass and band-reject filter for image signal rejection.

Figure 20-19 shows a transformer-coupled circuit that reduces the image-frequency signal that is similar to the Atwater Kent superheterodyne radio Models 70, 72, 74, and 76, which were made in the early or middle 1930s.

RF Transformer 1's secondary winding resonates with variable capacitor VC1A for tuning into the desired radio station frequency. A low-impedance tap from the secondary winding of RF Transformer 1 is coupled to a low-impedance winding of RF Transformer 2. The secondary winding of RF Transformer 2 is stepped up, which provides a higher impedance than its primary winding. The secondary winding of RF Transformer 2 then is connected to a series resonant circuit consisting of inductor L and variable capacitor VC1B. Preferably, VC1A and VC1B are identical. The inductor L generally has less inductance than the inductance of RF Transformer 1's secondary winding to provide a notch or band-reject characteristic at a higher frequency than the tuned RF signal.

For example, the notch or band-reject frequency is tuned RF frequency + 2 × IF, or

$$f_{\text{notch-band reject}} = f_{RF} + 2f_{IF} \qquad (20\text{-}18)$$

In the Atwater Kent Radio Model 70, IF = 130 kHz; thus the band-reject filter is tuned to the (tuned RF frequency + 2 × 130 kHz) = (tuned RF frequency + 260 kHz), or 260 kHz above the tuned RF frequency.

References

1. Arthur B. Williams and Fred J. Taylor, *Electronic Filter Design Handbook*, 2nd ed. New York: McGraw-Hill, 1988.
2. Class Notes EE140, Robert G. Meyer, UC Berkeley, Fall 1975.
3. Class Notes EE240, Robert G. Meyer, UC Berkeley, Spring 1976.
4. Paul R. Gray and Robert G. Meyer, *Analysis and Design of Analog Integrated Circuits*, 3rd ed. New York: John Wiley & Sons, 1993.
5. Kenneth K. Clarke and Donald T. Hess, *Communication Circuits: Analysis and Design*. Reading: Addison-Wesley, 1971.
6. Robert L. Shrader, *Electronic Communication*, 6th ed. New York: Glencoe/McGraw-Hill, 1991.
7. William G. Oldham and Steven E. Schwarz, *An Introduction to Electronics*. New York: Holt Reinhart Winston, 1972.
8. Allan R. Hambley, *Electrical Engineering Principles and Applications*, 2nd ed. Upper Saddle River: Prentice Hall, 2002.
9. Gene F. Franklin, J. David Powell, and Abbas Emami-Naeini, *Feedback Control of Dynamic Systems*, 3rd ed. Reading: Addison-Wesley, 1995.
10. Howard W. Sams & Co., *Sams Photofact Transistor Radio Series*, TSM-100. Indianapolis, May, 1969.
11. John F. Rider, *Servicing Superheterodynes*. New York: John F. Rider, 1934.
12. Lap-Tech Precision, Inc., *Quartz Products by Lap-Tech*. Bowmanville, Ontario, Canada: 1990.

Chapter 21

Image Rejection

One of the topics covered in this chapter is a quick review of image signals. These are radio-frequency (RF) signals within the band of interest that can mix into the intermediate-frequency (IF) stages and cause interference in the manner of having two radio stations' signals demodulated at the detector or having a single station that pops up in two places within the band. We will examine how image signals are reduced via tuned circuits from the beginning days of the superheterodyne radio to using I and Q signals in an image reject mixer. Finally, an image-rejection mixer using I and Q signals will be analyzed.

Therefore, the objectives of this chapter are

1. To explain image signals
2. To show methods to reduce image signals in a superheterodyne radio
3. To analyze image-reject mixers and the advantages of using I and Q signals to select either or both desired and image signals

Although image signals have been discussed previously, this chapter will show a very detailed analysis of the image-reject mixer with I and Q signals. In essence, the last portion of this chapter is a continuation of Chapters 15 and 16. For readers who did not go through the previous chapters that mention image signals, that's okay. This chapter is written to stand on its own.

What Is an Image Signal?

In a superheterodyne radio, an *image signal* is the signal whose frequency is the tuned RF signal frequency plus or minus twice the intermediate frequency. The local oscillator's frequency is halfway between the tuned RF signal and the image signal (Figures 21-1 and 21-2).

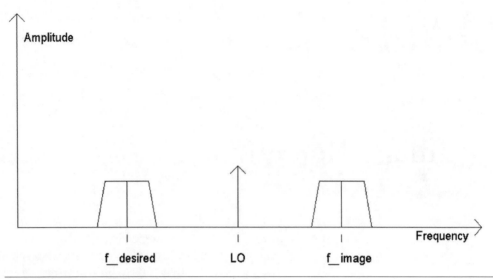

FIGURE 21-1 Spectrum of the image signal defined whose frequency is above the LO, local oscillator's frequency.

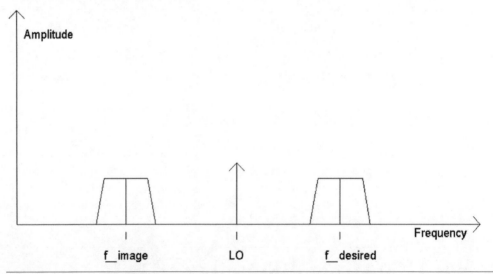

FIGURE 21-2 Spectrum of the image signal defined whose frequency is below the LO, local oscillator's frequency

In Figures 21-1 and 21-2, the image signal thus is down-converted to the same IF band pass as the (desired) tuned RF signal. For example, in this chapter, the local oscillator frequency will be higher than the frequency of the incoming RF signal. Thus, for analysis, Figure 21-1 "defines" the image signal that has a frequency above the local oscillator's frequency.

In the early days of frequency modulation (FM) radios, the frequency of the local oscillator (LO) was 10.7 MHz below the desired tuned RF signal, wherein the intermediate frequency was 10.7 MHz. Therefore, the image frequency would be located 10.7 MHz below the local oscillator's frequency. Figure 21-2 shows a general example of image frequencies located below the local oscillator's frequency. *However, for this chapter, the image frequency is defined to be above the local oscillator's frequency.*

In other radios, such as tuned radio-frequency (TRF), regenerative, and direct-conversion radios (zero-frequency IF), the image signal does not exist. The TRF and regenerative radios do not employ oscillators for mixing. In a direct-conversion radio, the incoming RF signal is mixed with exactly the same frequency as the carrier from the transmitter end, and thus the radio also does not have any image-signal problem. Therefore, image signals exist mainly in single-conversion superheterodyne receivers, where there is an RF signal that is multiplied or mixed by the local oscillator's signal.

Figure 21-3 shows the frequencies of "vulnerability" that result in image signals when the local oscillator's frequency is within the RF band of interest. The figure shows an example RF band of interest, such as the amplitude-modulation (AM) radio band from 540 kHz to 1,600 kHz. Also inserted is the local oscillator frequency, which can vary from about 995 KHz to 2,055 kHz. As the local oscillator's frequency is varied, as long as there is a frequency space that exists 455 kHz above the oscillator frequency within the RF band of interest, image signals will exist. Fortunately, these image signals are reduced in amplitude sufficiently by a single-inductor capacitor–tuned circuit prior to mixing or converting.

For those who have never heard image signals in an AM radio, one can try an experiment. In older 1960s to 1980s portable (e.g., shirt-pocket) AM/FM transistor radios where the components are exposed when the back is removed, there is access

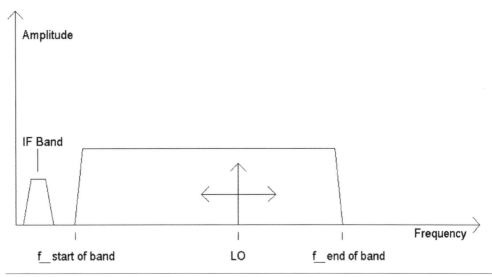

FIGURE 21-3   The RF band of interest and the local oscillator's frequency range.

to the primary winding of the AM ferrite antenna coil which is connected to the variable capacitor (Figure 21-4). With a soldering pencil, temporarily desolder the antenna coil lead to the tab of the variable capacitor. When the radio is turned back on, there is a good chance that stations that are broadcasting from 1,450 kHz to 1,600 kHz can be received when tuning to 540 kHz to 690 kHz on the radio dial. In addition, these stations can be received when tuned to 1,450 kHz to 1,600 kHz. Of course, the radio can be returned to normal operating condition by resoldering the antenna coil lead. When this is done, the tuned RF band-pass circuit/filter is activated again, and the image signals will no longer be received in the 540-kHz to 690-kHz range.

Desolder Antenna Coil Lead

FIGURE 21-4 An AM/FM radio with the antenna coil winding accessible for desoldering from its variable capacitor.

This experiment can be done with a single-band AM transistor radio. However, to desolder the antenna coil lead located on the bottom side of the printed circuit board, usually the printed circuit board must be removed from its case by taking out the mounting screws.

Methods to Reduce the Amplitude of the Image Signal

One or more band-pass variable-tuned filters were used originally to reduce the amplitude of the image signals in a superheterodyne receiver. As mentioned previously, in the 1930s or earlier with low IFs such as 130 kHz and below, multiple stages of RF band-pass and/or band-reject filters were used to remove the reception of image signals. Eventually, simply by raising the IF.intermediate frequency to 455 kHz, a single, tunable band-pass filter proved sufficient to solve the problem of image signals.

Therefore, reducing image signals requires one or more variable tuned circuits such as two sections of band-pass filtering. However, with multiple stages of variable-frequency band-pass filters, tracking throughout the tuning range between the variable capacitor sections becomes difficult. Another method was shown in Chapter 11 when the IF was raised to at least 600 kHz, and a steep roll-off low-pass filter at 1,600 kHz was used to prevent reception of image signals.

In Figure 21-3, it was shown that if the local oscillator's frequency falls within the RF band of interest, and if there is enough "room" frequency-wise in that RF band above the local oscillator's frequency, image signals can exist. Thus one method to remove image signals is to set the local oscillator signal beyond the highest RF signal of interest. This method includes double-conversion mixer circuits. A first mixer up-converts the incoming RF signal to a higher-frequency band beyond the original RF signal's spectrum. A second mixer then translates or mixes the output signal from the first mixer back down to an IF such as 450 kHz. Figure 21-5 shows a redrawn version of a double-conversion superheterodyne receiver from Robert L. Shrader's book, *Electronic Communication*, sixth edition (New York: Glencoe/McGraw-Hill, 1996). Also, Figure 21-6 shows a spectrum diagram of the up-conversion of the RF band and how image signals are handled.

In Figure 21-5, the RF spectrum of interest (e.g., the AM band, 540 kHz to 1,600 kHz) is up-converted to a high-frequency IF signal of 2 MHz. For example, the first local oscillator may have a range of 2.54 MHz to 3.6 MHz such that the difference frequency out of the first mixer is 2 MHz. When the local oscillator is at 2.54 MHz to 3.6 MHz, the image frequencies that are possible to mix or convert back to the 2-MHz IF are signals in the range of 4.54 MHz to 5.6 MHz (see Figure 21-6).

By up-converting the original RF band of interest to a higher IF frequency, the new image signals are at frequencies that are easily filtered from the front end of the receiver. For example, a low-pass filter at 1.7 MHz that does not have to be variably tuned can be used instead of the usual tunable band-pass filter. In any event, none of the image-frequency signals from the up-converted process are from the original RF band of interest.

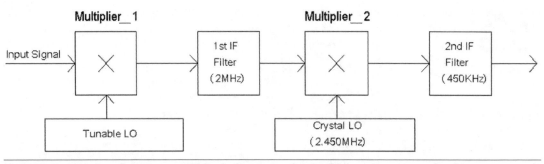

FIGURE 21-5 A redrawn partial block diagram of a double-conversion radio from *Electronic Communication*.

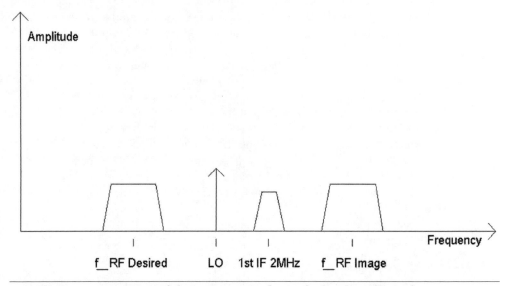

FIGURE 21-6 Spectrum of the up-converted signals, local oscillator frequencies, and possible image-signal frequencies.

Analysis of an Image-Rejection Mixer Using I and Q Signals

For IFs that are very low, such as near direct current (DC) to about 200 kHz, rejecting the image signal becomes a larger problem if variable-tuned band-pass filters are used. For example, having an IF of 20 kHz results in an image signal just 40 kHz away from the desired signal. If the RF signal is at 1,000 kHz and the local oscillator frequency is at 1,020 kHz, the image signal will be at 1,040 kHz. An LC band-pass filter with a 10-kHz bandwidth to recover 5 kHz of audio information will require a Q of 100, which is "difficult" but workable.

Recall from Equation (20-14) that the attenuation factor from the center or resonant frequency is

$$\frac{0.5BW_{-3\text{ db}}}{\Delta f}$$

At 1,000 kHz, a Q of 100 yields a bandwidth (BW) of 10 kHz. Therefore, 0.5BW = 5 kHz.

However, the attenuation factor of the image signal at 1,040 kHz is about 5 kHz/40 kHz = 0.125 = 12.5 percent, which is insufficient attenuation for a 20 kHz IF. In a system using a 455-kHz IF, the image signal has an attenuation factor of about 5 kHz/910 kHz = 0.0055 = 0.55 percent, which is acceptable. An attenuation factor of 2 percent or lower is sufficient.

Another way to attenuate image signals is to use an image-reject mixer that requires two IF signals. These two IF signals have the same amplitude but are phase-shifted in reference to each other by 90 degrees.

As stated previously, for the analysis on how an image-reject mixer works, the local oscillator frequency is always higher in frequency than the desired signal's frequency, and the image signal's frequency is above the local oscillator frequency (see Figure 21-1).

Analyzing the image-reject mixer requires three signals to be examined: the desired signal, the local oscillator signal, and the image signal. The question that summarizes the idea behind a image-signal-rejection mixer is simple: How does a circuit differentiate the image signal from the desired signal after both signals have been down-converted to the intermediate frequency?

If one looks at the IF signal in terms of the desired signal, the upper and lower sidebands have been swapped, whereas the image signal's sidebands are the same even when down-converted to the IF signal. So there is a difference between the two IF signals (Figures 21-7, 21-8, and 21-9). (*Note:* LSB = lower sideband and USB = upper sideband.)

Figure 21-7 shows an example spectrum of the three signals that are used in mixing down to an intermediate frequency. For now, a specific example is shown. For ease of arithmetic analysis, the intermediate frequency is set at or defined to be 450 kHz. (*Note:* Intermediate frequencies for standard AM radios were not always set to 455 kHz; there were slight variants at 450 kHz and 460 kHz.) As shown, the desired signal is centered at 550 kHz, with its lower-sideband limit at 545 kHz and its upper-sideband limit at 555 kHz. The local oscillator is set at 1,000 kHz, and the image signal's center frequency is 1,450 kHz, with the image signal's lower-sideband limit at 1,445 kHz and its upper-sideband limit at 1,455 kHz.

Figure 21-8 then shows what happens when the desired signal is mixed (multiplied) with the local oscillator's signal, and the resulting difference-frequency signal is taken as the IF signal. This IF signal has a range from 445 kHz to 455 kHz with the center at 450 kHz, the intermediate frequency (IF). The mixed-down IF signal is first phase-inverted, which will be shown later with trigonometric identities, but the IF signal also is inverted in spectrum. By spectrum inversion, it is meant that the upper-sideband signals and lower-sideband signals switch places. For example,

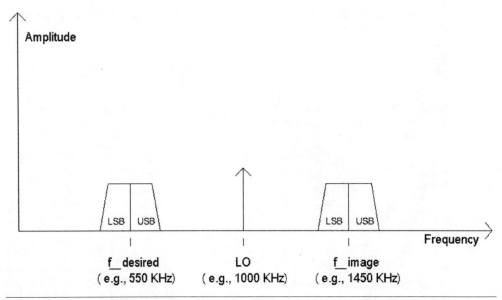

FIGURE 21-7 An example spectrum showing the frequencies of the desired signal, local oscillator signal, and image signal.

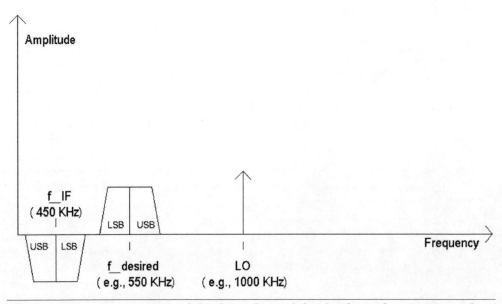

FIGURE 21-8 The spectrum of the desired signal that has been down-converted to or mixed to the intermediate frequency.

the upper-sideband frequency of the desired signal is 555 kHz. Therefore, with a 1,000-kHz oscillator signal, the 555-kHz upper-sideband signal is mapped to 1,000 kHz – 555 kHz = 445 kHz. Likewise, the lower-sideband signal at 545 kHz is mapped to 1,000 kHz – 545 kHz = 455 kHz. Thus the IF signal for the desired signal starts with the upper-sideband signal at 445 kHz and ends with the lower-sideband signal at 455 kHz. The swapping effect of the sidebands is called *spectrum inversion*.

Spectrum inversion is also used in simple audio scrambling processes. The audio signal's low frequencies are mapped to a higher frequency, and the audio signal's high frequencies are mapped to a lower frequency. In this type of audio scrambling, the signals for bass frequencies are converted (frequency translated) to treble frequencies and vice versa.

Figure 21-9 shows that the 450-kHz IF signal related to the image signal maintains lower- and upper-sideband integrity, so there is no sideband swapping. Also, there is no inversion, which is due to the trigonometric identities. For example, the lower-sideband frequency of the RF signal is at 1,445 kHz. When this signal is mixed with the oscillator signal, the resulting frequency in the IF band is 1,445 kHz – 1,000 kHz = 445 kHz. And similarly, the upper-sideband signal, when mixed with the oscillator signal, yields a frequency of 1,455 kHz – 1,000 kHz = 455 kHz. Thus the IF signal relating to the image signal has a frequency range of 445 to 455 kHz, with its lower-sideband frequency at 445 kHz and its upper-sideband frequency at 455 kHz, which then maintains spectrum integrity of the sidebands.

Because of trigonometric identities such as $\sin(-x) = -\sin(x)$ and $\cos(-x) = \cos(x)$, the other difference is that one of the IF signals will be "inverted" in phase relative to the other IF signal. Also, see Figures 15-15 and 15-17, which show an inverting effect

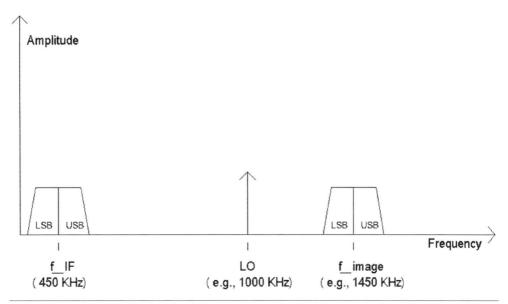

FIGURE 21-9   The spectrum of the image signal that has been down-converted to or mixed down to the intermediate frequency.

when (RF) signals are mixed or sampled at frequencies slightly above and below the sampling frequency.

Since the analysis of the image-reject mixer will involve a multitude of sine and cosines, Table 21-1 presents a list of trigonometric identities.

Figure 21-10 shows the spectrum of the three signals and the IF signal as tones (single-frequency signals). In Figure 21-10, with the local oscillator's frequency $= f_{LO}$, the desired signal's frequency will be $(f_{LO} - f_{IF})$, and the image signal's frequency will be $(f_{LO} + f_{IF})$.

TABLE 21-1   Trigonometric Identities

$\cos(\alpha + \beta) = [\cos(\alpha)][\cos(\beta)] - [\sin(\alpha)][\sin(\beta)]$
$\cos(\alpha - \beta) = [\cos(\alpha)][\cos(\beta)] + [\sin(\alpha)][\sin(\beta)]$
$\sin(\alpha + \beta) = [\sin(\alpha)][\cos(\beta)] + [\cos(\alpha)][\sin(\beta)]$
$\sin(\alpha - \beta) = [\sin(\alpha)][\cos(\beta)] - [\cos(\alpha)][\sin(\beta)]$
$[\sin(\alpha)][\cos(\beta)] = (^1/_2)\sin(\alpha + \beta) + (^1/_2)\sin(\alpha - \beta)$
$[\cos(\alpha)][\cos(\beta)] = (^1/_2)\cos(\alpha + \beta) + (^1/_2)\cos(\alpha - \beta)$
$[\sin(\alpha)][\sin(\beta)] = -(^1/_2)\cos(\alpha + \beta) + (^1/_2)\cos(\alpha - \beta)$
$\cos(\alpha - \beta) = \cos(\beta - \alpha)$
$\sin(\alpha - \beta) = -\sin(\beta - \alpha)$
$\sin(\alpha) - \sin(\beta) = 2[\cos(\frac{\alpha + \beta}{2})][\sin(\frac{\alpha - \beta}{2})]$

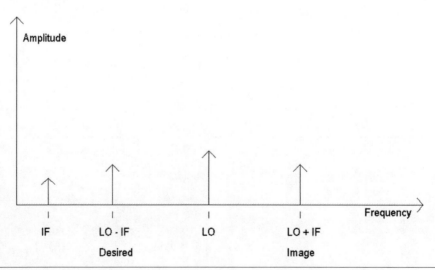

FIGURE 21-10   A general spectrum of signals needed for analyzing how image-reject mixers operate.

Analysis of the Image-Reject Mixer

The analysis of the image reject mixer will be explained in two parts. The first part will cover the outputs of I and Q channels, and the second part will combine the two channels together with one channel processed with a 90-degree phase shift to provide cancellation of the image signal.

Figure 21-11 shows the basic flow of signals into a "traditional" mixer to provide two IF signals. This system does not reject the image, but it illustrates the phases of the two IF signals.

In Figure 21-11, a desired signal S_desired(t) and an image signal S_image(t) are combined, usually by reception of the RF signals. The combination of S_desired(t) and S_image(t) are then mixed via a multiplier with a local oscillator signal LO(t). The output of the mixer then is filtered to pass on only the difference-frequency signal to provide the two IF signals. Let

$$S\_desired(t) = S\_des(t) \text{ and } S\_image(t) = S\_img(t)$$

Then the desired and image signals will be defined as follows:

$$S\_des(t) = K_{des} \sin[2\pi(f_{LO} - f_{IF})t] \tag{21-1}$$

$$S\_img(t) = K_{img} \sin[2\pi(f_{LO} + f_{IF})t] \tag{21-2}$$

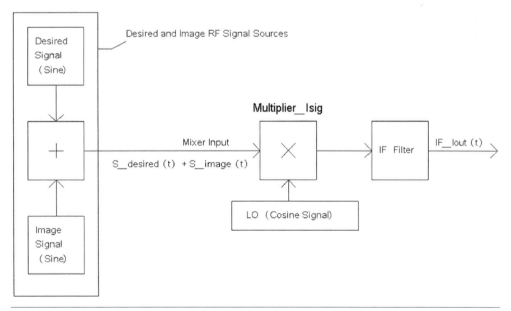

FIGURE 21-11   The image and desired signals are combined to a conventional mixer for providing IF signals.

The oscillator signal will be defined as

$$LO(t) = \cos[(2\pi f_{LO})t] \qquad (21\text{-}3)$$

The input to the mixer is then S_des(t) + S_img(t), and the output of the mixer is

$$[S\_des(t) + S\_img(t)] \times LO(t) =$$
$$\{K_{des} \sin[2\pi(f_{LO} - f_{IF})t] + K_{img} \sin[2\pi(f_{LO} + f_{IF})t]\} \cos[(2\pi f_{LO})t] \qquad (21\text{-}4)$$

$$[S\_des(t) + S\_img(t)] \times LO(t) =$$
$$\{K_{des} \sin[2\pi(f_{LO} - f_{IF})t]\} \cos[(2\pi f_{LO})t] +$$
$$\{K_{img} \sin[2\pi(f_{LO} + f_{IF})t]\} \cos[(2\pi f_{LO})t] \qquad (21\text{-}5)$$

There are two products that concern sines and cosines; thus the following trigonometric identity will be used:

$$[\sin(\alpha)][\cos(\beta)] = \tfrac{1}{2}\sin(\alpha + \beta) + \tfrac{1}{2}\sin(\alpha - \beta)$$

Also, since the IF filter will remove the summed frequency components $(\tfrac{1}{2}) \sin(\alpha + \beta)$, only the difference-frequency signals $(\tfrac{1}{2}) \sin(\alpha - \beta)$ will pass through the IF filter.

Thus, at the output of the IF filter, there are two IF signals:

$$\tfrac{1}{2}K_{des} \sin[2\pi(f_{LO} - f_{IF} - f_{LO})t] + \tfrac{1}{2}K_{img} \sin[2\pi(f_{LO} + f_{IF} - f_{LO})t] =$$

$$\tfrac{1}{2}K_{des} \sin[2\pi(- f_{IF})t] + \tfrac{1}{2}K_{img} \sin[2\pi(f_{IF})t]$$

Note that $\sin(-x) = -\sin(x)$; therefore, at the output of the mixer IF_Iout(t), the two IF signals are

$$-\tfrac{1}{2}K_{des} \sin[2\pi(f_{IF})t] + \tfrac{1}{2}K_{img} \sin[2\pi(f_{IF})t] = IF\_Iout(t) \qquad (21\text{-}6)$$

From the output of the mixer and IF filter, there is a sign change in the IF signal related to the desired signal compared with the IF signal related to the image signal. Can we somehow use this sign change advantageously to null one of the IF signals? For example, could a simple phase shifter as shown in Figure 21-12 work?

In Figure 21-12, unfortunately, when one of the IF signals is canceled out, so is the other. Thus the phase-shifting network does *not* work because we only want one of the signals canceled, not both.

However, we can take advantage of the two following trigonometric identities:

$$[\sin(\alpha)][\sin(\beta)] = -\tfrac{1}{2}\cos(\alpha + \beta) + \tfrac{1}{2}\cos(\alpha - \beta) \qquad \text{and} \qquad \cos(-x) = \cos(x)$$

Note that there is no sign change in the cosine function compared with $\sin(-x) = -\sin(x)$.

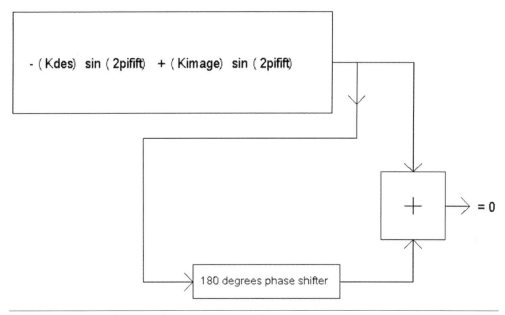

FIGURE 21-12   **Phase-shifting network and summing circuit.**

Now let's take the same two signals below and use them in Figure 21-13:

$$S\_des(t) = K_{des} \sin[2\pi(f_{LO} - f_{IF})t] \tag{21-1}$$

$$S\_img(t) = K_{img} \sin[2\pi(f_{LO} + f_{IF})t] \tag{21-2}$$

The circuits in Figure 21-13 are the same as the ones in Figure 21-11 except that the local oscillator signal now is a sine function instead of a cosine function. Thus the oscillator signal in Figure 21-13 is phase-shifted by 90 degrees when compared with the one in Figure 21-11.

Therefore, the output of the mixer in Figure 21-13 is

$$\{K_{des} \sin[2\pi(f_{LO} - f_{IF})t] + K_{img} \sin[2\pi(f_{LO} + f_{IF})t]\} \sin[(2\pi f_{LO})t] =$$
$$\{K_{des} \sin[2\pi(f_{LO} - f_{IF})t]\} \sin[(2\pi f_{LO})t] + \{K_{img} \sin[2\pi(f_{LO} + f_{IF})t]\} \sin[(2\pi f_{LO})t]$$

With the previously mentioned trigonometric identity, that is,

$$[\sin(\alpha)][\sin(\beta)] = -\tfrac{1}{2}\cos(\alpha + \beta) + \tfrac{1}{2}\cos(\alpha - \beta)$$

the output of the IF filter IF_Qout(t), which passes only the difference-frequency signals, is

$$\tfrac{1}{2}K_{des} \cos[2\pi(f_{LO} - f_{IF} - f_{LO})t + \tfrac{1}{2}K_{img} \cos[2\pi(f_{LO} + f_{IF} - f_{LO})t = \text{IF\_Qout(t)}$$

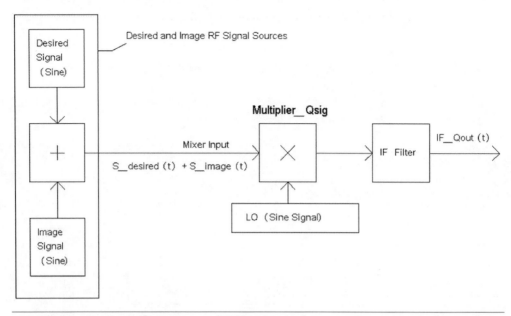

FIGURE 21-13 A similar mixer and IF circuit as shown in Figure 21-11, but with a phase-shifted local oscillator signal.

$$\tfrac{1}{2}K_{des} \cos[2\pi(-f_{IF})t + \tfrac{1}{2}K_{img} \cos[2\pi(f_{IF})t] = IF\_Qout(t)$$

and because $\cos(-x) = \cos(x)$, we have

$$\tfrac{1}{2}K_{des} \cos[2\pi(f_{IF})t + \tfrac{1}{2}K_{img} \cos[2\pi(f_{IF})t] = IF\_Qout(t) \qquad (21\text{-}7)$$

Now let's compare the signals in Equation (21-6):

$$-\tfrac{1}{2}K_{des} \sin[2\pi(f_{IF})t] + \tfrac{1}{2}K_{img} \sin[2\pi(f_{IF})t] = IF\_Iout(t) \qquad (21\text{-}6)$$

Also let

$$IF\_Iout(t) = I \text{ Signal}$$

$$IF\_Qout(t) = Q \text{ Signal}$$

At this point in the analysis, by phase shifting either signal, IF_Qout(t) or IF_Iout(t), by 90 degrees and combining the two signals, extraction of the IF signal related just to the desired signal or just to the image signal can be implemented. Actually, if two combiners are used, one as a summer and another as a subtractor, both IF signals can be separated as an IF signal related to the desired signal and another separate IF signal related to the image signal.

Before the complete image mixer is shown, let's take a look at a phase-shifted cosine function. For any cosine function such as $\cos(x)$, a phase shift by a lagging 90 degrees = $-\pi/2$ results in a sine function, that is, $\cos[x - (\pi/2)] = \sin(x)$; if the cosine function is phase-shifted by a leading +90 degrees = $\pi/2$, then $\cos[x + (\pi/2)] = -\sin(x)$ (see Figures 21-14A, 21-14B, 21-14C, and 21-14D).

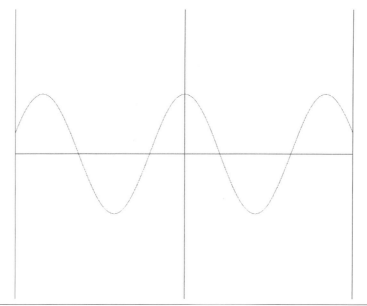

FIGURE 21-14A   A cosine signal.

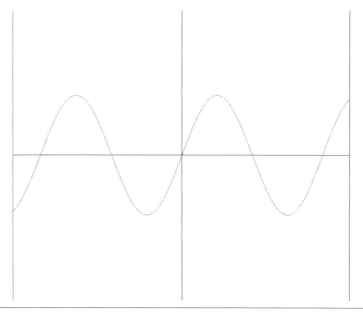

FIGURE 21-14B   A phase-shifted cosine signal by –90 degrees.

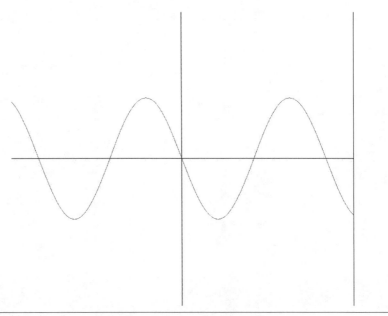

FIGURE 21-14C A phase-shifted cosine signal by +90 degrees.

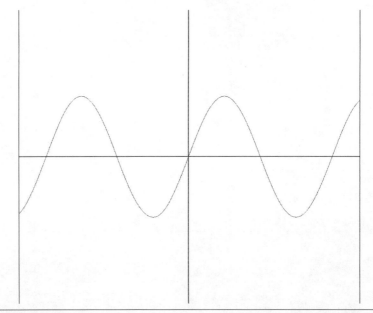

FIGURE 21-14D A sine signal.

Now let's combine the circuits from Figures 21-11 and 21-13 and include phase-shifting circuits or functions as shown in Figure 21-15 to form the complete image-reject mixer.

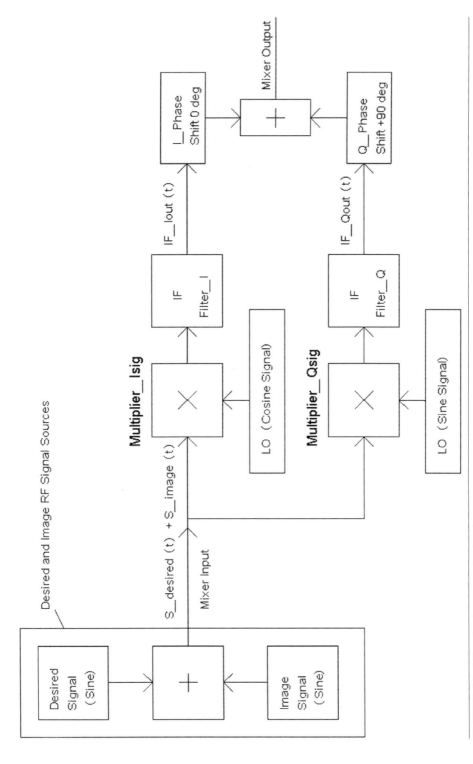

FIGURE 21-15 An image-reject mixer.

Referring to Figure 21-15, there is the signal IF_Iout(t) that is connected to a 0-degree phase shifter (I_Phase), which outputs the same signal IF_Iout(t) to the first input of the combiner or summer. The combiner provides the Mixer Output (see Figure 21-15) by summing the outputs from the phase shifters, I_Phase and Q_Phase.

$$-\tfrac{1}{2}K_{des}\sin[2\pi(f_{IF})t] + \tfrac{1}{2}K_{img}\sin[2\pi(f_{IF})t] = IF\_Iout(t) \qquad (21\text{-}6)$$

The other signal, IF_Qout(t), is connected to a +90-degree phase-shifter circuit (Q_Phase):

$$\tfrac{1}{2}K_{des}\cos[2\pi(f_{IF})t] + \tfrac{1}{2}K_{img}\cos[2\pi(f_{IF})t] = IF\_Qout(t) \qquad (21\text{-}7)$$

And the output of the +90-degree phase-shifter circuit provides a phase-shifted version of $+\pi/2$ of the signal IF_Qout(t), which is then

$$\tfrac{1}{2}K_{des}\cos[2\pi(f_{IF})t + \tfrac{\pi}{2}] + \tfrac{1}{2}K_{img}\cos[2\pi(f_{IF})t + \tfrac{\pi}{2}] = IF\_Qout(t + \tfrac{\pi}{2})$$

However, $\cos[x + (\pi/2)] = -\sin(x)$. Therefore,

$$-\tfrac{1}{2}K_{des}\sin[2\pi(f_{IF})t] - \tfrac{1}{2}K_{img}\sin[2\pi(f_{IF})t] = IF\_Qout(t + \tfrac{\pi}{2}) \qquad (21\text{-}8)$$

Combining Equations (21-6) and (21-8) results in

$$IF\_Iout(t) + IF\_Qout(t + \tfrac{\pi}{2}) = -\tfrac{1}{2}K_{des}\sin[2\pi(f_{IF})t] + \tfrac{1}{2}K_{img}\sin[2\pi(f_{IF})t] +$$

$$-\tfrac{1}{2}K_{des}\sin[2\pi(f_{IF})t] - \tfrac{1}{2}K_{img}\sin[2\pi(f_{IF})t] = -K_{des}\sin[2\pi(f_{IF})t]$$

$$= Mixer\ Output \qquad (21\text{-}9)$$

The result of adding the outputs of the 0- and +90-degree phase-shifting circuits provides an IF signal related to the desired signal at the output of the image-reject mixer.

Note the cancellation of the IF signals related to the image signal:

$$+\tfrac{1}{2}K_{img}\sin[2\pi(f_{IF})t]$$

and

$$-\tfrac{1}{2}K_{img}\sin[2\pi(f_{IF})t]$$

Given the two signals,

$$-\tfrac{1}{2}K_{des}\sin[2\pi(f_{IF})t] + \tfrac{1}{2}K_{img}\sin[2\pi(f_{IF})t] = IF\_Iout(t) \qquad (21\text{-}6)$$

$$-\tfrac{1}{2}K_{des}\sin[2\pi(f_{IF})t] - \tfrac{1}{2}K_{img}\sin[2\pi(f_{IF})t] = IF\_Qout(t + \tfrac{\pi}{2}) \qquad (21\text{-}8)$$

If we subtract the two equations by using a subtractive circuit or function, as shown in Figure 21-16, the IF signal related to the desired signal will be canceled out, and the IF signal related to the image signal will be provided instead.

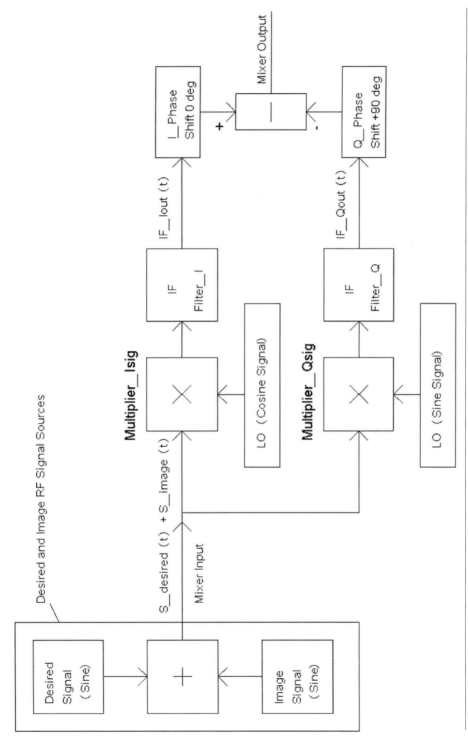

FIGURE 21-16 A subtractive circuit following the phase-shifting networks provides an IF signal related to the image signal.

385

$$\text{IF\_Iout(t)} - \text{IF\_Qout}(t + \tfrac{\pi}{2}) = -\tfrac{1}{2}K_{des} \sin[2\pi(f_{IF})t] + \tfrac{1}{2}K_{img} \sin[2\pi(f_{IF})t] -$$

$$[-\tfrac{1}{2}K_{des} \sin[2\pi(f_{IF})t] - \tfrac{1}{2}K_{img} \sin[2\pi(f_{IF})t] = \tfrac{1}{2}K_{img} \sin[2\pi(f_{IF})t] - -\tfrac{1}{2}K_{img} \sin[2\pi(f_{IF})t]$$

$$= K_{img} \sin[2\pi(f_{IF})t] = \text{Vout'}(t) \qquad (21\text{-}9)$$

Two minuses = a plus! That is,

$$- -\tfrac{1}{2}K_{img} \sin[2\pi(f_{IF})t] = +\tfrac{1}{2}K_{img} \sin[2\pi(f_{IF})t]$$

With a summing and subtracting circuit, the image-reject mixer can separately retrieve or extract RF signals above and below the local oscillator's frequency (see Figure 20-17).

What is more serendipitous about retrieving signals from both sides of the local oscillator signals is that if the IF filter is a low-pass filter instead of a typical band-pass filter, the image-reject mixer can output two signals, each with a spectrum as wide as an IF low-pass filter. For example, suppose that both IF filters in Figure 20-17 are 470-kHz low-pass filters and the local oscillator is set at 1,000 kHz. Out of the summing combiner will come an IF signal covering a bandwidth of 470 kHz, which contains signals of frequency-translated versions of the "desired" RF signals from 1,000 kHz to 530 kHz. Then, out of the subtracting circuit will come an IF signal also covering a bandwidth of 470 kHz, which contains signals of frequency-translated versions of the "image" RF signal from 1,000 kHz to 1,470 kHz (see Figure 20-18).

In Figure 21-18, the desired signals below the local oscillator's frequency of 1,000 kHz, 550 kHz to 1,000 kHz, are mapped to the IF band from frequencies F1 to F2. Frequency F1 represents the down-converted desired RF signal frequency near (or just below) 1,000 kHz, whereas frequency F2 represents the down-converted desired RF signal at the bottom end of the RF band at 550 kHz. Similarly, the image signals are retrieved separately, and frequency F3 represents the down-converted image RF signals near (or just above) 1,000 kHz, whereas frequency F4 represents the down-converted image RF signal at the top of the RF band at 1,450 kHz.

Even though in this example the IF bandwidth is 450 kHz, the actual bandwidth of the recovered signals is twice 450 kHz because the information from 550 kHz to 1,000 kHz is extracted in one IF channel and the information from 1,000 to 1,450 kHz is retrieved in the second IF channel. Therefore, by using I and Q signals IF_Iout(t) and IF_Qout(t) in the image-reject mixer, twice the bandwidth of RF information can be recovered.

This feature of recovering twice the IF bandwidth of information is useful in software-defined radios (SDRs). For example, suppose that the IF filters are low-pass filters at 96 kHz each, and the I and Q signals, IF_Iout(t) and IF_Qout(t), are instead connected to the two audio channels of a computer's 192-kHz sampling-rate sound card. A 192-kHz sampling rates results in maximum bandwidth of 96 kHz for each of the two sound channels. The software in the computer can emulate the phase shifter and combining functions to extract two separate 96-kHz IF signals to demodulate, which allows demodulating down-converted RF signals over a 192-kHz range. This

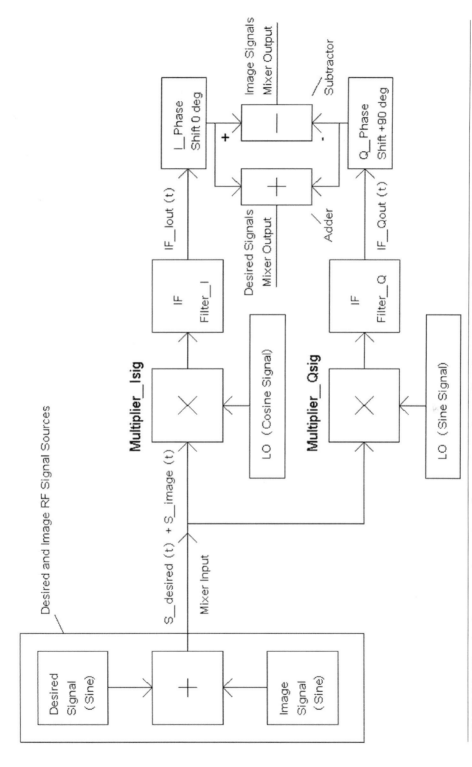

FIGURE 21-17 An image reject mixer that extracts desired and image signals separately.

387

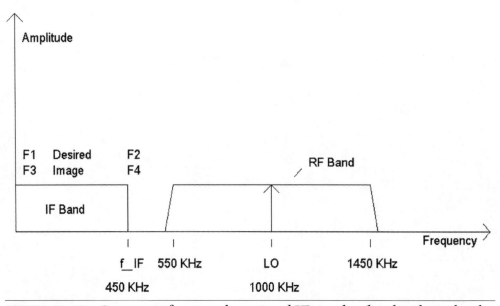

FIGURE 21-18 Spectrum of separately retrieved IF signals related to desired and image signals.

192-kHz range then includes related RF signals above and below the local oscillator's frequency.

Consequences of an Imperfect 90-Degree Phase Shifter on Reducing the Image Signal

In practice, the 90-degree phase shifter is not always 90 degrees. The phase shifter may be 89 degrees or 91 degrees. So let's examine what happens when there is a small phase error $\Delta\varphi$ from the 90-degree phase shifter.

In the perfect case,

$$\cos(x + \tfrac{\pi}{2}) = \sin(x)$$

But suppose the cosine function is phase-shifted by $(\pi/2) + \Delta\varphi$ instead? Then

$$\cos(x + \tfrac{\pi}{2} + \Delta\varphi) = -\sin(x + \Delta\varphi) \tag{21-10}$$

How did Equation (21-10) come about? Let $x' = x + \Delta\varphi$, and we know that

$$\cos(x + \tfrac{\pi}{2}) = -\sin(x)$$

so for any x, even x',

$$\cos(x' + \tfrac{\pi}{2}) = -\sin(x')$$

and by substitution of x' = x + $\Delta\varphi$, we get

$$\cos(x + \Delta\varphi + \tfrac{\pi}{2}) = -\sin(x + \Delta\varphi)$$

or

$$\cos(x + \tfrac{\pi}{2} + \Delta\varphi) = -\sin(x + \Delta\varphi)$$

In Figure 21-15, suppose that the +90-degree phase shifter (Q_Phase) has a slight phase error of $\Delta\varphi$. The output of the phase shifter then will change from

$$\tfrac{1}{2}K_{des}\cos[2\pi(f_{IF})t + \tfrac{\pi}{2}] + \tfrac{1}{2}K_{img}\cos[2\pi(f_{IF})t + \tfrac{\pi}{2}] = \text{IF\_Qout}(t + \tfrac{\pi}{2})$$

to

$$\tfrac{1}{2}K_{des}\cos[2\pi(f_{IF})t + \tfrac{\pi}{2} + \Delta\varphi] + \tfrac{1}{2}K_{img}\cos[2\pi(f_{IF})t + \tfrac{\pi}{2} + \Delta\varphi] =$$

$$\text{IF\_Qout}(t + \tfrac{\pi}{2} + \Delta\varphi) = -\tfrac{1}{2}K_{des}\sin[2\pi(f_{IF})t + \Delta\varphi] - \tfrac{1}{2}K_{img}\sin[2\pi(f_{IF})t + \Delta\varphi] =$$

$$\text{IF\_Qout}(t + \tfrac{\pi}{2} + \Delta\varphi) \tag{21-11}$$

The output of the 0-degree phase shifter in Figure 21-11 is

$$-\tfrac{1}{2}K_{des}\sin[2\pi(f_{IF})t] + \tfrac{1}{2}K_{img}\sin[2\pi(f_{IF})t] = \text{IF\_Iout}(t) \tag{21-6}$$

When Equation (21-6) is added to Equation (21-11) to yield a signal with a reduction in the IF signal related to the image signal, the result is

$$-\tfrac{1}{2}K_{des}\sin[2\pi(f_{IF})t] + \tfrac{1}{2}K_{img}\sin[2\pi(f_{IF})t] + -\tfrac{1}{2}K_{des}\sin[2\pi(f_{IF})t + \Delta\varphi] -$$

$$-\tfrac{1}{2}K_{img}[2\pi(f_{IF})t + \Delta\varphi] \approx -K_{des}\sin[2\pi(f_{IF})t] + \tfrac{1}{2}K_{img}\{\sin[2\pi(f_{IF})t] - \sin[2\pi(f_{IF})t + \Delta\varphi]\}$$

$$= \text{Mixer Output signal} \tag{21-12}$$

However, the trigonometric identity

$$\sin(\alpha) - \sin(\beta) = 2[\cos(\tfrac{\alpha + \beta}{2})][\sin(\tfrac{\alpha - \beta}{2})]$$

says that

$$\sin[2\pi(f_{IF})t] - \sin[2\pi(f_{IF})t + \Delta\varphi] = 2[\cos(\tfrac{2\pi(f_{IF})t + 2\pi(f_{IF})t + \Delta\varphi}{2})][\sin(\tfrac{2\pi(f_{IF})t - (2\pi(f_{IF})t + \Delta\varphi)}{2})] =$$

$$2[\cos(2\pi(f_{IF})t + \tfrac{\Delta\varphi}{2}][\sin(\tfrac{-\Delta\varphi}{2})] = \sin[2\pi(f_{IF})t] - \sin[2\pi(f_{IF})t + \Delta\varphi] \tag{21-13}$$

Substituting Equation (21-13) into Equation (21-12) yields

$$-K_{des}\sin[2\pi(f_{IF})t] + \tfrac{1}{2}K_{img}\,2[\cos(2\pi(f_{IF})t + \tfrac{\Delta\varphi}{2}][\sin(\tfrac{-\Delta\varphi}{2})] = \text{Vout}$$

The amplitude to the IF signal related to the image signal is then

$$\tfrac{1}{2}K_{img}\,2[\sin(\tfrac{-\Delta\varphi}{2})] = K_{img}\sin(\tfrac{-\Delta\varphi}{2})$$

But for small angles in radians, $\Delta\varphi \ll 1$,

$$\sin(\tfrac{-\Delta\varphi}{2}) = \tfrac{-\Delta\varphi}{2}$$

Thus the amplitude related to the image signal is

$$K_{img}\tfrac{-\Delta\varphi}{2}$$

and the magnitude related to the image signal is $K_{img}|-\Delta\varphi/2|$. Thus, the attenuation factor of the image signal is $|-\Delta\varphi/2|$.

Given 57.3 degrees = 1 radian, then 1-degree-error $\Delta\varphi$ is 0.0174 radian, which would mean that the IF signal–related image signal is multiplied by $|-\Delta\varphi/2| = 0.0174/2 = 0.0087 = 0.87$ percent, a reduction of the (image) signal by a factor 114.

For small errors in the phase-shifting network, the errors amount to an attenuation factor of approximately 1 percent per 1 degree of error. This means that if the error $\Delta\varphi = 2$ degrees (e.g., a 92-degree or 88-degree phase shifter instead of 90 degrees), the attenuation factor is 2 percent, or a reduction of 50 for the IF signal related to the image signal.

References

1. U.S. Patent 1,855,576, Clyde R. Keith, "Frequency Translating System," filed on April 9, 1929.
2. U.S. Patent 2,086,601, Robert S. Caruthers, "Modulating System," filed on May 3, 1934.
3. U.S. Patent 2,025,158, Frank Augustus Cowan, "Modulating System," filed on June 7, 1934.
4. U.S. Patent 5,058,159, Ronald Quan, "Method and System for Scrambling and Descrambling Audio Information Signals," filed on June 15, 1989.
5. U.S. Patent 5,159,531, Ronald Quan and Ali R. Hakimi, "Audio Scrambling System Using In-Band Carrier," filed on April 26, 1990.
6. U.S. Patent 5,471,531, Ronald Quan, "Method and Apparatus for Low-Cost Audio Scrambling and Descrambling," filed on December 14, 1993.

7. U.S. Patent 6,091,822, Andrew B. Mellows, John O. Ryan, William J. Wrobleski, Ronald Quan, and Gerow D. Brill, "Method and Apparatus for Recording Scrambled Video Audio Signals and Playing Back Said Video Signal, Descrambled, Within a Secure Environment," filed on January 8, 1998.

8. Kenneth K. Clarke and Donald T. Hess, *Communication Circuits: Analysis and Design*. Reading: Addison-Wesley, 1971.

9. Thomas H. Lee, *The Design of CMOS Radio-Frequency Integrated Circuits*, 2nd ed. Cambridge: Cambridge University Press, 2003.

10. Allan R. Hambley, *Electrical Engineering Principles and Applications*, 2nd ed. Upper Saddle River: Prentice Hall, 2002.

11. Robert L. Shrader, *Electronic Communication*, 6th ed. New York: Glencoe/McGraw-Hill, 1991.

12. Mischa Schwartz, *Information, Transmission, and Noise*. New York: McGraw-Hill, 1959.

13. Harold S. Black, *Modulation Theory*. Princeton: D. Van Nostrand Company 1953.

14. B. P. Lathi, *Linear Systems and Signals*. Carmichael: Berkeley Cambridge Press, 2002.

15. Alan V. Oppenheim and Alan S. Wilsky, with S. Hamid Nawab, *Signals and Systems*, 2nd ed. Upper Saddle River: Prentice Hall, 1997.

16. Howard W. Coleman, *Color Television*. New York: Hastings House Publishers, 1968.

17. Geoffrey Hutson, Peter Shepherd, and James Brice, *Colour Television System Principles*. 2nd ed. London: McGraw-Hill, 1990.

18. Mary P. Dolciani, Simon L. Bergman, and William Wooton, *Modern Algebra and Trigonometry*. New York: Houghton Mifflin, 1963.

19. Taiwa Tomiyama and Hiroshi Arimoto, "Advanced Low-Voltage Single Chip Radio IC," *IEEE Transactions on Consumer Electronics* 38(3), August 1992.

20. John F. Rider, *Servicing Superheterodynes*. New York: John F. Rider, 1934.

Chapter 22

Noise

This chapter covers the basic concepts behind random noise (e.g., hiss). In radio circuits, the noise in a front-end circuit hampers the ability to pick out weak radio-frequency (RF) signals. In audio circuits, the noise sources degrade the signal-to-noise ratio in mostly preamp circuits such as microphone preamplifiers. However, as mentioned in Chapter 12, where op amps and mixers are used in software-defined radio (SDR) front-end circuits, the noise contribution not only from the mixer circuits but also from the intermediate-frequency (IF) amplifiers can hamper reception of weak signals. In this chapter, the examination of noise will be limited to amplifier circuits and not mixers.

Therefore, the objectives for this chapter are

1. To identify sources of electronic random noise
2. To present some basic noise theory or noise models, including analyzing noise in a single-transistor amplifier (including referred input noise), lowering noise via paralleling transistors, and analyzing noise in a differential-pair amplifier
3. To select op amps for noise performance

For more details on noise, several books are recommended:

1. National Semiconductor, *Audio/Radio Handbook*. Santa Clara: National Semiconductor Corporation, 1980.
2. Paul R. Gray and Robert G. Meyer, *Analysis and Design of Analog Integrated Circuits*, 3rd ed. New York: John Wiley & Sons, 1993.
3. R. E. Zeimer and W. H. Tranter, *Principles of Communications*. Boston: Houghton Mifflin Company, 1976.

In the *Audio/Radio Handbook*, the subject of noise is excellently written about in the section on preamplifiers. *Analysis and Design of Analog Integrated Circuits*, by Gray and Meyer, superbly covers the subject of noise in integrated circuits, which is also applicable to discrete devices. Actually, this book is a standard textbook on analog integrated circuits in many university classes. And in *Principles of Communications*,

Appendix A, "Physical Noise Sources and Noise Calculations in Communication Systems," a very good discussion of noise is provided.

Sources of Random Electronic Noise and Some Basic Noise Theory

Random electronic noise is not like periodic noise sources, such as power-supply noise, hum, or interfering noise from external signal generators. Instead, there are two main sources of random electronic noise: thermal noise and shot noise.

Thermal noise is present in all resistors no matter what type. That is, metal-film, carbon-film, carbon-composition, wire-wound, and oxide resistors all have thermal noise. Also, field-effect transistors (FETs) also generate thermal noise. Shot noise arises from current flow in diodes and bipolar-junction transistors. Older devices such as vacuum tubes also generate shot noise.

 Random noise signals do not add or subtract the same way as the familiar voltage sources we use. When independent random noise signals are added in terms of current or voltage, the total noise voltage is the sum of the squares of the random noise voltages followed by taking the square root. For example, three independent noise voltages V_{n1}, V_{n2}, and V_{n3} are added in the following manner:

$$V_{n1} + V_{n2} + V_{n3} = \sqrt{(V_{n1})^2 + (V_{n2})^2 + (V_{n3})^2}$$

Similarly, independent random noise current sources I_{n1}, I_{n2}, and I_{n3} add in the same manner:

$$I_{n1} + I_{n2} + I_{n3} = \sqrt{(I_{n1})^2 + (I_{n2})^2 + (I_{n3})^2}$$

It should be noted that for most practical purposes, inductors and capacitors do not generate random noise. Figures 22-1A, 22-1B, and 22-1C show sources of thermal and shot noise.

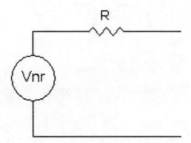

FIGURE 22-1A A noise model for resistors generating thermal noise.

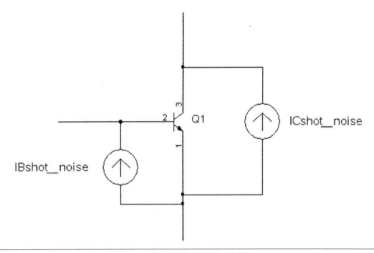

FIGURE 22-1B   A noise model for shot-noise generators in a transistor.

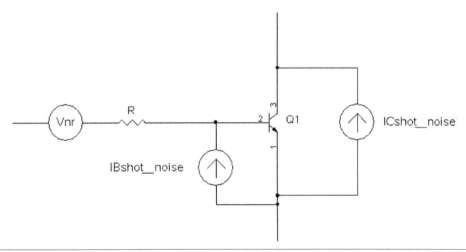

FIGURE 22-1C   Thermal- and shot-noise components in a transistor.

Before the actual noise of the resistors and shot-noise generators is calculated, we need a rough estimate of actual signal-to-noise ratio (SNR) for different types of products. SNR is the ratio of the signal in root-mean-square (RMS) voltage over the RMS noise voltage. In terms of sinusoidal signals, the RMS voltage is 0.707 × peak voltage = $0.707V_p$. For example, a test sine-wave signal is given as $V_p \sin(2\pi ft)$. Table 22-1 lists various devices and their SNRs.

TABLE 22-1   Signal-to-Noise Ratios of Various Devices

Device	SNR	Bandwidth	Input Voltage	Output Voltage
Phase Linear 700 power amplifier	108.5 dB	20 kHz	1.3 volts RMS	53 volts RMS
Rotel RA-610 integrated amplifier	68 dB	20 kHz	0.125 volt RMS at the aux. input	15 volts RMS
Onkyo TX-NR807 A/V receiver	101.81 dB	~10 kHz via an A-weighting filter	—	2.83 volts RMS
Mission PCM-4000 CD player	111.5 dB	~10 kHz via an A-weighting filter	—	2.08 volts RMS
Luxman T-110 FM tuner	50 dB	~15 kHz	2.2-µV RMS RF signal	—
JVC HR-D140U video cassette recorder	45 dB	4 MHz	1-volt peak-to-peak video	1-volt peak-to-peak video

This table shows the SNRs of electronic devices ranging from audio amplifiers to a video recorder. What is important to note is that the SNR is usually measured at some input or output level for the sine-wave signal, and even more important, the noise is measured over a specific bandwidth. That is, the noise is usually measured from the device through a band-pass or low-pass filter of specified bandwidth.

Note that all the measurements in Table 22-1 refer to voltages, so

$$\text{SNR (in dB)} = 20 \log \frac{V_{signal}}{V_{noise}} \tag{22-1}$$

or, equivalently,

$$10^{SNR/20} = \frac{V_{signal}}{V_{noise}} \tag{22-2}$$

Therefore, when a device is measured and a noise measurement is given in an RMS voltage, one of the most important things to find out is: What is the bandwidth that the noise measurement is made under?

Let's start off with the noise voltage caused by a resistor $V_{nr} = \sqrt{4kTBR}$, which is measured in RMS voltage, and where k is Bolzmann's constant ($= 1.38 \times 10^{-23}$), T is temperature in degrees kelvin ($= 298$ degrees K at room temperature), B is bandwidth in hertz, and R is resistance in ohms. That is,

$$V_{nr} = \sqrt{4kTBR} \tag{22-3}$$

Thus, for a 1,000-Ω resistor with a bandwidth of 1 Hz,

$$V_{nr\,(1\,k\Omega)} = \sqrt{4kT(1\text{ Hz})(1{,}000\ \Omega)} = 4\text{ nV (nanovolts)}$$

If the bandwidth is an audio bandwidth such as 10 kHz,

$$V_{nr\,(1\,k\Omega)} = \sqrt{4kT(10{,}000\text{ Hz})(1{,}000\ \Omega)} = 406\text{ nV}$$

A normalized noise voltage expression is to set the bandwidth B to 1 Hz to provide a noise density voltage of

$$\frac{V_x}{\sqrt{\text{Hz}}} = \text{noise density voltage} \tag{22-4}$$

where V_x is noise voltage when bandwidth B = 1 Hz. For the 1,000-Ω resistor, V_x = 4.06 nV.

To find the noise voltage, simply multiply the noise density voltage by the square root of the bandwidth. Noise voltage then can be expressed as

$$\frac{V_x}{\sqrt{\text{Hz}}}\sqrt{B} = V_{nr} \tag{22-5}$$

Therefore, from the preceding example for a 1,000-Ω resistor, the noise voltage is

$$\frac{4.06\text{ nV}}{\sqrt{\text{Hz}}}\sqrt{B} = V_{nr}$$

And for B = 100 Hz, $V_{nr} = (4.06\text{ nV}/\sqrt{\text{Hz}}) \times \sqrt{100\text{ Hz}} = 40.6\text{ nV} = V_{nr\,(1\,k\Omega)}$.

For B = 20,000 Hz, $V_{nr} = (4.06\text{ nV}/\sqrt{\text{Hz}}) \times \sqrt{20\text{ kHz}} = 573\text{ nV} = V_{nr\,(1\,k\Omega)}$.

Also, it should be noted bandwidth B is defined by the difference of an upper frequency and a lower frequency. For example, for a bandwidth B of 20,000 Hz, the 1,000-Ω resistor produces 573 nV. Thus, if the noise is measured from 20,000 and 40,000 Hz or from 3,000 Hz to 23,000 Hz, then the noise voltage from a 1,000-Ω resistor is still 573 nV.

In one of the example devices listed in Table 22-1, the Phase Linear 700 amplifier had an SNR of 108.5 dB for 1.3 volts at an input. Figure 22-2 provides a model for equivalent input noise. The figure shows how the SNR of an amplifier can be modeled as a perfect noiseless amplifier with noise added to the input signal.

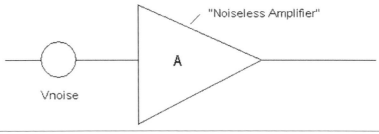

FIGURE 22-2   A model showing equivalent noise input voltage with a noisy input source connected to a "noiseless" amplifier.

This then means that to find an equivalent noise voltage at the input of the Phase Linear 700 amplifier, Equation (22-2) is used:

$$10^{SNR/20} = \frac{V_{signal}}{V_{noise}} \qquad (22\text{-}2)$$

$$10^{108.5/20} = \frac{1.3\,V}{V_{noise}} = 255{,}070$$

or

$$V_{noise} = \frac{1.3\,V}{255{,}070} = 4.88 \ \mu V$$

Figure 22-3 shows how to relate the equivalent input noise voltage into an equivalent input noise resistance, where the resistor generates V_{noise} via $V_{nr} = \sqrt{4kTBR}$ at the input of the amplifier. The equivalent noise resistance at the input is calculated by

$$(V_{noise})^2 = 4kTBR$$

and then rearranging the terms to

$$\frac{(V_{noise})^2}{4kTB} = \frac{4.88 \ \mu V^2}{4kTB} = R$$

for B = 20,000 Hz, as shown in Table 22-1 for the Phase Linear 700 amplifier, R = 72 kΩ = equivalent input noise resistance (as shown in Figure 22-3).

 Although the input noise is a little high by today's standards, the noise voltage at the output of the amplifier is really a combination of noise from resistors and transistors inside the amplifier. But still, the Phase Linear 700 amplifier's SNR is 108.5 dB, and the SNR exceeds the dynamic range of 16-bit CD audio players today.

Another convenient way of calculating noise is to first set V_{nr} to $\sqrt{4kTBR} =$ 1 nV for B = 1 Hz and find R. After some calculations, R = 61 Ω for 1 nV of noise for B = 1 Hz. Then one can find the equivalent noise resistance for a given noise voltage

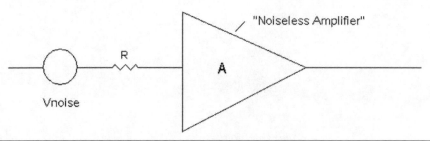

FIGURE 22-3 An equivalent input noise resistance model.

density specification in Y nV/$\sqrt{\text{Hz}}$, where Y is the scalar for nanovolts per (square) root hertz. Then

$$\text{Equivalent noise resistance} = (Y^2)61 \ \Omega$$

For example, a FET such as the 2N3382 has an equivalent input noise voltage density of 10 nV/$\sqrt{\text{Hz}}$; thus Y = 10, and the equivalent noise resistance is $(Y^2)61 \ \Omega$ = (100)61 Ω = 6,100 Ω. In another example, a TL072 op amp has an equivalent input noise of 18 nV/$\sqrt{\text{Hz}}$; thus the equivalent noise resistance that the equivalent voltage generated by a resistor to provide 18 nV/$\sqrt{\text{Hz}}$, is $(18^2)61 \ \Omega$ = 19.76 kΩ.

For active components such as transistors, the shot-noise generators exist as current noise generators across the collector and emitter and the base and the emitter. Figure 22-1B shows the shot-noise generators ICshot_noise and IBshot_noise. These two current-source generators are characterized by

$$\text{ICshot\_noise} = \sqrt{2(I_{CQ})qB} \tag{22-6}$$

where I_{CQ} is the quiescent collector current, q is electron charge (= 1.6×10^{-19} Coulomb) and B is bandwidth in hertz.

$$I_{Bshot\_noise} = \sqrt{2(\tfrac{I_{CQ}}{\beta})qB} \tag{22-7}$$

where β = the current gain of the transistor and $\frac{I_{CQ}}{\beta}$ = IB is the base current.

For a transistor direct-current (DC) collector current of 1 mA, and for a bandwidth of 1 Hz, the collector shot-noise current is $\sqrt{2(0.001)1.6 \times 10^{-19}}$ Amp = 17.9 pA, and for a current gain β of 100, the base-current shot noise is 1.79 pA. Therefore, at a base current of 10 μA = I_C/β = 0.001 mA/100, the noise current density is

$$\frac{1.79 \text{ pA}}{\sqrt{\text{Hz}}}$$

A way to reflect the collector shot-noise current to the input is to find an equivalent input voltage that is amplified by the transconductance of the transistor to provide the same amount of collector noise current (Figure 22-4).

So what we need is a noise voltage, which is the shot-noise current from the collector to the emitter expressed as a referred input voltage. This referred input noise voltage source is defined as Vcol_ref_in_noise such that

$$[\text{Vcol\_ref\_in\_noise}][g_m] = \sqrt{2(I_{CQ})qB} =$$
$$\text{collector-to-emitter shot-noise current source generator}$$

Let's square both sides for now:

$$[(\text{Vcol\_ref\_in\_noise})(g_m)]^2 = 2(I_{CQ})qB$$

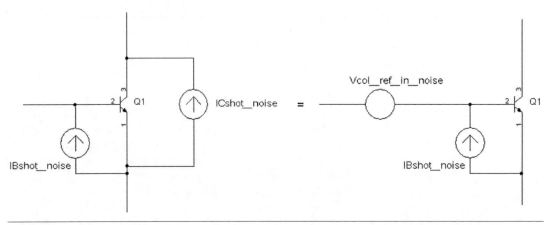

FIGURE 22-4 "Reflecting" the collector noise current source back to the input via an equivalent noise source.

$$(Vcol\_ref\_in\_noise)^2[g_m]^2 = 2(I_{CQ})qB$$

$$(Vcol\_ref\_in\_noise)^2 = [2(I_{CQ})qB]/[g_m]^2$$

However,

$$g_m = \frac{I_{CQ(q)}}{kT}, \text{ and } [g_m]^2 = [g_m]g_m = [g_m]\frac{I_{CQ(q)}}{kT}, \text{ and therefore,}$$

$$(Vcol\_ref\_in\_noise)^2 = \frac{2(I_{CQ})qB}{[g_m]^2} = \frac{2(I_{CQ})qB}{[g_m]\frac{I_{CQ(q)}}{kT}} = \frac{2BkT}{[g_m]}$$

$$(Vcol\_ref\_in\_noise)^2 = \frac{2BkT}{[g_m]} \qquad (22\text{-}8)$$

Recall that the resistor noise is related to 4kTBR. If we use $1/g_m$ as a resistor, Equation (22-8) almost fits the 4kTBR factor. So we just multiply the numerator and denominator by $1 = 2/2$ to provide the following:

$$(Vcol\_ref\_in\_noise)^2 = \frac{2BkT}{g_m} \times \frac{2}{2} = \frac{4BkT}{2g_m} = \frac{4kTB}{2g_m} = 4kTB\frac{1}{2g_m}$$

$$Vcol\_ref\_in\_noise = \sqrt{4kTB\frac{1}{2gm}} \qquad (22\text{-}9)$$

From Equation (22-9), the referred input noise voltage, Vcol_ref_in_noise looks like a noise voltage from a resistor of resistance value $1/2g_m$.

We can further express equivalently that there is an input referred noise resistance $1/2g_m$. In the analysis of noise, sometimes calculating the total noise resistance referred to the input allows the designer to evaluate the values of feedback resistors or voltage-source resistors.

Thus the equivalent input noise resistance is $1/2g_m$, where

$$g_m = \frac{I_{CQ(q)}}{kT} = g_m = \frac{I_{CQ}}{0.026\ V}$$

where I_{CQ} is the DC collector current. For a DC collector current of 1 mA, the equivalent input noise resistance that is referenced from the collector current shot noise generator then is

$$1/2g_m = \frac{1}{2} \times \frac{1}{g_m} = \frac{1}{2} \times 26\ \Omega = 13\ \Omega$$

For 100 µA of collector current, the equivalent noise resistance is 130Ω, and for 10 µA of collector current, the equivalent noise resistance is $1{,}300\ \Omega$.

Accordingly, from the preceding three example calculations, lowering the collector current actually results in more equivalent noise referred to the input. And increasing the collector current results in less equivalent noise referred to the input. However, with bipolar transistors, increasing the collector current also increases the base-current shot-noise-generator current source. Therefore, there is a tradeoff in terms of setting the collector current based on the source-driving resistance. At 1 mA of collector current and a β of 100, the base shot-noise current density is IBshot_noise_density = 1.79 pA/$\sqrt{Hz}$. The noise voltage generated by the base noise current is just the source resistance multiplied by the base noise current, or the noise voltage density is equal to the base shot-noise current density multiplied by the source resistance (Figure 22-5).

For example, if the signal source resistance Rsource is less than a few hundred ohms, such as an RF 50-Ω or 75-Ω source or a low-impedance microphone ($<$300 Ω),

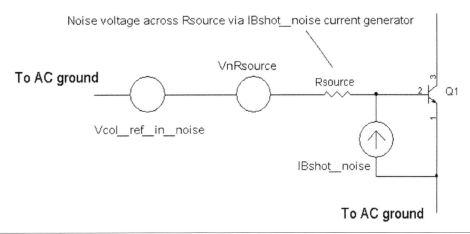

FIGURE 22-5   Base current shot-noise generator contributing to noise voltage via the input source resistor, Rsource.

most likely the collector current can be set up to about 1 mA for a transistor current gain of 100. If the current gain is very high, such as a β of 300 or more, then the collector current in this example can be set higher to a few milliamps.

At 1 mA with a β of 100, the base noise current has an IBshot_noise_density of 1.79 pA/$\sqrt{\text{Hz}}$, and with Rsource at 300 Ω, the noise voltage density caused by base noise current is then 1.79 pA/$\sqrt{\text{Hz}}$ × 300 Ω = 0.537 nV/$\sqrt{\text{Hz}}$ or the equivalent noise voltage generated from a 17.5-Ω resistor [(0.537²)61 Ω], which is almost comparable with ½g_m = 13 Ω, the equivalent input noise resistance from the collector shot-noise current generator.

Thus, with a perfect transistor, the total noise resistance referred to the input is the source resistance R_{source} = 300 Ω + 17.5 Ω + 13 Ω. The noise level is contributed substantially by the source resistance and not the transistor.

However, if the source resistance is increased to 10,000 Ω, collector current is 1 mA, and a β is 100, the base current shot noise will provide 1.79 pA/$\sqrt{\text{Hz}}$ × 10,000 Ω = 17.9 pA/$\sqrt{\text{Hz}}$ or the equivalent noise resistance of (17.9)² = 19.5 kΩ. Therefore, the total input-referred resistance is 10,000 Ω + 19,500 Ω + 13 Ω. From this example, we can see that the base-current shot noise into the 10-kΩ source resistance generates most of the noise voltage. One solution is to select a transistor with a higher β (>100) such as a β = 300 to 800.

However, a better alternative to designing a low-noise preamplifier with a medium to high signal source resistance (e.g., a parallel-tank circuit) is to use a low-noise JFET such as the 2SK152, J110, or LSK170. These JFETs have an equivalent input-noise voltage density of about 1.0 nV/$\sqrt{\text{Hz}}$ with negligible gate current.

In the preceding example, the transistor was deemed to be perfect in the sense that there was no internal base series resistor Rbb (see Figure 22-1C). However, commercially made transistors include an internal base series resistor as a result of the process of making transistors. In general, typical base series resistors can be in the range of several hundred ohms to a thousand or more ohms. Certain low-noise, switching, or power transistors have Rbb < 50 Ω. For low-noise applications, the commonly available PNP transistor 2N4403 has an Rbb in the 40-Ω range according to low-noise-amplifier expert John Curl. Note that PNP transistor 2N4403 has a little less noise than NPN versions 2N4400 and 2N4401.

Another very low-noise bipolar transistor is the Toshiba 2SA1316, rated at 0.60 nV/$\sqrt{\text{Hz}}$, which has an equivalent input noise resistance of 22 Ω, including the Rbb and the shot-noise components. The specification sheet states that the Rbb is about 2 Ω for the 2SA1316. Some of the earlier "low-noise transistors," such as NPN transistor 2N5089, have an Rbb in the area of about 1,000 Ω, whereas its PNP counterpart, the 2N5087, does much better with an Rbb in the range of 200 to 300Ω. Other common low-noise NPN transistors, such as the PN2222 and PN2222A, have specified characteristics with comparable Rbb values (<50 Ω) to the 2N2N4401. And the data sheet for the PN2907A, its PNP transistor counterpart, has a similar noise performance as the 2N4403. However, the noise performance of transistors may vary from the specification sheet depending on the manufacturer, so transistors such as the PN2222, PN2222A, and PN2907A should be tested and verified for noise performance, according to John Curl.

Note The PN2222(A) and PN2907A have their pin-outs reversed from 2N part-number transistors such as the 2N3904. Be careful to verify the pin-outs for the transistors starting out with PNxxxx.

Typical general-purpose transistors such as the 2N3904 or 2N4126 NPN transistors have Rbb values on the order of more than 500 Ω, and most likely Rbb ≈ 1,000 Ω. Their complementary PNP transistors, 2N3906 and 2N4126, fared much better, with Rbb ≈ 100 to 150 Ω or so. In general, the dominant noise source for a bipolar transistor results from the series base resistor. For example, the equivalent input noise resistor pertaining to collector current shot noise at 1 mA is 13 Ω, whereas the typical base series resistor, Rbb is typically greater, such as at least 50 Ω.

Paralleling Transistors for Lower Noise

An alternative to lowering the noise effects of the series base resistor Rbb is to parallel transistors. I learned this paralleling transistor technique from John Curl and from his design of the JC-1 moving-coil pre-preamp. Generally, if the transistors are on the same package or die, paralleling them is not a problem. For discrete transistors to be paralleled, some combination of matching for the same V_{BE} turn-on voltage between each transistor and/or using emitter resistors to help stabilize the biasing must be chosen. Otherwise, there will be an imbalance of collector currents, and one transistor can hog most of the desired collector current, leaving the other transistors at very low collector currents. For an example circuit that shows paralleling transistors for low-noise applications, go to www.google.com/patents and type in 4035737, which is U.S. Patent no. 4,035,737, inventor John Curl, filed on March 5, 1976. Also see Figure 22-6, a preamp circuit from U.S. Patent no. 4,035,737.

For audio situations, paralleling transistors for low-noise applications works very well. The "side effect" is increasing capacitance across the collector, base, and emitter terminals, but for audio, this is generally fine. The paralleled transistors also can be coupled at their collectors to a common-base amplifier to form a cascode circuit. In this way, the Miller effect is minimized, and for RF applications, the input capacitances of the paralleled transistors can be tuned or resonated with an inductor.

Note that paralleling JFETs also can lower noise as well. For example, in a 1973 Ampex BC-230 studio color television camera, three JFETs were paralleled for the camera tube preamplifier. The drains of the three FETs then were connected to the emitter of a grounded base amplifier to form a cascode preamplifier. Figure 22-7 shows a simplified circuit diagram of n bipolar transistors connected in parallel. The transistors are all matched in terms of V_{BE} turn-on voltage.

Figure 22-7 shows n transistors connected in parallel. For this analysis, the series base resistors Rbb1, Rbb2,...Rbbn all have the same resistance value. Thus Rbb1 = Rbb2... = Rbbn.

For a total DC collector current of I_{cq} where I_{cq} is the sum of the quiescent collector currents for all the transistors Q1 to Qn, each transistor is operating at a DC collector current of I_{cq}/n because there are n transistors, and the collector currents are

United States Patent [19]

Curl

[11]   **4,035,737**

[45]   **July 12, 1977**

[54]   **LOW NOISE AMPLIFIER**

[76]   Inventor: **John J. Curl,** c/o Ihem, 1820
Montreaux-Vaude, Switzerland

[21]   Appl. No.: **664,108**

[22]   Filed: **Mar. 5, 1976**

Related U.S. Application Data

[63]   Continuation of Ser. No. 499,608, Aug. 22, 1974,
abandoned.

[51]   **Int. Cl.²** **H03F 3/18**
[52]   **U.S. Cl.** .... **330/13**; 179/1 A;
179/100.4 A; 330/15; 330/19
[58]   **Field of Search** .... 179/100.4 A, 100.41 M,
179/100.41 T, 1 A; 330/13, 15, 17, 18, 19

[56]   **References Cited**

U.S. PATENT DOCUMENTS

3,418,590   12/1968   Rongen et al. .... 330/15 X
3,500,221   3/1970   Mercier .... 330/22 X

Primary Examiner—Rudolph V. Rolinec
Assistant Examiner—Lawrence J. Dahl
Attorney, Agent, or Firm—De Lio and Montgomery

[57]   **ABSTRACT**

A low noise amplifier for amplifying a signal from a
moving coil phono cartridge which comprises a plural-
ity of pairs of complimentary first and second transistors in push-pull and each pair in parallel, with a resistance of relatively low-value in the emitter circuit of
each transistor and all collectors connected to an output terminal.

2 Claims, 1 Drawing Figure

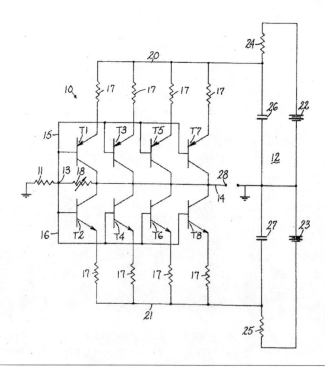

FIGURE 22-6   A low-noise preamplifier with paralleled transistors.

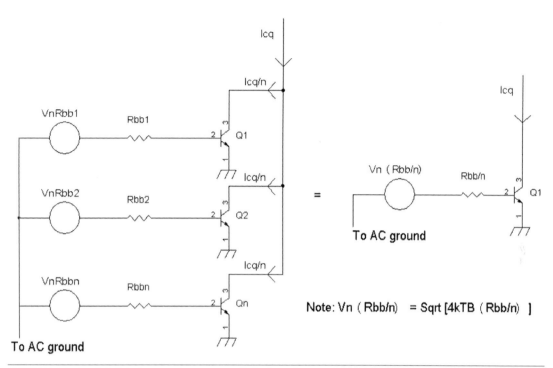

FIGURE 22-7   Multiple transistors paralleled with series base resistors to effectively reduce base resistance.

evenly divided. Therefore, the transconductance of each transistor is g_m/n, where g_m is the tranconductance of a transistor operating at a collector current of I_{cq}.

Also, for the analysis, the small signal input resistance of each transistor is $\beta/(g_m/n) = n\beta/g_m >> Rbb1$. For example, if $\beta = 100$ and $n = 2$ with $I_{cq} = 1$ mA, then $n\beta/g_m = 200/0.0384$ mho $= 5,200\ \Omega$. Typically, Rbb $< 500\ \Omega$, so the noise voltage formed by Rbb with Rbb in series is applied to the base emitter junction of the transistor with negligible loss. That is, the input resistance at the base of the transistor is typically much larger than Rbb.

The total noise from the n transistors thus is

$$\text{Iout\_noise} = \frac{g_m}{n}\sqrt{4kTB(Rbb1)} + \frac{g_m}{n}\sqrt{4kTB(Rbb2)} + \cdots + \frac{g_m}{n}\sqrt{4kTB(Rbbn)}$$

$$(22\text{-}10)$$

$$\text{Iout\_noise} = \frac{g_m}{n}\left(\sqrt{4kTB(Rbb1)} + \sqrt{4kTB(Rbb2)} + \cdots + \sqrt{4kTB(Rbbn)}\right)$$

$$\frac{\text{Iout\_noise}}{\frac{g_m}{n}} = \sqrt{4kTB(Rbb1)} + \sqrt{4kTB(Rbb2)} + \cdots + \sqrt{4kTB(Rbbn)}$$

$$(22\text{-}11)$$

Because the noise voltages

$$\sqrt{4kTB(Rbb1)} + \sqrt{4kTB(Rbb2)} + \cdots + \sqrt{4kTB(Rbbn)}$$

from each series base resistor are independent random noise sources, squaring the sum of the noise voltages is just the sum of the squares of each noise voltage. That is, the independent sources are *orthogonal* such that each noise source can only multiply with itself to provide a finite number; other noise terms that multiply one noise source with a different independent noise source will amount to zero. And the square of

$$\sqrt{4kTB(Rbb1)} = 4kTB(Rbb1)$$

Therefore,

$$\left[\frac{\text{Iout\_noise}}{\frac{g_m}{n}}\right]^2 = 4kTB(Rbb1) + 4kTB(Rbb2) + \cdots + 4kTB(Rbbn) \qquad (22\text{-}12)$$

Let $Rbb = Rbb1 = Rbb2 = \ldots Rbbn$ because these are all the same resistance values. And since there are n series resistors,

$$4kTB(Rbb1) + 4kTB(Rbb2) + \ldots 4kTB(Rbbn) = n4kTB(Rbb) \qquad (22\text{-}13)$$

Thus

$$\left(\frac{\text{Iout\_noise}}{\frac{g_m}{n}}\right)^2 = n4kTB(Rbb) \qquad (22\text{-}14)$$

$$[\text{Iout\_noise}]^2 = [\tfrac{g_m}{n}]^2\, n4kTB(Rbb) = [g_m]^2 \tfrac{1}{n}\tfrac{1}{n} n4kTB(Rbb) = [g_m]^2 \tfrac{1}{n} 4kTB(Rbb) \qquad (22\text{-}15)$$

$$[\text{Iout\_noise}]^2 = [g_m]^2\, 4kTB\tfrac{Rbb}{n} \qquad (22\text{-}16)$$

$$\sqrt{[\text{Iout\_noise}]^2} = \sqrt{[g_m]^2\, 4kTB\tfrac{Rbb}{n}} \qquad (22\text{-}17)$$

$$\text{Iout\_noise} = \sqrt{[g_m]^2\, 4kTB\tfrac{Rbb}{n}} \qquad (22\text{-}18)$$

$$\text{Iout\_noise} = g_m \sqrt{4kTB\tfrac{Rbb}{n}} \qquad (22\text{-}19)$$

Note that $g_m = I_{cq}/0.026$ volt, where I_{cq} is the summed DC collector currents of all n transistors.

Therefore, paralleling n transistors with each transistor having a series base resistor is equivalent to a single transistor with the series base resistor's resistance divided by n.

Equation (22-19) thus represents the output noise current of a transistor that has a thermal noise voltage related to a series base resistor Rbb/n. See the right-hand side circuit of Figure 22-6. Therefore, the (series base resistor) noise voltage contributed by paralleling n transistors results in the noise of a single transistor multiplied by a factor of $1/\sqrt{n}$. For example, regarding noise from Rbb, paralleling four transistors or $n = 4$ results in $1/\sqrt{4}$, the noise of a single transistor, or one-half the noise of a single transistor.

 Note that the collector shot-noise current does not change when the transistors are paralleled, and it is still $= \sqrt{2(I_{CQ})qB}$.

Differential-Pair Amplifier Noise

A differential-pair amplifier with no series emitter resistors gives about 3 dB ($1.41 = \sqrt{2}$) more noise than a grounded emitter amplifier. The differential-pair amplifier compared with a grounded emitter amplifier has 1.41 more input referred noise voltage from the collector-to-emitter shot-noise current generators. If one factors in the series base resistors, in a differential-pair amplifier there is twice the series base resistance, which comes from a series base resistor from each of the two transistors. As a result of the two series base resistors in the differential-pair amplifier, the thermal noise generated by them yields 1.41 times the noise voltage from a single series base resistor of a grounded emitter amplifier. Recall that the thermal noise voltage has a square-root function of resistance (see Figures 22-8, 22-9, and 22-10).

Figure 22-8 shows a differential-pair amplifier with Q1 and Q2 transistors that have series base resistors Rbb1 and Rbb2 along with collector-to-emitter shot-noise generators In1 and In2. For the noise analysis, the base-current shot-noise components will be considered "small." The signal source across the bases of Q1 and Q2 has sufficiently low source resistance that the base-current shot-noise generators are insignificant in the noise performance of the differential-pair amplifier.

Since Q1 and Q2 are matched transistors, the total series base resistance is just $2 \times$ Rbb1 or $2 \times$ Rbb2, twice the series base resistance, thereby increasing the thermal noise voltage by a single series base resistor Rbb, where Rbb $=$ Rbb1 $=$ Rbbb2. Thus

$$\sqrt{4kTB(2)(\text{Rbb})} = \sqrt{2}\ \sqrt{4kTB(\text{Rbb})} \qquad (22\text{-}20)$$

Thus, for noise owing to series base resistance, the differential-pair amplifier has twice the noise resistance of a single-ended grounded emitter amplifier.

Also seen in Figure 22-8 is the emitter DC current source IEE, that evenly divides the emitter currents for Q1 and Q2 when both bases are tied to the same voltage, such as a ground (0 V). For that matter, any current source connected to the emitters of Q1

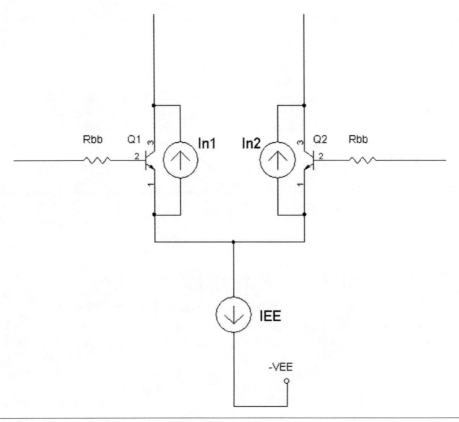

FIGURE 22-8   A differential-pair amplifier with collector-to-emitter shot-noise generators and series base resistors.

and Q2 will have its current evenly split to half the current flowing out of the collector of Q1 and the other half flowing out of Q2, for a current gain of the transistors of $\beta >> 10$. Recall that for large β ($>>10$), the collector current is equal to the emitter current.

In Figure 22-8, the collector-to-base shot-noise generators In1 and In2 are "floating" in a sense that unlike the grounded emitter amplifier, where the collector-to-emitter noise current source is grounded at one end, the noise sources are not grounded in the differential-pair amplifier. For each transistor Q1 and Q2, there is a noise current source summing with the collector and the same noise current source flowing into the emitter of the respective transistor in an opposite direction (see Figure 22-9). For an analysis of the collector shot-noise current generators, the series base resistors have been omitted in Figures 22-9 and 22-10.

Thus Figure 22-9 shows each "floating" noise source as a combination of two current noise sources, one flowing noise current with the collector and another current noise source flowing in the opposite direction into the emitters. These two

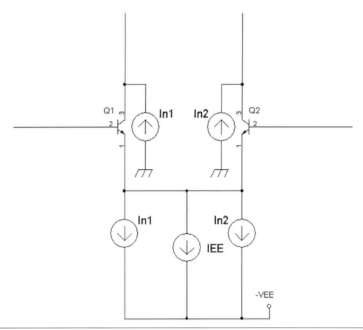

FIGURE 22-9 Collector-to-emitter shot-noise generators redrawn as current-source generators at the emitters and collectors of Q1 and Q2.

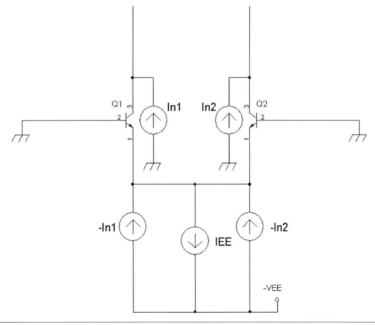

FIGURE 22-10 Collector-to-emitter shot-noise generators shown as opposing current sources at the emitters of Q1 and Q2.

current sources are correlated, which means that they are not independent of each other, and their noise voltages add like other voltages. For example, The In1 noise current source at the collector will be partially canceled out by the In1 noise current source flowing into the emitter because the In1 noise current source flowing into the emitter is flowing in the opposite direction as the noise current source In1 at the collector. However, In1 and In2 are still independent of each other and add by the sum of the squares and then taking the square root (see Figure 22-10).

Figure 22-10 thus shows noise current sources flowing into the emitters of Q1 and Q2 as –In1 and –In2. The differential-pair amplifier is configured to evenly split both the emitter noise current sources (–In1 and –In2) in half going to the collectors of Q1 and Q2. To analyze shot-noise components and refer them back to the input of the differential-pair amplifier as an equivalent input noise voltage, see Figure 22-10.

Therefore, the total noise current at the collector of Q1 is

$$\text{In1} + -\tfrac{1}{2}[\text{ In1} + \text{In2}] = \tfrac{1}{2}[\text{In1} - \text{In2}] = I_{\text{shot\_ noise total col\_Q1}} \tag{22-21}$$

and the total noise current at the collector of Q2 is

$$\text{In2} + -\tfrac{1}{2}[\text{ In1} + \text{In2}] = \tfrac{1}{2}[\text{In2} - \text{In1}] = I_{\text{shot\_ noise total col\_Q2}} \tag{22-22}$$

Because the shot-noise currents In1 and In2 are independent sources,

$$(I_{\text{shot\_ noise total col\_Q1}})^2 = [\tfrac{1}{2}]^2([\text{In1}]^2 + [-\text{In2}]^2) = \tfrac{1}{4}([\text{In1}]^2 + [\text{In2}]^2) \tag{22-23}$$

and,

$$(I_{\text{shot\_ noise total col\_Q2}})^2 = [\tfrac{1}{2}]^2([\text{In2}]^2 + [-\text{In1}]^2) = \tfrac{1}{4}([\text{In1}]^2 + [\text{In2}]^2) \tag{22-24}$$

$$[\text{In1}]^2 = 2(I_{CQ})qB \tag{22-25}$$

and

$$[\text{In2}]^2 = 2(I_{CQ})qB \tag{22-26}$$

I_{CQ} = quiescent collector current of Q1 = quiescent collector current of Q2.

$$(I_{\text{shot\_ noise total col\_Q1}})^2 = \tfrac{1}{4}([\text{In1}]^2 + [\text{In2}]^2) = \tfrac{1}{4}[2(I_{CQ})qB + 2(I_{CQ}qB] = \tfrac{1}{2}[2(I_{CQ})qB] \tag{22-27}$$

and

$$(I_{\text{shot\_ noise total col\_Q2}})^2 = \tfrac{1}{4}([\text{In1}]^2 + [\text{In2}]^2) = \tfrac{1}{4}[2(I_{CQ})qB + 2(I_{CQ}qB] = \tfrac{1}{2}[2(I_{CQ})qB] \tag{22-28}$$

To find the equivalent input noise owing to shot-noise current generators from Q1 and Q2, we have

$$V_{\text{noise\_diff\_pair\_reflected input}} = \text{equivalent input noise owing to collector-to-emitter shot-noise generators In1 and In2}$$

Then

$$\frac{(V_{\text{noise\_diff\_pair\_reflected input}})g_{\text{m\_diff\_pair\_1\_out}}}{\sqrt{[I_{\text{shot\_ noise total col\_Q1}}]^2}} = \sqrt{\tfrac{1}{2}[2(I_{CQ})qb]} \tag{22-29}$$

However, $g_{\text{m\_diff\_pair\_1\_out}} = (\tfrac{1}{2})g_{\text{m\_Q1}}$ because the transconductance of a differential-pair amplifier is reduced by half as a result of the output signal being taken at one collector output.

Note   If the output signal is taken differentially from the collector of Q1 and the collector of Q2, then $g_{\text{m\_diff\_pair\_1\_out}} = g_{\text{m\_Q1}}$.

Therefore,

$$(V_{\text{noise\_diff\_pair\_reflected input}})\tfrac{1}{2}g_{\text{m\_Q1}} = \sqrt{\frac{(V_{\text{noise\_diff\_pair\_reflected input}})g_{\text{m\_diff\_pair\_1\_out}}}{(\tfrac{1}{2})2(I_{CQ})qB]}} \tag{22-30}$$

Squaring both sides of the equation yields

$$(V_{\text{noise\_diff\_pair\_reflected input}})^2 [\tfrac{1}{2}g_{\text{m\_Q1}}]^2 = \tfrac{1}{2}2(I_{CQ})qB \tag{22-31}$$

$$(V_{\text{noise\_diff\_pair\_reflected input}})^2 \tfrac{1}{4}[g_{\text{m\_Q1}}]^2 = \tfrac{1}{2}2(I_{CQ})qB \tag{22-32}$$

Multiplying by 4 on both sides, we get

$$(V_{\text{noise\_diff\_pair\_reflected input}})^2 [g_{\text{m\_Q1}}]^2 = [\tfrac{4}{2}]2(I_C)qB = 4(I_{CQ})qB \tag{22-33}$$

$$(V_{\text{noise\_diff\_pair\_reflected input}})^2 = \frac{4(I_{CQ})qB}{(g_{\text{m\_Q1}})^2} \tag{22-34}$$

But

$$g_{\text{m\_Q1}} = \frac{(I_{CQ})kT}{q}$$

so

$$1/(g_{\text{m\_Q1}})^2 = \frac{1}{(I_{CQ})q/kT} \frac{1}{g_{\text{m\_Q1}}}$$

$$\left(V_{\text{noise\_diff\_pair\_reflected input}}\right)^2 = \frac{4(I_{CQ})qB}{(I_{CQ})q/kT} \frac{1}{g_{m\_Q1}} = \frac{4kTB}{1} \frac{1}{g_{m\_Q1}} = 4kTB \frac{1}{g_{m\_Q1}} \quad (22\text{-}35)$$

When the square root is taken on both sides,

$$V_{\text{noise\_diff\_pair\_reflected input}} = \sqrt{4kTB \frac{1}{g_{m\_Q1}}} \quad (22\text{-}36)$$

Equation (22-36) relates to an equivalent input noise resistance of $1/g_{m\_Q1}$, which is twice the input noise resistance of a single-ended ground emitter amplifier whose equivalent input noise resistance is $1/2g_m$, as shown in Equation (22-9). Thus, for the same collector current, a differential-pair amplifier has twice the equivalent noise resistance at the input than a single-ended grounded emitter amplifier pertaining to shot-noise currents across the collector to the emitter. This twice noise resistance thus relates to an equivalent input noise voltage of 1.41 or $\sqrt{2}$ of the equivalent input noise voltage of the grounded emitter amplifier in terms of shot noise.

In practice, since there is also an extra series base resistor in the differential-pair amplifier, the equivalent input noise in a differential amplifier is increased by 1.41 owing to both shot noise and series base resistors over a single-ended grounded emitter amplifier.

Before we leave the noise analysis for a differential-pair amplifier, what if we take the output differentially from the collectors of Q1 and Q2 in Figure 22-10? Would the equivalent input noise due to the shot noise currents In1 and In2 for a differential output be different from a single ended output? The answer is no, the equivalent input noise is still the same for both cases. That is, the noise currents taken differentially at the outputs will be

$$\tfrac{1}{2}\left[\text{In1} - \text{In2}\right] - \tfrac{1}{2}\left[\text{In2} - \text{In1}\right] = I_{\text{shot\_ noise total col\_Q1}} - I_{\text{shot\_ noise total col\_Q2}}$$

or

$$I_{\text{shot\_ noise total col\_Q1}} - I_{\text{shot\_ noise total col\_Q2}} = \text{In1} - \text{In2}$$

Thus the noise current is increased by a factor of 2 when taken differentially as opposed to being taken from a single-ended output. The reason why the output noise current is increased by 2 and not by $\sqrt{2}$ is because the collector noise currents from Q1 and Q2 are correlated but out of phase. Also see equations (22-21) and (22-22) for the shot noise from the collector of Q1 and Q2, single ended output.

However, the transconductance also increases by a factor of 2 in a differential-pair amplifier when the two output currents from Q1 and Q2 are taken differentially as opposed to an output current taken in a single-ended manner (e.g., collector current from either Q1 or Q2). Note that the equivalent input noise voltage is the ratio of the output noise current to the transconductance of the differential-pair amplifier. So, therefore, although the noise increases by a factor of 2 (in differential mode), so does the transconductance (in differential mode), which results in no change in equivalent input noise between the differential mode output and the single ended output.

Typically, the main contribution to noise in a differential-pair amplifier is the series base resistor for Q1 and Q2. Therefore, to lower noise in a differential-pair amplifier, one can use the technique of paralleling transistors. For example, Q1 and Q2 would consist of multiple paralleled, matched transistors.

Cascode Amplifier Noise

In a cascode amplifier, most of the noise is contributed by the common-emitter stage or bottom transistor (Q1). The top transistor or common-base transistor amplifier (Q2) contributes negligible amounts of noise. Figure 22-11 shows the noise sources for a cascode amplifier.

Figure 22-11 shows series base resistors Rbb1 and Rbb2 for the common-emitter amplifier Q1 and common-base amplifier Q2. Because the emitter of Q2 is loading into a current source Q1 with very high output resistance, the series base resistor of Q2 does generate a noise signal at the collector of Q2. When there is a noise voltage at the base of Q2, the emitter of Q2 is loaded to ground by the output resistance of Q1 r_{01}, which does not contribute noise. Thus r_{01} is an emitter resistor to ground for Q2, and r_{01} is a local feedback resistor that lowers the transconductance in terms of Q2 for voltage at the base of Q2 and output current at the base of Q2. Typically, $r_{01} > 10$ kΩ.

The transconductance from the viewpoint of a noise voltage at the base of Q2 then is

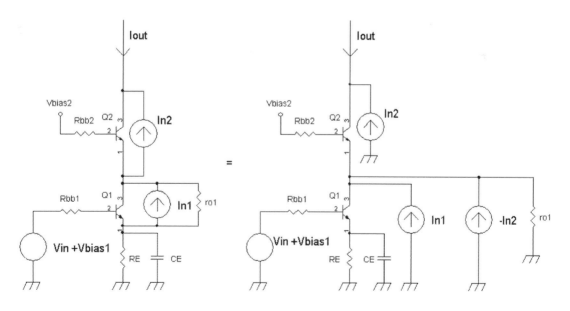

Note: Capacitor CE is an AC short circuit.

FIGURE 22-11   Cascode amplifier with thermal and shot-noise sources.

$$g'_{m\_Q2} = \frac{g_{m\_Q2}}{1 + g_{m\_Q2}(r_{01})} \approx \frac{1}{r_{01}} \approx 0$$

Therefore, noise voltages at the base of Q2 do not appear at the output of Q2's collector terminal. If there are appreciable capacitances from the emitter of Q2 to ground, then there will be a noise contribution from the noise voltage at the base of Q2. But generally at lower frequencies, this is not the case.

In terms of shot-noise contribution from the collector to the emitter of Q2, the shot-noise current generator In2 cancels out at the output of Q2's collector. That is, at Q2's collector, there is In2, but at Q2's emitter, there is –In2. Since the emitter current is the same as the collector current for a large current gain β (>>1) of Q2, the –In2 noise-current source appears at the collector of Q2 and sums with the In2 noise source to cancel noise related to Q2.

Therefore, most of the noise in a cascode amplifier comes from the noise sources of the bottom transistor Q1. The series base resistor noise and the shot noise from the collector to the emitter of Q1 appear at the output of Q2 because the collector current of Q1 equals the collector current of Q2 for the current gain β >> 1 of Q2. Intuitively, it makes sense that the main noise contributions in a cascode amplifier come from the bottom transistor because it is the collector current of Q1 that "controls" the collector current of Q2.

Selecting Op Amps

In terms of selecting for low-noise to ultralow-noise op amps, see Table 22-2.

TABLE 22-2   Equivalent Input Noise Densities for Low Noise Op Amps

Device	Noise Density in $\frac{nv}{\sqrt{Hz}}$	Type
TL072/TL082	18	Dual JFET
RC4558	8	Dual bipolar
RC4136	8	Quad bipolar
NE5532	5	Dual bipolar
RC4560	~ 3	Dual bipolar
LM4562	2.7	Dual bipolar
LT1115	0.9	Single bipolar
AD797	0.9	Single bipolar

Table 22-3 illustrates the noise performance of some general-purpose op amps.

TABLE 22-3   Equivalent Noise Densities for General Purpose Op Amps

Device	Noise Density in $\frac{nv}{\sqrt{Hz}}$	Type
LM741	45	Single bipolar
LM1458	45	Dual bipolar
TL062	42	Dual JFET
LM358	40	Dual bipolar
LM324	35	Quad bipolar
TLC272	30	Dual JFET

To maintain noise performance with op amps, the selection of feedback resistor values is important. High-value resistors generally degrade noise performance (see Figure 22-12).

Generally, a rule of thumb is to have the parallel combination of R1 and RF = (R1 × RF)/(R1 + RF) maintain the low noise capability of the op amp by having

$$\frac{R1 \times RF}{R1 + RF} < 0.5(\text{equivalent input noise resistance}) \tag{22-37}$$

where Y nV/$\sqrt{Hz}$ and Y is the scalar for nanovolts per square root of hertz. Then

$$\text{Equivalent noise resistance} = (Y^2)61\ \Omega$$

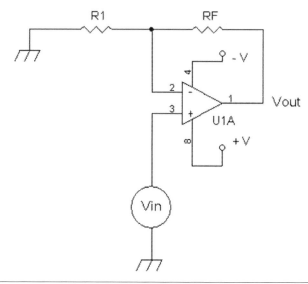

FIGURE 22-12   A noninverting amplifier.

For example, if the gain is greater than 10 and RF >> R1, then (R1 × RF)/ (R1 + RF) ≈ R1, and if an AD797 op amp is chosen, then the equivalent noise resistance is $(0.9^2)61\,\Omega = 49\,\Omega$, which means that R1 < 24.5 Ω. Once the gain is known, the value of RF can be determined. For example, for a noninverting gain of 100, RF = 99(R1), where the gain = (R1 + RF)/R1.

Under the same conditions of gain > 10, if an RC4558 is selected, then the equivalent noise resistance is $(8^2)61\,\Omega = 3,900\,\Omega$, which means that R1 < 1,850 Ω. Again, the value of RF can be determined when the gain = (R1 + RF)/R1 is known.

References

1. Class Notes EE141, Paul R. Gray and Robert G. Meyer, UC Berkeley, Spring 1975.
2. Class Notes EE241, Paul R. Gray and Robert G. Meyer, UC Berkeley, Fall 1975.
3. Class Notes EE140, Robert G. Meyer, UC Berkeley, Fall 1975.
4. Class Notes EE240, Robert G. Meyer, UC Berkeley, Spring 1976.
5. Paul R. Gray and Robert G. Meyer, *Analysis and Design of Analog Integrated Circuits*, 3rd ed. New York: John Wiley & Sons, 1993.
6. Kenneth K. Clarke and Donald T. Hess, *Communication Circuits: Analysis and Design*. Reading: Addison-Wesley, 1971.
7. National Semiconductor, *Audio/Radio Handbook*. Santa Clara: National Semiconductor Corporation, 1980.
8. R. E. Zeimer and W. H. Tranter, *Principles of Communications*. Boston: Houghton Mifflin Company, 1976.
9. William G. Oldham and Steven E. Schwarz, *An Introduction to Electronics*. New York: Holt, Reinhart and Winston, Inc., 1972.
10. Allan R. Hambley, *Electrical Engineering Principles and Applications*, 2nd ed. Upper Saddle River: Prentice Hall, 2002.
11. Texas Instruments, *Linear Circuits Amplifiers, Comparators, and Special Functions, Data Book 1*. Dallas: Texas Instruments, 1989.
12. Motorola Semiconductor Products, Inc., *Linear Integrated Circuits*, Vol. 6, Series B., 1976.
13. Thomas M. Frederiksen, *Intuitive IC Op Amps from Basics to Useful Applications* (National's Semiconductor Technology Series). Santa Clara: National Semiconductor Corporation, 1984.
14. Alan B. Grebene, *Bipolar and MOS Analog Circuit Design*. New York: John Wiley & Sons, 1984.
15. U.S. Patent 4,035,737, "Low-Noise Amplifier," John J. Curl, filed on March 5, 1976.
16. U.S. Patent 5,471,531, "Method and Apparatus for Low-Cost Audio Scrambling and Descrambling," Ronald Quan, filed on December 14, 1993.
17. U.S. Patent 4,963,764, "Low-Noise Current Mirror Active Load Circuit," Ronald Quan, filed on March 23, 1989.
18. U.S. Patent 6,617,910, "Low-Noise Analog Multiplier Utilizing Nonlinear Local Feedback Elements," Ronald Quan, filed on August 10, 2001.

19. *High Fidelity's Test Reports*. Great Barrington: Billboard Publications, Inc., 1973.
20. *High Fidelity*. New York: ABC Consumer Magazines, Inc. May 1987, pgs. 20–21.
21. *Home Theater*. New York: Source Interlink Media,), December 2009, pgs. 60–62.
22. Victor Company of Japan, Ltd., *JVC Service Manual HR-D140U*. Elmwood Park: JVC Company of America, 1985.

Chapter 23

Learning by Doing

In the last 10 chapters, Chapters 13 through 22, quite a lot of math was involved in explaining how signals and circuits work. However, for this last chapter, I will return to a more "hands-on" approach with experimentation with simple circuits. Before these elementary circuits are explored, I will revisit two previous circuits, the one-transistor superheterodyne radio and the software-defined radio (SDR) front-end circuit.

Therefore, the objectives of this chapter are

1. To show an updated improvement to the one-transistor radio
2. To comment on the 40-meter-band SDR front-end cicuit
3. To experiment with mixers and using Spectran, a free spectrum analyzer software program for the personal computer (PC)
4. To run experiments on op amps and current sources
5. To observe how an LC circuit works
6. To learn the Thevenin equivalent circuit
7. To analyze a bridge circuit
8. To give my thoughts on electronics, writing this book, and learning

The test equipment required in this chapter includes two signal generators with adjustable amplitude. The frequency range of one of the generators should provide up to a 1-MHz sine-wave signal. Also, a dual-trace oscilloscope of 10-MHz bandwidth or more is required. A four-trace scope is preferred.

Update on the One-Transistor Superheterodyne Radio

The first version of the one-transistor superheterodyne radio had a few drawbacks; most notably, it had insufficient volume (Figure 23-1). The original one-transistor schematic diagram is shown as Figure 23-1.

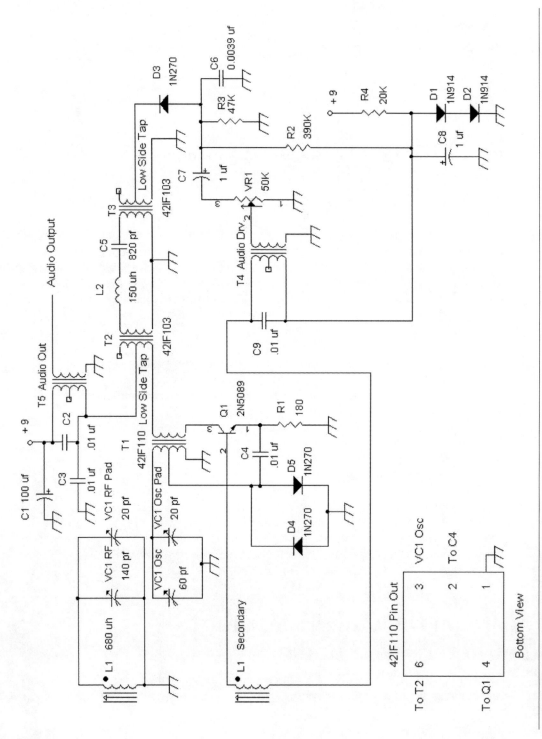

FIGURE 23-1 Original one-transistor radio design.

In Figure 23-1, the audio signal from the audio driver transformer T4 is fairly small. The emitter resistor R1 reduces the transconductance of Q1, thereby reducing the audio gain. When C4 was increased to a larger capacitance, such as 1,000 µF, to increase the audio gain by using the tapped winding of T1 to ground the emitter, squegging resulted. *Squegging* is a condition where an oscillator's signal turns on and off at a low frequency rate and is caused by the large coupling capacitors in the signal path.

One possible solution is to replace R1 with an inductor with a value of 470 µH to 1,000 µH; at audio frequencies, the impedance of the inductor is the direct-current (DC) coil resistance, and at the oscillator frequency, the impedance of the inductor is high. The resistance of these inductors typically will run about 5 Ω to 10 Ω, and therefore, one has to measure the resistance of the inductor first before replacing R1. The DC bias line to the base of Q1 via C8 will have to be removed because it now requires a variable bias voltage to set a new bias voltage to the base of Q1 such that the emitter voltage divided by the (470-µH) coil resistance is about 10 to 20 mA.

In Figure 23-2, the collector current of Q1 is set by VR2. The coil resistance of L3 is measured, and VR2 is varied until the emitter voltage across the coil resistance results in an emitter current of about 15 mA.

Alternatively, see Figure 23-3, a modified schematic with changes in the oscillator configuration and the IF filter, T2, T6, and T3.

In Figure 23-3 the emitter resistor R1 is reduced dramatically from the previous design from 180 Ω to 5.1 Ω. This reduction allows for higher audio gain to provide a sufficient loudness from a speaker. Although T5 has a primary impedance of 1,000 Ω and 8 Ω for the secondary, it was found that replacing T5 with a 500-Ω primary impedance and an 8-Ω secondary impedance rating allowed for a higher audio level.

The adjustment for VR2, the bias control, requires some care. After the initial adjustment of VR2, the transistor warms up. As the transistor warms up, the collector current changes. This collector-current change then causes the oscillator's frequency to shift somewhat. Therefore, VR2 may need readjustment. VR2 is set to provide 0.070 volt DC at the emitter of Q1. Then, with an oscilloscope, measure the oscillator signal at the base of Q1 for 200-mV peak-to-peak amplitude by adjusting VR3.

Please note that the oscillator circuit in Figure 23-3 is different from the circuits shown in Figures 23-1 and 23-2. Therefore, notice that pins 4 and 6 are reversed in oscillator coil T1 (42IF110). See the notes in Figure 23-3.

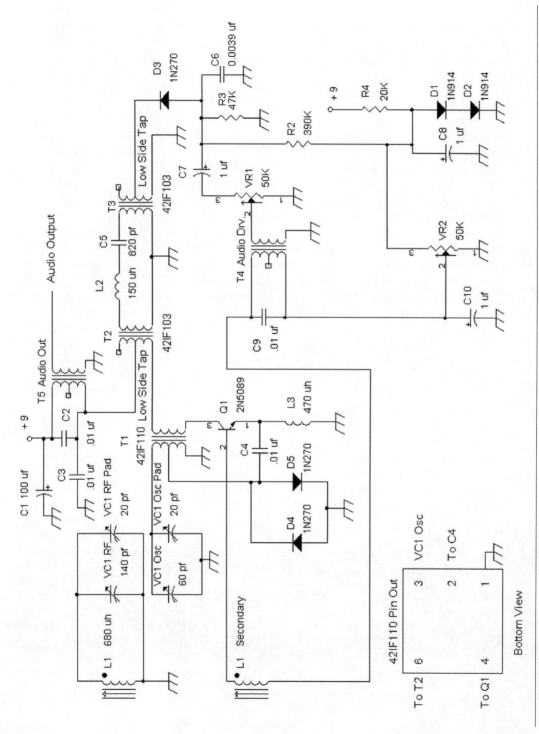

FIGURE 23-2 One-transistor radio with a "possible" modification for higher audio volume.

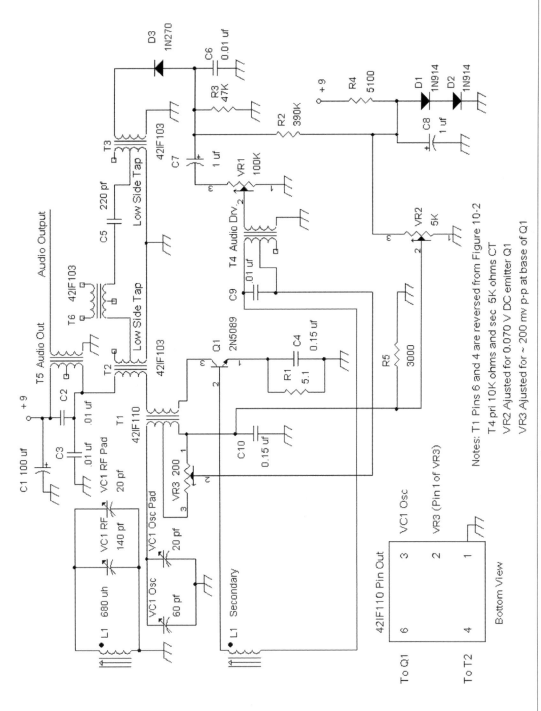

FIGURE 23-3 One-transistor superheterodyne radio version 2.0.

423

Comments on SDR 40-Meter Front-End Circuit

In Chapter 12, the front-end circuit from Figure 12-11B proved satisfactory for receiving amateur radio signals on the 40-meter band. The figure is shown again in Figure 23-4.

This front-end circuit supplied the I and Q low-frequency IF signals to a 192-kHz sampling-rate sound card and then was compared against a Yaesu FT-707 high-frequency (HF) band receiver. The FT-707 was better overall, but this front-end circuit managed to pull in some weak stations fine with a 25-foot-long wire antenna. If I had to redesign the circuit, I would make one small change. I would replace the bipolar transistor Q1, an MPSH10, with a low-noise junction field effect transistor (JFET), such as a 2SK152 or a J110, to reduce noise further, but the input biasing circuits consisting of R1, R2, R7, and R4 would have to be changed. By using a JFET, input bias noise current is not a problem compared with a bipolar transistor. Thus the radio-frequency (RF) transformer T1 can be configured to provide a higher signal level from the high-side tap (instead of signal from the low-side tap) to the gate of the JFET. Other ways to reduce noise would involve replacing the op amps with ultralow-noise types (<1 nV/$\sqrt{\text{Hz}}$) or replacing Q1 and Q2 with a voltage-gain stage. However, these changes may not be worth the trouble because it is always easier to use a better antenna, which would overcome the noise caused by the front-end circuit. One of my amateur radio friends told me that when a good 40-meter antenna is connected to the receiver, the atmospheric (electrical) noise dominates over any noise caused by the front-end circuits, or alternatively, he recommends a digital signal processing (DSP) audio filter from AM-COM, Inc., that reduces hiss by at least 3- to 10-fold voltage-wise (10 dB to 20 dB).

Experimenting with Mixers and Using the Spectran Spectrum Analyzer Program

An audio spectrum analyzer used to cost $5,000 twenty or thirty years ago. However, today one can obtain an audio spectrum analyzer with a computer and a free program, Spectran. The minimum sampling rate for most sound cards today is at least 48,000 samples per second, which allows processing audio signals up to about 24 kHz.

A doubly balanced mixer is a mixer that ideally provides only the product of the two input signals, without the passage of either or both input signals to the output of the mixer. We can use Spectran to observe the performance of a doubly balanced mixer, as shown in Figure 23-5.

To ensure that the input signal is not leaked through into the output, the commutating switcher U1 is provided with a square-wave switching signal (50 percent duty cycle) via binary divider U3. And to ensure that the square-wave switching signal is not leaked through to the output, the DC bias points are identical at both inputs of

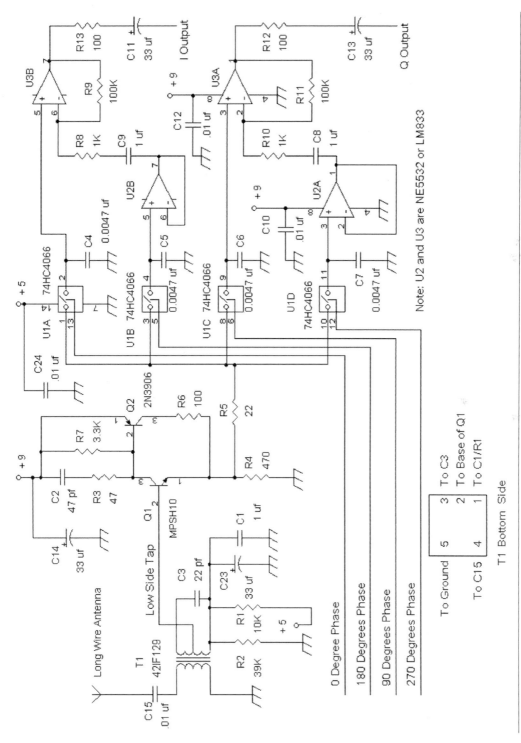

FIGURE 23-4 A 40-meter band front-end circuit.

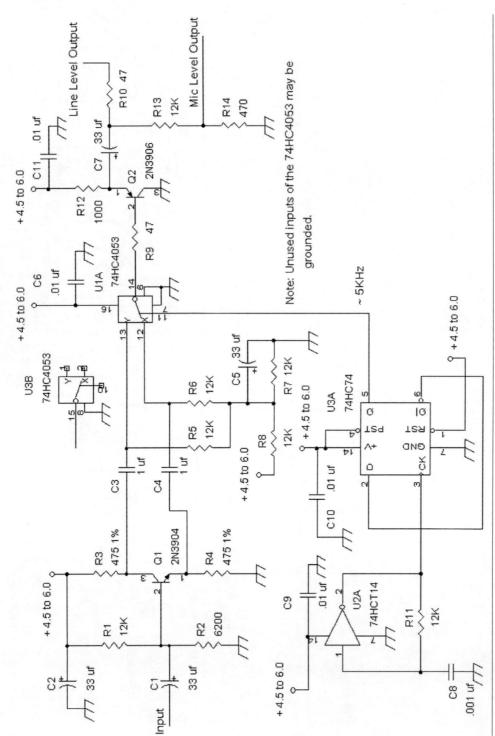

FIGURE 23-5 A doubly balanced mixer for spectrum analysis.

the analog switch U1A and there is also equal amplitude but opposite phases of the input signal provided to the inputs of U1A. Phase-splitter circuit Q1 provides nearly equal amplitude and opposite phases when R3 and R4 are identical in value (e.g., 1 percent or better tolerance resistors) and when the β of Q1 is high (>100). In practice, R4 is typically 1 percent larger in value than R3, or R3 is 1 percent smaller in value than R4. For R4 = 475 Ω, the preferred resistance value is about 480 Ω, whereas R3 = 475 Ω.

The supply voltage is at 4.5 volts to 6 volts DC to avoid overvoltage on the logic chips and analog switch. With the components R11 and C8 selected, the output of pin 9 of U3 (74HC74) is a square wave of 5 kHz. The input terminal provides a 4-kHz sine-wave signal of about 0.5 volt peak to peak. Couple the output of the mixer circuit via R10 to the left or right channel of the computer's sound card line input. If a microphone input is used, a lower-amplitude signal from the mixer can be taken from the microphone level output terminal, where R13 and R14 are connected.

Download the Spectran program from the Web via http://digilander.libero.it/ i2phd/spectran.html. The version 2 spectrum analyzer file is about 600 kB, and there is a 252-kB PDF operating manual that also can be downloaded. See Figure 23-6 for a

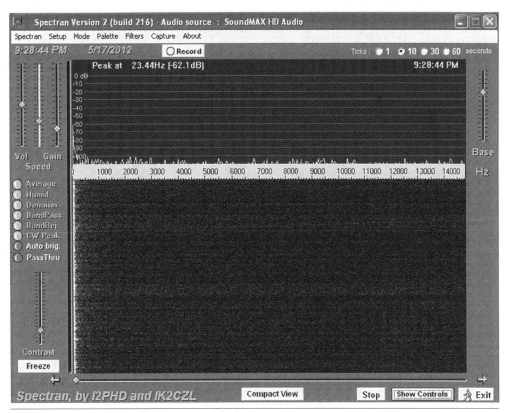

FIGURE 23-6 A sample Spectran display.

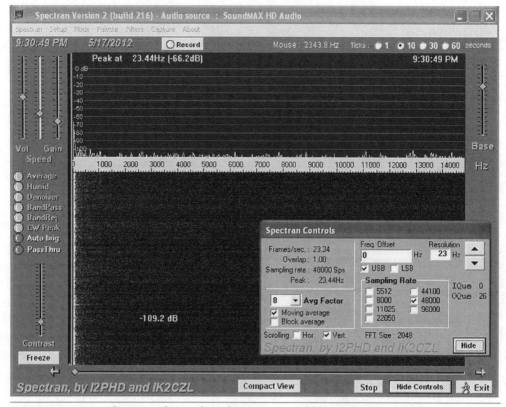

FIGURE 23-7 Selection of sampling frequency and bin width.

sample Spectran display. Click on "Show Controls" to select the sampling frequency and resolution, as shown in Figure 23-7.

The sampling rate of 48 kHz is checked because this sampling rate should be available to all sound cards using the Window XP operating system. A resolution of 23 Hz is chosen such that a span from 0 Hz to almost 15 kHz is available without having to use the frequency-offset slider control ← and →. A finer resolution such as 12 Hz can be chosen, but the span will be cut in half to about 7.5 kHz of bandwidth, and the frequency-offset slider control must be used. See Figure 23-8.

Figure 23-8 shows the frequency-offset slider control displaying a span of frequencies from about 5,800 Hz to about 13,100 Hz.

With a 4-kHz sine-wave signal of about 0.5 volt peak to peak (0.25 volt peak) at the input, the spectrum of the doubly balanced mixer typically should have a spectrum displayed as shown in Figure 23-9. Figure 23-9 shows that the switching signal at about 5.2 kHz and the input signal at 4 kHz are suppressed by at least 40 dB or to 1 percent of the output signals whose frequencies are at 1.2 kHz (e.g., an IF signal or difference-frequency signal) and at 9.2 kHz, the summed frequency signal.

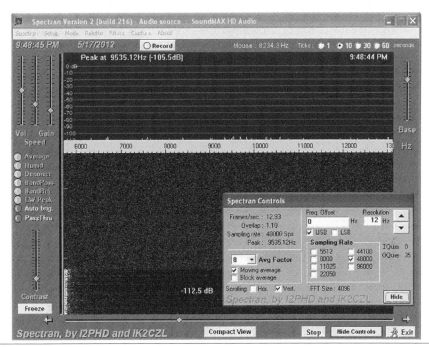

FIGURE 23-8   Using the frequency-offset slider control on a finer resolution mode.

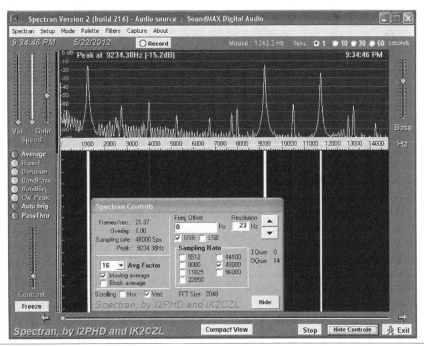

FIGURE 23-9   Output spectrum for a 4-kHz input signal.

Also observe that there is mixing action around 15 kHz by using the frequency-offset slider control (Figure 23-10). Figure 23-10 shows harmonic mixing at around 15.6 kHz, the third harmonic of the 5.2-kHz switching signal. The difference frequency is 15.6 kHz – 4 kHz = 11.6 kHz, and the sum-frequency component is at 15.6 kHz + 4 kHz = 19.6 kHz.

Spectran is also great for analyzing the Fourier transform of pulse waveforms. For example, in Chapter 15, a pulse of 25 percent duty cycle produced frequencies at the fundamental frequency plus all multiples of harmonics except every fourth harmonic (e.g., fourth, eighth, etc.). Figure 23-11 provides a spectrum of a 25 percent duty-cycle pulse for a 1-kHz signal with a 250-μs pulse, a 1,000-μs period, and 0.380 volt peak-to-peak amplitude (into the line input of a sound card).

In Figure 23-11, the spectrum of a 1-kHz 25 percent duty-cycle pulse signal shows frequency components at all frequencies with every fourth harmonic nulled. Thus signals at 4 kHz, 8 kHz, and 12 kHz and so on are reduced substantially in amplitude. The 1-kHz 25 percent duty-cycle pulse waveform was provided by a function generator with a duty-cycle adjustment. Alternatively, one can use a counter and logic gates to provide the 25 percent duty-cycle waveform, as shown in Chapter 12 with the 40-meter-band SDR front-end circuit.

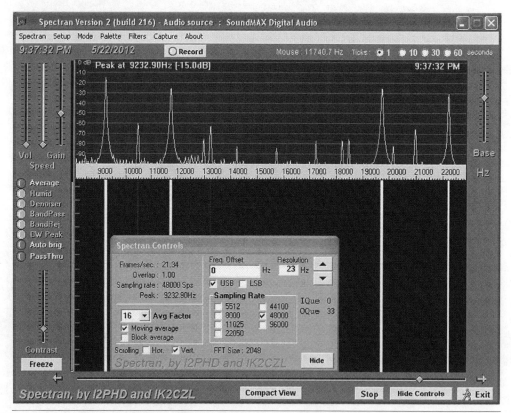

FIGURE 23-10 Harmonic mixing at around 15.6 kHz.

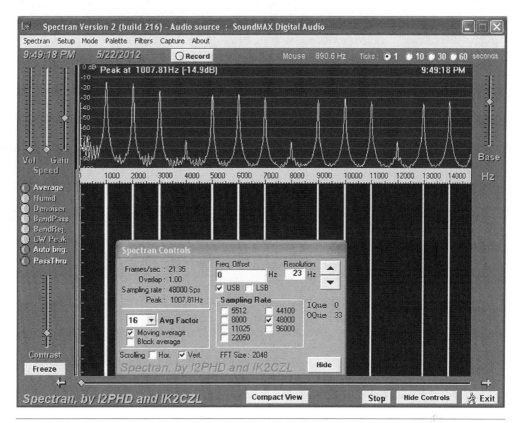

FIGURE 23-11   Spectrum analysis via Spectran of a 25 percent duty-cycle pulse-train waveform.

Now we will turn to a basic one-transistor mixer, as shown in Figure 23-12. The figure looks more like a common-emitter or grounded-emitter amplifier, which is for small signals at the base of less than 10 mV peak to peak. However, when a large sinusoidal signal of more than 200 mV peak to peak is combined with a small signal of a different frequency, a mixing or multiplying effect occurs.

Look at Figure 23-12. At the base of Q1, resistors R1 and R3 attenuate the input signal Vin1 by 100-fold, and resistors R2 and R3 attenuate the input signal Vin2 by 11-fold. With a sine-wave generator at 11 kHz with 4 volts peak of amplitude connected to input terminal Vin2 of the mixer, connect a 1-volt peak-to-peak sine-wave signal at 6.5 kHz into input terminal Vin1. Now connect the output of the mixer via R8 or R9/R10 to the sound card of the computer with the Spectran program running. Observe that there is a difference-frequency signal of 4.5 kHz (an IF signal) and that its amplitude is almost the same as the 6.5-kHz signal at the output of the mixer. This "equality" in amplitude level confirms that when a large-amplitude sinusoidal signal of a first frequency is combined with a small-amplitude signal of a second frequency, the output of the mixer provides that intermodulation distortion signal, which is the IF signal at about the same amplitude as the amplified small-amplitude signal

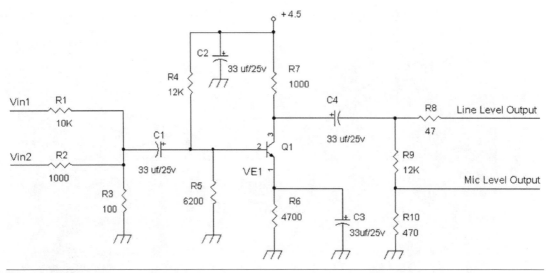

FIGURE 23-12   A one-transistor mixer that is typically used in radios.

of the second frequency. And this "equality" in amplitude level confirms that the conversion transconductance of a bipolar mixer is just the same as the small-signal transconductance when a large-amplitude signal is involved (see Figure 23-13).

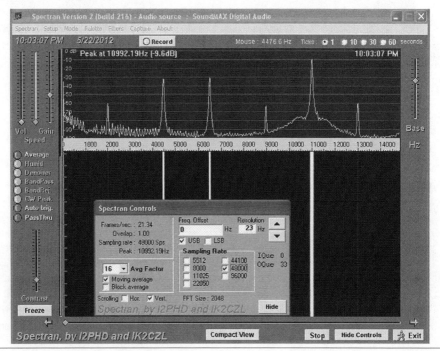

FIGURE 23-13   Spectrum of the one-transistor mixer via Spectran.

Lowering the large-amplitude 11 kHz signal will lower the conversion transconductance and thus lower the intermodulation distortion signal at 4.5 kHz.

In Figure 23-13, note at the output of the mixer that the IF signal at about 4.5 kHz has roughly the same amplitude as the input signal at 6.5 kHz. This confirms that a large oscillator voltage of greater than 200 mV peak to peak into the base emitter junction of a mixer transistor provides a conversion gain roughly equal to the small-signal gain.

With the 11-kHz signal removed and disconnected from the input terminal V_in2, one can confirm that second-harmonic distortion is about 1 percent per 1-mV peak sine wave. Therefore, a 1-V peak-to-peak 6.5-kHz sine-wave signal at V_in1 provides about 10 mV peak to peak or 5 mV peak sine wave into the base of Q1. Thus the second-harmonic distortion signal at 13-kHz should be about 5 percent (–26 dB) of the fundamental-frequency signal at 6.5 kHz. Spectran should confirm that the second harmonic is then about 26 dB down from the fundamental-frequency signal (see Figure 23-14).

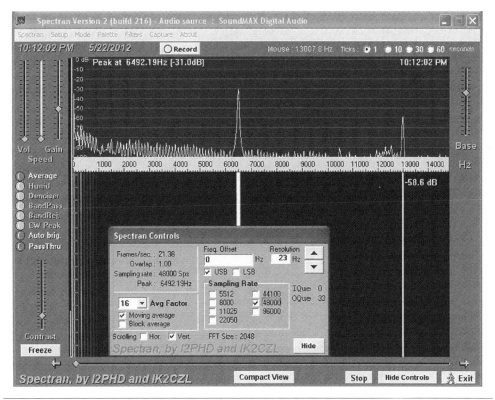

FIGURE 23-14 Spectrum of second-harmonic distortion via Spectran.

As seen in Figure 23-14, with the fundamental-frequency 6.5-kHz signal at –31.0 dB, the second-harmonic signal at 13 kHz is at –58.6 dB, which is about 27.6 dB lower than the fundamental-frequency signal. The calculated second harmonic is about 26 dB, so theory and practice match up well. Also note that the cable length of the audio cable to the sound card should be 3 feet or less if possible. The preceding measurement was done with a 15-foot-long cable, whose capacitance slightly rolls off the frequency response at about 13 kHz.

Another experiment one can try is to connect a series 200-Ω variable resistor to C3 such that the local feedback can be controlled at the emitter of Q1. By noting the emitter voltage VE1 and dividing it by the emitter resistor R6 = 4,700 Ω, the collector current is known. Then find the transconductance $I_{CQ}/0.026$ volt, and set the series variable resistor to (½) × 0.026 volt/I_{CQ} to observe whether the third-harmonic distortion is nulled or minimized.

 One can test harmonic and intermodulation distortion of audio devices (e.g., high-fidelity receivers and amplifiers) via Spectran.

Oh, just one more thing (as Lieutenant Columbo would say), the common-emitter amplifier shown in Figure 23-12 generates lots of distortion and is not usually used in practice for providing high-quality audio signal amplification. This type of amplifier is commonly shown in beginning electronic circuits classes, but what the textbook or the instructor does not tell the student is that the grounded-emitter amplifier distorts heavily with normal line-level audio signals (e.g., 100 mV RMS). In order to improve the distortion performance of this amplifier, some type of negative feedback is required.

Conducting Experiments on Op Amps and Amplifiers

This section explores frequency response of operational amplifier circuits. In Chapter 19 it was shown that a multiple-channel input linear mixing amplifier results in lower bandwidth than expected. Again, by linear mixing, I refer to additive mixing (e.g., Figure 23-15) instead of multiplicative mixing in RF circuits.

With about a 1-volt peak-to-peak sine-wave signal, measure the frequency response of the inverting gain (–1) amplifier at Vout1 and the voltage follower amplifier at Vout2. The –3-dB bandwidth happens when the 1-V peak-to-peak signal drops to 0.707 volt peak to peak. Typically, the –3-dB bandwidth of an LM1458 for a unity-gain follower via Vout2 is about 1 MHz and 500 kHz at Vout1, where resistors R2, R3, R4, and R5 are not connected to ground yet. The inverting amplifier configuration for a gain of –1 should have half the bandwidth of the voltage follower.

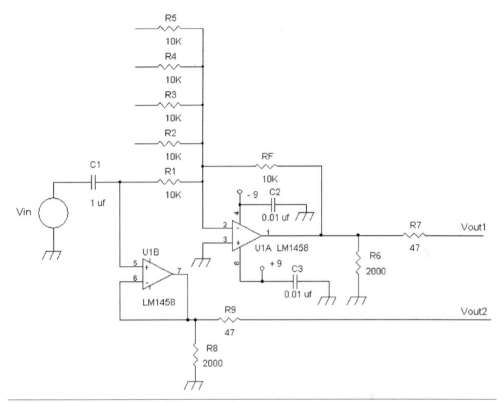

FIGURE 23-15   A five-input mixer and a voltage follower circuit with four unused input resistors floating.

Now short resistors R2, R3, R4, and R5 to ground and remeasure the bandwidth at Vout1, which should be about one-third the previous measurement, or about 166 kHz (see Figure 23-16). Thus one needs to be careful in terms of bandwidth reduction caused by a multiple-input summing amplifier, as shown in Figure 23-15. When more inputs are added, the bandwidth shrinks further.

Recall that the –3-dB response is given by bandwidth $\approx$ GBWP[$R_{par}/(RF + R_{par})$], where R_{par} is the parallel combination of all input resistors, and GBWP is usually the unity-gain bandwidth of the op amp.

We now turn to observing how an op amp provides flat frequency response for closed-loop gains set by the feedback resistors. An op amp in general without any negative feedback has an open-loop bandwidth of less than 1 kHz and typically less than 100 Hz. The question one would ask is: How does feedback provide a flat frequency response beyond the open-loop bandwidth?

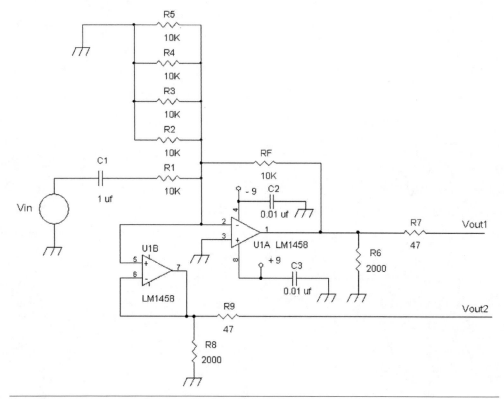

FIGURE 23-16   Unused inputs tied to ground for remeasurement of frequency response at Vout1.

To uncover this "mystery," one way to find out is to build the circuit as seen in Figure 23-17. Figure 23-17 shows an inverting amplifier set for a gain of about –100. To prevent overloading the output, the input signal is attenuated by 10-fold such that a 1-kHz, 1-volt peak-to-peak sinusoidal waveform at input resistor R7 produces a 10-volts peak-to-peak signal at Vout1. The error signal $V_{error\ U1A}$ at pin 2 of U1 (LM1458) is amplified by about 100 times via U2, a wider bandwidth op amp (a TL082). The (+) terminal of the LM1458 is grounded to provide a convenient way to measure the differential voltage across the (–) and (+) inputs of the 1458 op amp.

Take a look at Figure 23-17. The experiment is as follows: Confirm that a 10-volt peak-to-peak signal appears at Vout1 at frequencies from 100 Hz to 5,000 Hz. The response at Vout1 should be flat from 100 Hz to 5,000 Hz with a small drop-off at 5,000 Hz. However, the frequency response at Vout2 should show a rising amplitude as frequency is increased (see Figure 23-18). Vout2 represents an amplified version of of the error signal across the (–) and (+) inputs of U1A, an LM1458 op amp, in Figure 23-17. Note that the (+) input of U1A is grounded.

If the error signal at the (–) terminal of the LM1458 has a flat frequency response, then the output signal at the LM1458 will have a rolled-off response at the open-loop

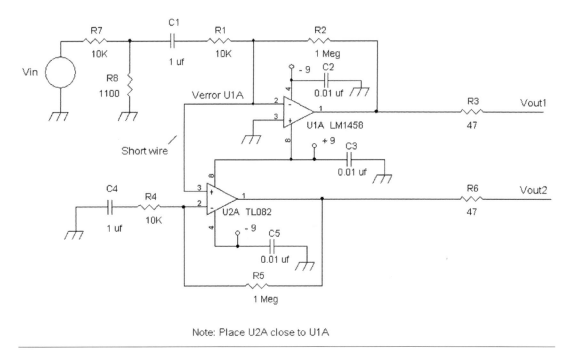

FIGURE 23-17   Analyzing the frequency response of the input terminal of an op amp.

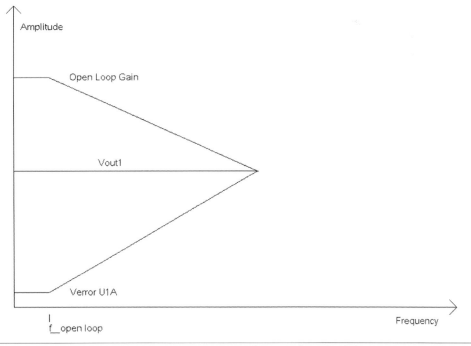

FIGURE 23-18   Frequency response at open loop and at the (–) input terminal.

bandwidth frequency such as 10 Hz. At 10 Hz, the gain is 0.707 of the gain at DC or at 1 Hz. The gain falls off after 10 Hz at approximately 10 Hz/f, where $f > 10$ Hz. For example, at 100 Hz, the gain is one-tenth the gain at 1 Hz. At 1,000 Hz, the gain is one-hundredth or 1 percent of the gain at 1 Hz. Therefore, to provide a flat frequency response beyond 10 Hz, the error signal at the (–) input of the LM1458 must be increasing in a complementary manner or at approximately the reciprocal of the fall-off response. Thus, at the (–) terminal of U1A, it has a response of about $f/10$ Hz. For example, the signal at 100 Hz at the (–) input terminal of the LM1458 must be 10 times higher than at 1 Hz.

This makes sense in that the open-loop frequency response has the characteristic of an integrator, and the frequency response at the (–) input terminal must have the characteristics of a differentiator. When a signal is integrated and then differentiated, or vice versa, the frequency response is flat.

So the next question is, is there a limit as to how high in frequency the input signal can go before something else happens at the (–) input? The answer is that there is a high-frequency limit if the output wants to maintain a flat frequency response. What happens eventually is that the alternating-current (AC) voltage amplitude across the (–) and (+) input terminals of U1A becomes too high and starts to overload the input stage of the op amp. This input overload condition is also known as *slewing*, which provides a slew-rate limitation on the op amp. Thus the design of the input stage of an op amp often dictates the slew rate at the output of the op amp. If the input stage is designed in a very robust manner, able to take on large input signals, then the slew rate will be very high. Op amps such as the LM318 use emitter series resistors for the input differential-pair amplifier to increase the dynamic range. Other op amps may use junction field effect (JFET) or metal-oxide semiconductor field-effect transistor (MOSFET) input stages that have an inherently greater dynamic range than bipolar transistors.

In an LM1458, when the error voltage across the (–) and (+) input terminals approaches 0.5 volt peak to peak, the input stage is starting to limit or is at limiting. As a result, the limiting is similar to the limiting characteristic in Figure 13-10 in Chapter 13 except that each transistor in the differential-pair amplifier input stage of an LM1458 is connected to the emitter of a common-base amplifier that results in twice the dynamic range over a simple differential-pair amplifier. For the overload characteristics of the LM1458 op amp, see Figure 23-19.

To observe the effects of slew-rate limiting, connect a sine-wave generator of amplitude 5 volts peak to peak to the input of the circuit in Figure 23-15. Start at 1 kHz and increase the frequency until the output signal at Vout1 changes from a sine wave to a triangle waveform. As the frequency increases, the sinusoidal signal at the (–) input of U1A in Figure 23-12 will increase sufficiently to cause limiting in the input stage and thus provide a square-wave current signal into the second stage of the LM1458 op amp. This square-wave signal then is fed to the second stage of the op amp that acts as a low-pass filter or integrator (Miller capacitance multiplier), which then results in the output waveform looking like a triangle wave.

We will now turn to a shunt feedback circuit for dividing resistance. The circuit in Figure 23-20 will demonstrate how shunt negative feedback provides an equivalent

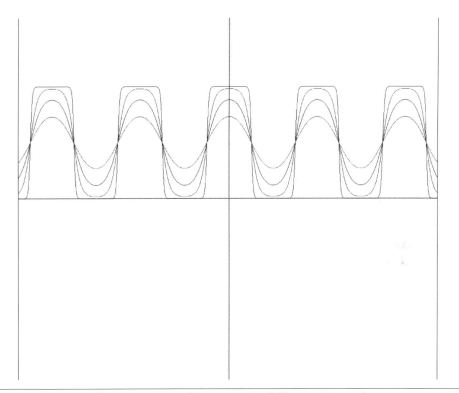

FIGURE 23-19   Collector output of an LM1458 differential pair for various input sine-wave input levels at 0.052, 0.104, 0.208, and 0.832 volt peak.

resistance that is lower than "expected." High-impedance buffer amplifier U1B ensures current from the input voltage source only flows into R1. By setting the gain of U1A via the feedback resistor's Rin = R1/(1 – – A) = R1/(1 + A) = R1/(1 + R3/R2).

See Figure 23-20. In this case, R3 = R2; therefore,

$$\text{Rin} = \frac{R1}{1+1} = \frac{R1}{2}$$

With the voltage source a 1.5-volt battery and milliamp meter in series, as shown in Figure 23-20, measure the current flowing into R1. The current flowing into R1 should be about 3 mA. Therefore, by Ohm's law, the equivalent resistance Rin, the resistance from R1 at the input to ground, should be 1.5 volts/3 mA = 500 Ω.

Alternatively, if the milliamp meter and 1.5-volt battery are removed, then Rin can be measured directly with an ohmmeter [e.g., a digital VOM (volt-ohm-milliammeter) or an analog VOM] at the R1–pin 5 U1B junction and ground.

If the gain of the inverting amplifier is increased, such as by replacing R3 with a 20-kΩ resistor instead, then A = 2, and the current drain will increase and provide an equivalent Rin = R1/(1 + A) = 1,000/3 = 333 Ω.

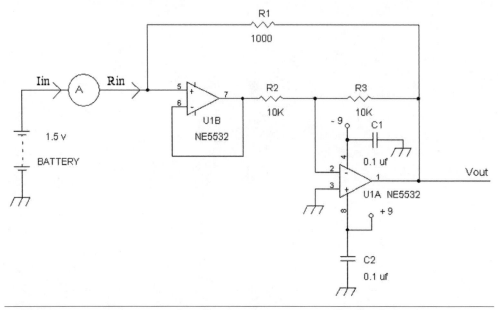

FIGURE 23-20 Resistance-value division circuit using a feedback resistor.

Similarly, this circuit works with AC signals as well (see Figure 23-21). Set an oscillator at 1 kHz and 1 volt peak to peak sine wave to Vin. Then measure the AC signal at R4/R1 or at the (+) input of U1B, which should read 0.5 volt peak to peak. By using "shunt" feedback to synthesize a resistor to ground via feedback resistor R1, a lower than expected resistance value is achieved because the output of U1A is providing a negative or out-of-phase signal to pull more current through R1.

Figure 23-22 shows another application of shunt feedback using a capacitor instead of a resistor. The result is again more than "expected" current flowing into the feedback capacitor, which mimics a larger capacitor from the input of U1B to ground. As a result of this larger than expected capacitor current via an out-of-phase signal at the output of U1A, the effective capacitance is increased. Therefore, $Cin = (1 + A)C1$.

To ensure a DC bias potential at the (+) input of U1A with an AC coupled signal source, a large-valued resistor R1 = 2.2 MΩ is connected to R_FIL and ground. See Figure 23-22. With the shunt feedback capacitor C1 = 0.001 µF, and R_Fil = 10 kΩ, feed a sinusoidal signal of 1 V peak to peak at 100 Hz, and connect the scope probe to R_FIL/C1. With R3 = R2, the closed-loop gain A = 1; thus $Cin = (1 + 1)C1 = 2(0.001 \text{ µF}) = 0.002 \text{ µF} = Cin$.

The –3-dB (0.707 volt peak to peak) frequency then should be

$$\frac{1}{2\pi(10k\Omega)(0.002\ \mu F)} = 7.96 \text{ kHz} \approx 8 \text{ kHz}$$

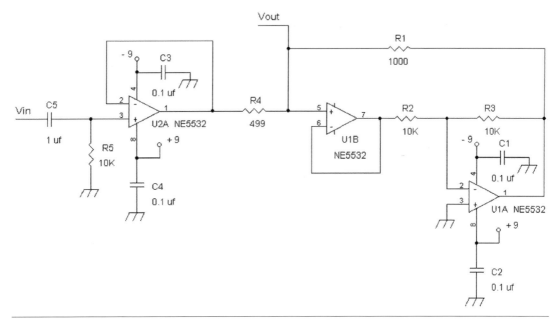

FIGURE 23-21   AC signal into a divided resistance value.

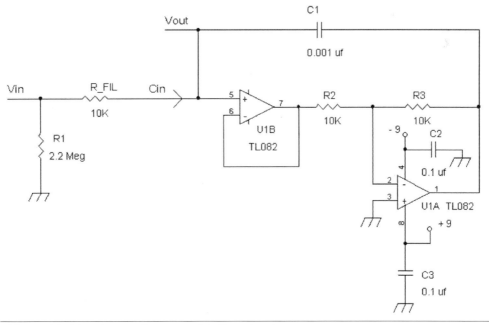

FIGURE 23-22   A capacitance multiplier circuit.

With the oscillator's frequency initially at 500 Hz, increase the oscillator's frequency and confirm that the 1 volt peak to peak signal dropped to 0.707 volt peak to peak at around 8 kHz. One can experiment by replacing R3 with a 30-kΩ resistor to multiply the C1 capacitance by $1 + 3 = 4$, which then results in $C_{in} = 0.004$ μF, or a –3-dB frequency at about 4 kHz.

In operational amplifiers, usually the second stage includes a capacitance multiplier circuit (Miller multiplier) that multiplies the feedback capacitance value greatly with $A >> 10$ to ensure a unity-gain stable op amp. More information on Miller multipliers and compensating op amps can be found in the book, *Analysis and Design of Analog Integrated Circuits*, by Paul R. Gray and Robert G. Meyer (third edition, New York: John Wiley & Sons, 1993).

It should be noted that if R2 or R3 is replaced with a variable resistor, then Cin can be varied in capacitance values. Furthermore, if a voltage-controlled amplifier replaces R2, R3, and U1A, the capacitance value of Cin can be varied as a function of voltage. An example of a voltage-controlled variable capacitor using a variable-gain amplifier can be found in U.S. Patent 4,516,041.

And now for something completely different, well, sort of. Figures 23-23, 23-24, and 23-25 take me back to my college days when I had no proto boards to conveniently build or test circuits. But I did the next best thing. I used the old IBM 80 column computer programming cards as substrates or bases for holding down the electronic components. Since these IBM cards are not easily obtainable anymore, the circuits will be "taped" down on index cards instead.

A simple current source is shown in Figure 23-23. The object of this experiment is to show that the current source changes very little in current even when the supply voltage is varied threefold. When the voltage is changed threefold across a resistor,

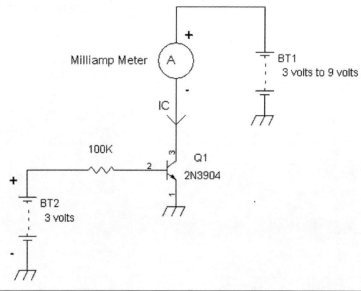

FIGURE 23-23   Simple NPN current-source circuit.

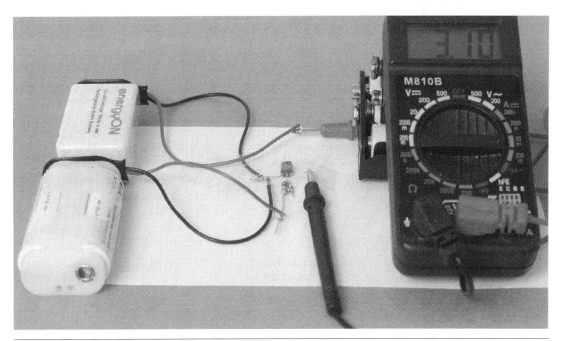

FIGURE 23-24   NPN current-source circuit assembled on an index card.

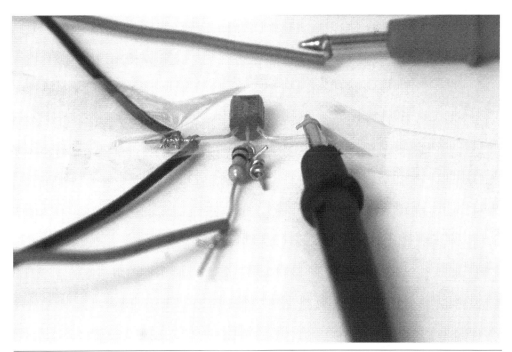

FIGURE 23-25   Close-up picture of index-card circuit board.

the current changes by threefold, but this threefold increase is not so with a constant current source.

As mentioned earlier in this book, most people can relate to a constant voltage source such as a battery or power supply, which delivers a constant voltage "regardless" of the load current, within reason, of course. A constant current source delivers a constant current "regardless" of the voltage across it. In Figure 23-23, measure the collector current via a digital VOM at a supply voltage BT1 of 3 volts DC; then remeasure the collector current at 9 volts DC for BT1. Note that the current readings are almost the same at 3 volts and 9 volts for BT1, with very little increase in current when the supply voltage is raised. Also, the output resistance of this circuit can be determined by noting the change in voltage divided by the change in collector current. R_{out} is the resistance from collector to the (grounded) emitter. Thus

$$R_{out} = \frac{\Delta VCE}{\Delta IC}$$

where ΔVCE = 9 volts − 3 volts = 6 volts, and ΔIC = IC at 9 volts − IC at 3 volts.

Depending on the transistor, even with the same part number (e.g., 2N3904), the collector current can vary better than 2:1 owing to the current gain β of Q1. The collector current also affects the output resistance R_{out} because higher collector currents usually result in lower output resistance.

For this experiment, the following measurements were recorded:

IC at 9 volts = 3.10 mA
IC at 3 volts = 2.92 mA

Thus

$$R_{out} = \frac{9V - 3V}{3.10\ mA - 2.92\ mA} = \frac{6V}{0.180\ mA} = 333\ k\Omega$$

However, because of the variation in current gain and other parameters among transistors, the reader should expect to see different results.

Figure 23-26 demonstrates another concept of a current source and its relationship with a load resistor. Thus Figure 23-26 shows a PNP constant current source with a resistive load. In this experiment, the DC current is essentially constant such that the voltage across the load resistor is determined by IC × RL = VCollector. First insert RL = 1 kΩ, and measure the voltage across the 1-kΩ resistor. Then replace RL with a 3-kΩ resistor and note that the voltage across the 3-kΩ resistor has now increased threefold.

Experiments with a Resonant Circuit

Figure 23-27 shows a series resonant circuit that is used as a notch or band-reject filter. The AC voltages across the capacitor and inductor are out of phase by 180 degrees but are of the same magnitude at the resonant frequency

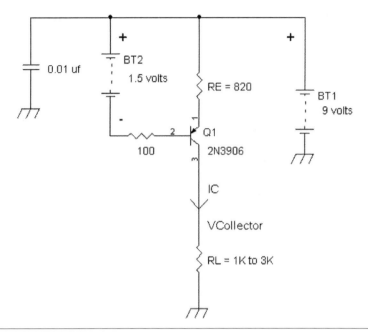

FIGURE 23-26   PNP constant-current-source experiment.

$$f_{res} = 1/[(2\pi)\sqrt{L1C1}]$$

With L1 = 470 µH and C1 = 220 pF,

$$f_{res} \approx 495 \text{ kHz}$$

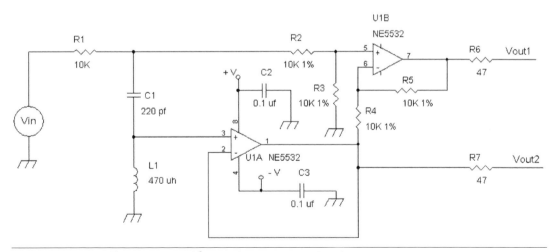

FIGURE 23-27   Experiment with a series resonant circuit.

Thus the series capacitor C1 and inductor L1 voltages add to zero (volt) at resonance (see Figure 23-27). This experiment will show that the voltage across capacitor C1 is indeed 180 degrees out of phase with the voltage across inductor L1. A voltage follower amplifier provides a voltage identical to the voltage across L1. A differential amplifier consisting of R2, R3, R4, R5, and U1B provides a signal indicative of the voltage across capacitor C1 at the output of U1B. With the values chosen for C1 and L1, the resonant frequency is about 500 kHz. Set the signal generator for 100 mV peak to peak at 500 kHz. Place a scope probe on C1/R1, and vary the frequency of the generator for minimum signal amplitude at C1/R1 to determine the resonant frequency, which should be near 500 kHz.

With a dual-trace display, Vout1 is on channel 1 and Vout2 is on channel 2 of the oscilloscope. Triggering of the signal can be chosen at channel 1 of the scope. Notice that the signals at channels 1 and 2 are about the same magnitude but opposite in phase. Therefore, there are AC voltages across the capacitor and inductor, but when these voltages are added in series, the resulting voltages cancel out (see Figure 23-28).

As seen in Figure 23-28, the top trace represents the voltage across capacitor C1, and the bottom trace represents the voltage across the inductor L1. Notice that the amplitudes are essentially the same but 180 degrees apart. The middle trace shows that summing the signals from the top and bottom traces results in zero or nearly zero.

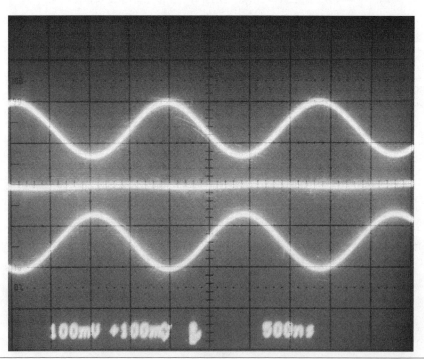

FIGURE 23-28 Dual-trace scope waveforms of voltages across C1 and L1 shown with the top and bottom traces.

For another experiment, one can sweep the signal with frequencies from 1 kHz to 1 MHz and observe the outputs at Vout1, which will provide a low-pass filter characteristic, and at Vout2, which has a high-pass-filter frequency response. With R1 = 10 kΩ, the Q of the circuit is generally too low for a low-pass or high-pass filter.

$$Q = \frac{2\pi f_{res} L}{R1}$$

where $f_{res} = \dfrac{1}{2\pi\sqrt{L1C1}} = \dfrac{1}{2\pi\sqrt{470~\mu H \times 220~pF}} \approx 495~kHz$

$$Q = \frac{2\pi(495~kHz)}{R1} = \frac{1,461~\Omega}{R1}$$

With R1 = 10 kΩ,

$$Q = \frac{1,461~\Omega}{R1} = \frac{1,461~\Omega}{10~k\Omega} = 0.146$$

which is a very low Q because the Q for a two-pole low-pass filter typically is

- ~0.577 for a Bessel filter, a more constant group delay filter for a linear phase response
- ~0.707 for a Butterworth filter, a maximally flat frequency-response filter
- ~1.0 for a slightly peaked filter

For example, change R1, which includes the output resistance of the generator, from 10 kΩ to 2,000 Ω for a Butterworth filter. If the generator has a 600-Ω output resistance, then R1 should be about 1,400 Ω to provide a total resistance of 2,000 Ω.

One may ask why such a low Q for low-pass or high-pass two-pole filters? The reason is that generally low-pass and high-pass filters provide a wide bandwidth frequency response as opposed to band-pass filters, which have a much narrower bandwidth. Typical band-pass two-pole filters require a Q > 2 (e.g., for video chroma subcarrier band-pass filters centered at 3.58 MHz or 4.43 MHz, the Q is typically between 2 and 5.) and Q > 10 for RF/IF filter circuits. Recall that high-Q circuits provide narrow bandwidth, and low-Q circuits result in wide bandwidth.

Thevenin-Equivalent Circuit

One of the most common circuits is a voltage-divider circuit using two resistors R1 and R2 that attenuates or divides the amplitude of a signal. A Thevenin-equivalent circuit of a voltage-divider circuit uses the values of the two resistors and the original voltage source to "transform" the voltage-divider circuit to a lower-voltage source and a single resistor (Figures 23-29 and 23-30).

One characteristic of reducing a voltage-divider circuit to a Thevenin-equivalent circuit is that the source resistance or Thevenin resistance is now known as the parallel combination of R1 and R2. Therefore, one can almost use the resistive divider

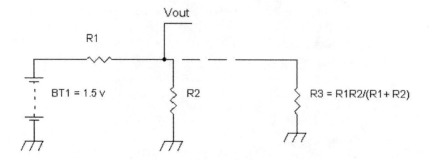

Note: For example, R1 = R2 = 20K ohms, R3 =10K ohms

FIGURE 23-29   **A simple two-resistor voltage-divider circuit.**

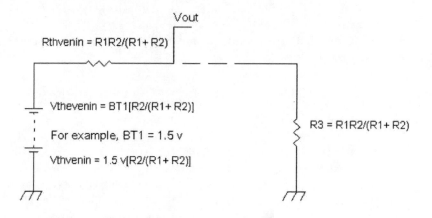

Note: For example, R1 = R2 = 20K ohms, R3 =10K ohms

FIGURE 23-30   **A Thevenin-equivalent circuit to the voltage-divider circuit.**

circuit like a step-down transformer. But unlike a transformer that preserves power transfer, a resistive divider circuit creates losses in power.

The divided down-voltage via voltage source BT1, loading into an open circuit then is

$$V_{Thevenin} = BT1 \frac{R2}{R1 + R2} = \text{Thevenin voltage}$$

In general BT1 = an arbitrary voltage V_{in}, and so

$$V_{Thevenin} = V_{in} \frac{R2}{R1 + R2} = \text{Thevenin voltage}$$

For example, dividing down the voltage allows for providing a low-impedance source. Therefore, to confirm that the Thevenin-equivalent circuit really works in that

$$R_{Thevenin} = \frac{R1\,R2}{R1 + R2} = \text{Thevenin resistance}$$

let R1 = R2 = 20 kΩ so that

$$R_{Thevenin} = \frac{20 \text{ k}\Omega \times 20 \text{ k}\Omega}{20 \text{ k}\Omega + 20 \text{ k}\Omega} = 10 \text{ k}\Omega$$

A cell or battery, such as a 1.5-volt battery = V_{in}, provides the voltage to R1 = 20 kΩ as shown in Figures 23-29 and 23-30. The voltage at R2 = 20 kΩ then should measure as

$$V_{Thevenin} = V_{in} \frac{R2}{R1 + R2} = 1.5 \text{ V} \frac{20 \text{ k}\Omega}{20 \text{ k}\Omega + 20 \text{ k}\Omega} = (1.5)\frac{1}{2} = 0.75 \text{ V} = \text{voltage at R2}$$

Now take a 10-kΩ resistor for R3 and connect it in parallel with R2. If the Thevenin resistance is indeed 10 kΩ, which is now the source resistance, paralleling R3 = 10 kΩ should drop the 0.75 volt to half, or 0.375 volt.

One can try different voltages and values for R1 and R2 to confirm the Thevenin voltage and resistance. For example, use a 9-volt battery for Vin, and let R1 = 47 kΩ and R2 = 2.7 kΩ. Then

$$V_{Thevenin} = V_{in} \frac{R2}{R1 + R2} = 9V \frac{2.7 \text{ k}\Omega}{47 \text{ k}\Omega + 2.7 \text{ k}\Omega} = 9V \frac{2.7 \text{ k}\Omega}{49.7 \text{ k}\Omega}$$

$$= 0.489 \text{ V} = \text{voltage at R2 without R3 connected}$$

$$R_{Thevenin} = \frac{47 \text{ k}\Omega \times 2.7 \text{ k}\Omega}{47 \text{ k}\Omega + 2.7 \text{ k}\Omega} = 2.55 \text{ k}\Omega$$

If 2.55 kΩ = R3, now connect R3 across R2, and the 0.489 volt should drop by half to 0.2445 volt.

The Thevenin-equivalent circuit also is applicable to AC voltage sources. With AC signals, one can use either a resistive or a capacitive voltage divider. Therefore, the voltage-divider circuit sometimes can be used as a minimum-loss pad circuit for impedance matching (e.g., from high to low impedance).

Analyzing a Bridge Circuit

Figure 23-31 shows a bridge circuit with current flowing through R3 as the particular DC current of interest. R3 may represent an ammeter such as a galvanometer's coil resistance, and R1 and R2 are the same resistance. In a Wheatstone bridge circuit, normally an unknown resistor such as R5 is inserted and a known calibrated variable resistance in R4 is varied until there is no current flowing into R3. When the null in current through R3 is achieved, the calibrated resistance value is read off R4.

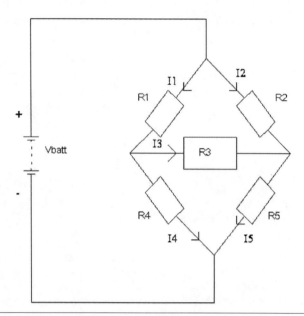

FIGURE 23-31   A bridge circuit.

For this experiment, suppose that we want to find the current through R3 for any value of R1, R2, R3, R4, and R5 for a given voltage supply. How would one start to solve the equations? Five currents are identified as I1, I2, I3, I4, and I5, and the five resistors, R1, R2, R3, R4, and R5, are shown, along with a known voltage Vbatt. We have five unknown currents and need to set up five equations to solve for the current through R3. Or do we?

Is there a faster way to solve for I3, and if so, how is it done? These and other questions enquiring minds would like know. Therefore, the strategy to solve this problem is to "Get Smart."

Would you believe that this problem can be solved in 99 seconds? Or how about 86 seconds? (See the TV show *Get Smart*.) Actually, this problem can be solved in 20 seconds or less. Of course, one may say, "Surely you are joking!" and my response would be, "I am not joking . . . and don't call me Shirley!" (See the movie *Airplane!*).

Okay, kidding aside, the bridge circuit can be split and analyzed by converting the four outer resistors into two Thevenin-equivalent circuits (see Figures 23-32 and 23-33).

The voltage divider consisting of R1 and R4 with Vbatt is equivalent to a Thevenin-equivalent circuit with

$$V_{\text{Thevenin\_R1R4}} = V\text{batt} \frac{R4}{R1 + R4}$$

and the Thevenin resistor is just the parallel combination of R1 and R4; that is,

$$R_{\text{Thevenin\_R1R4}} = \frac{R1 R4}{R1 + R4}$$

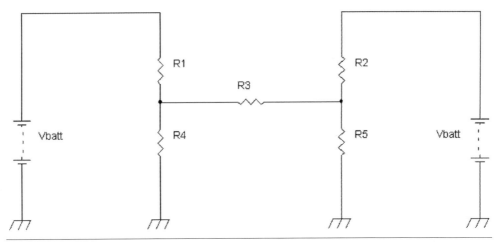

FIGURE 23-32   The bridge circuit is redrawn with a ground reference and two voltage-divider circuits.

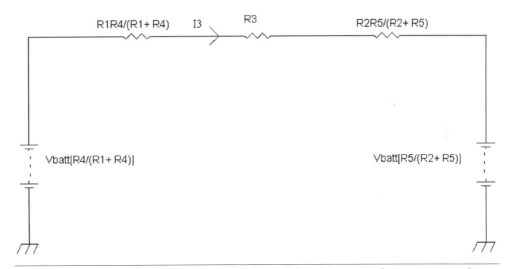

FIGURE 23-33   The circuit in Figure 23-32 is redrawn as two Thevenin-equivalent circuits.

Similarly for the voltage divider consisting of R2 and R5, the Thevenin voltage and resistor are

$$V_{Thevenin\_R2R5} = Vbatt \frac{R5}{R2 + R5}$$

$$R_{Thevenin\_R2R5} = \frac{R2R5}{R2 + R5}$$

With R3 bridging or connecting the two Thevenin circuits, the current through R3 is reduced to one equation and one unknown. Thus

$$I_3 = \frac{V_{Thevenin\_R1R4} - V_{Thevenin\_R2R5}}{R_{Thevenin\_R1R4} + R3 + R_{Thevenin\_R2R5}} = \frac{Vbatt\frac{R4}{R1 + R4} - Vbatt\frac{R5}{R2 + R5}}{\frac{R1R4}{R1 + R5} + R3 + \frac{R2R5}{R2 + R5}} = I_3$$

One can further simplify I_3 as

$$I_3 = \frac{Vbatt(\frac{R4}{R1 + R4} - \frac{R5}{R2 + R5})}{\frac{R1R4}{R1 + R4} + R3 + \frac{R2R5}{R2 + R5}}$$

So there you have it, the answer for I_3, which is the current through resistor R3 solved within 20 seconds if your handwriting is quick enough. And if you solved the problem in 21 seconds instead, a famous TV (or movie) secret agent would say to you, "Missed it by that much!" Actually I *am* kidding now, and solving the problem in 21 seconds to a few minutes is fine.

Of course, one actually can see if this analysis works. Set Vbatt = 9 volts; R1 = R2 = R3 = R4 = 20 kΩ; and R5 = 10 kΩ. The current through R3 should be

$$\frac{9V(\frac{20\ k\Omega}{20\ k\Omega + 20\ k\Omega} - \frac{10\ k\Omega}{20\ k\Omega + 10\ k\Omega})}{[\frac{20\ k\Omega(20\ k\Omega)}{20\ k\Omega + 20\ k\Omega} + 20\ k\Omega + \frac{20\ k\Omega(10\ k\Omega)}{20\ k\Omega + 10\ k\Omega}]} = \frac{9V(\frac{1}{2} - \frac{1}{3})}{10\ k\Omega + 20\ k\Omega + 6.67\ k\Omega} = \frac{9V(\frac{1}{6})}{36.67\ k\Omega} = \frac{1.5V}{36.67\ k\Omega} = 40.9\ \mu A = I_3$$

With 5 percent resistors for R1 = R2 = R3 = R4 = 20 kΩ, R5 = 10 kΩ 1 percent, and Vbatt = 9 volts, the measured I_3 = 39.5 μA, which is very close to the calculated value of 40.9 μA.

Some Final Thoughts on the Book

When I worked at Ampex, the company that invented the first practical videotape recorder, my mentor told me that in school you learn how to learn, and after you have completed your studies, then comes the real learning. Learning electronics or almost anything, for that matter, is a lifelong process. What I have found is that we all become to some degree self-taught in our subsequent studies. For instance, I learned the subject of noise from reading books and asking questions of experts in the field. I also had two or three great mentors who showed me the ropes in designing circuits. So, if you can find a mentor, please learn as much as you can from him or her. Experience counts a lot!

So, for me, I am always exploring electronics and related subjects such as math. And although readers may think after reading Chapters 13 through 22 that I am too math-centric, these particular chapters were written to show how the math is used to explain signals and circuits but not how to necessarily design circuits. However, I also will take knowledge from history and music as well for finding a solution to an engineering problem. One key to invention is the integration or merging and picking and choosing from the various knowledge bases. For example, one of my audio scrambling patents utilized timing signals in a form of "Rubato," or stolen time. Rubato is a way of playing music where the tempo speeds up and slows down for emotional effect.

There are many approaches to designing analog circuits. And the readers should know that I am a circuit designer first, as can be seen in Chapters 4 through 12. That is my usual "state." One approach is to scope out the analog design or problem mathematically and then design, which can be verified by a simulation program such as PSPICE and/or by building the circuits and testing them. The other approach is to design by intuition based on lots of experience and judgment and then later to analyze mathematically/scientifically why the design worked. I tend to switch back and forth between these two methods, but at the end of the day, I want to ensure that the design works both by practice and by theory. Just having an analog design work without full scrutiny sometimes is not good enough in terms of reliability or repeatability of the circuit.

This book is really meant for self-teaching by anyone who is interested in electronics. The math, although voluminous, really never gets too complicated beyond algebra and trigonometry. When I explain technology to people, I generally do not use too much math at all, but if I know that the person has some math, even algebra, I can include "simple" mathematics in my tutorials.

What the reader also should take away from this book is that designs do not always work the first time, even when "every" detail had been taken into account. But perseverance is one of the keys to successful problem solving. For example, one of my first full-blown video designs involved designing a miniature TV monitor, a cathode-ray tube (CRT) viewfinder for a TV camera. The problem I was having was that the horizontal scanning system was nonlinear, causing people to literally change width or weight as they moved across the displayed TV frame. I tried various conventional approaches that got close but not good enough. And because of a space constraint, I could not use a linearity coil. So I was stuck for two weeks not producing a solution. Finally, I solved the problem by applying negative feedback in an unconventional manner and sending a correction waveform via the damper diode.

There are times when a solution to a particular problem is not accessible through the Internet, books, or papers. Thus the particular problem has to be reduced to: What is the real problem? Finding out the main problem takes time. For example, in Chapter 7, the project was to show a transistorized regenerative radio. I thought it would be easy because my tube version performed straight from the get-go. What I found, however, was that the first try ended up with very poor sensitivity. I had to literally get out and take a hike for about half an hour before I realized why the radio was going into oscillation too soon, and by the time I finished the hike, I figured out a solution. During the hike, I figured out that there was insufficient radio-frequency (RF) gain for demodulation without applying regeneration, and I was trying to apply regeneration to bring up or compensate for a weak RF signal. However, because the RF signal was much too low in strength, the radio broke into oscillation instead when the regeneration control was turned up. Therefore, the solution was to first provide enough RF signal and then devise a way to control the regeneration (positive-feedback signal) in a manner that allowed the demodulation of the amplitude-modulated (AM) signal.

Therefore, designs do not always work as planned. But keep at it, and if you are stuck, take a shower, go to the park, get out of the office, and/or get some sleep.

Writing this book has been a fantastic journey for me, as well as a great learning experience. Well, that is all I have to say—except thank you for your interest. And after writing over a 100,000 words, I think I will watch some cartoons and listen to "The Merry Go-Round Broke Down." So . . . *That's all folks!*

References

1. Class Notes EE141, Paul R. Gray and Robert G. Meyer, UC Berkeley, Spring 1975.
2. Class Notes EE241, Paul R. Gray and Robert G. Meyer, UC Berkeley, Fall 1975.
3. Class Notes EE140, Robert G. Meyer, UC Berkeley, Fall 1975.
4. Class Notes EE240, Robert G. Meyer, UC Berkeley, Spring 1976.
5. Paul R. Gray and Robert G. Meyer, *Analysis and Design of Analog Integrated Circuits*, 3rd ed. New York: John Wiley & Sons, 1993.
6. Kenneth K. Clarke and Donald T. Hess, *Communication Circuits: Analysis and Design*. Reading: Addison-Wesley, 1971.
7. Mischa Schwartz, *Information, Transmission, and Noise*. New York: McGraw-Hill, 1959.
8. William G. Oldham and Steven E. Schwarz, *An Introduction to Electronics*. New York: Holt, Reinhart and Winston, Inc., 1972.
9. Allan R. Hambley, *Electrical Engineering Principles and Applications*, 2nd ed. Upper Saddle River: Prentice Hall, 2002.
10. Murray H. Protter and Charles B. Morrey, Jr., *Modern Mathematical Analysis*. Reading: Addison-Wesley, 1964.
11. Texas Instruments, *Linear Circuits Amplifiers, Comparators, and Special Functions*, Data Book 1. Dallas: Texas Instruments, 1989.
12. Motorola Semiconductor Products, Inc., *Linear Integrated Circuits*, Vol. 6, Series B., 1976.
13. Thomas M. Frederiksen, Intuitive IC Op Amps from Basics to Useful Applications (National's Semiconductor Technology Series). Santa Clara: National Semiconductor Corporation, 1984.
14. Alan B. Grebene, *Bipolar and MOS Analog Circuit Design*. New York: John Wiley & Sons, 1984.
15. U.S. Patent 4,516,041, Ronald Quan, "Voltage Controlled Variable Capacitor," filed on November 22, 1982.
16. U.S. Patent 5,058,159, Ronald Quan, "Method and Apparatus for Scrambling and Descrambling Audio Information Signals," filed on June 15, 1989.

Appendix 1

Parts Suppliers

Oscillator Coils, Intermediate-Frequency (IF) Transformers, Audio Transformers

1. Mouser Electronics: www.mouser.com (Xicon transformers and coils).
2. Centerpointe Electronics: www.cpcares.com (Xicon transformers and coils; for IF transformers, search under "Xicon I.F. Transformer.")

Antenna Coils

1. Scott's Electronics (Hard to Find Electronics Parts): www.angelfire.com/electronic2/index1/loopstick.html (for ferrite antenna coils).
2. MCM Electronics: www.mcmelectronics.com (Search for "loop antenna," the AM radio type. This type of loop antenna will require an RF transformer as described in Chapter 3. For example, see the link http://electronics.mcmelectronics.com/?N = &Ntt = loop + antenna.)
3. Antique Electronics Supply: www.tubesandmore.com (The search term is "antenna coil." As of 2012, the only antenna coil available has an inductance of 240 μH, which will match with a 330- to 365-pF variable capacitor.)
4. eBay (Search under "antenna coil," "loopstick antenna," "ferrite bar antenna," or "ferrite antenna." The typical inductances are from 330 to 788 μH. For antenna coils with inductances > 600 μH, one can unwind some of the wire and measure for the desired inductance using an inductance meter.)
5. eBay (Search under "ferrite bar" or "ferrite rod." These blank rods or bars can be wound with wire such as Litz wire to the desired inductance. Typically, the length is at least 2 inches or at least 50 millimeters.)

Variable Capacitors

1. Scott's Electronics (Hard to Find Electronics Parts): www.angelfire.com/electronic2/index1/index.html (For a 140-pF/60-pF dual variable capacitor, www.angelfire.com/electronic2/index1/Variable141pf.htm; for a twin

266-pF/266-pF variable capacitor, www.angelfire.com/electronic2/index1/
266pf.html; for a twin 335-pF/335-pF variable capacitor, www.angelfire.com/
electronic2/index1/335PF-Polyvaricon.html.)

2. eBay (Search under the terms "polyvaricon," "variable capacitor," "tuning
   capacitor," or "crystal radio parts." Note that some of the variable capacitors
   on eBay are trimmer or padding capacitors, which do not have sufficient
   capacitance, e.g., <100 pF, for the radios in this book.)

Crystal Earphones

1. Electronics (Hard to Find Electronics Parts): www.angelfire.com/electronic2/
   index1/index.html
2. eBay (Search under the terms "crystal earphone," "crystal radio," or "crystal radio
   parts.")

Passive Components, Resistors, Capacitors, Fixed-Valued Inductors

1. Mouser Electronics: www.mouser.com
2. Digi-Key Corporation: www.digikey.com
3. Anchor-Electronics: www.anchor-electronics.com
4. HSC Electronic Supply: www.halted.com/commerce/index.jsp?czuid=
   1338011465538 (May have to purchase items at their stores.)

Crystals

1. Mouser Electronics: www.mouser.com
2. Digi-Key Corporation: www.digikey.com
3. Anchor-Electronics: www.anchor-electronics.com
4. HSC Electronic Supply: www.halted.com/commerce/index.jsp?czuid=
   1338011465538 (May have to purchase items at their stores; call to confirm.)

Ceramic Resonators (~455 kHz)

1. Mouser Electronics: www.mouser.com
2. Digi-Key Corporation: www.digikey.com
3. HSC Electronic Supply: www.halted.com/commerce/index.jsp?czuid=
   1338011465538 (May have to purchase items at their stores; call to confirm.)
4. eBay (Search term is "ceramic resonator.")

Ceramic Filters (~455 kHz)

1. Mouser Electronics: www.mouser.com (The search term is "455-kHz ceramic filter.")
2. HSC Electronic Supply: www.halted.com/commerce/index.jsp?czuid = 1338011465538 (May have to purchase items at their stores; call to confirm.)
3. eBay (The search term is "ceramic filter 455 kHz.")

Transistors, Diodes, and Integrated Circuits

1. Mouser Electronics: www.mouser.com
2. Digi-Key Corporation: www.digikey.com
3. Anchor-Electronics: www.anchor-electronics.com
4. HSC Electronic Supply: www.halted.com/commerce/index.jsp?czuid = 1338011465538 (May have to purchase items at their stores; call to confirm.)
5. MCM Electronics: www.mcmelectronics.com
6. Jameco Electronics: www.jameco.com

Low-Noise Transistors and JFETs

1. Linear Integrated Systems, Inc.: www.linearsystems.com

Loudspeakers

1. MCM Electronics: www.mcmelectronics.com (Search terms are "cone speaker," "Mylar cone speaker," "50-mm speaker," "66-mm speaker," or "85-mm speaker.")
2. Jameco Electronics: www.jameco.com (Search term is "speakers.")
3. HSC Electronic Supply: www.halted.com/commerce/index.jsp?czuid = 1338011465538 (May have to purchase items at their stores; call to confirm.)

Appendix 2

Inductance Values of Oscillator Coils and Intermediate-Frequency (IF) Transformers

42IFxxx Pin Out

Bottom View

FIGURE A2-1 Pin-out of coil.

TABLE A2-1 AM Radio Oscillator Coils

Part Number	Impedance Ratio	Turns Ratio	Inductance Pins 1 & 3	Inductance Pins 1 & 2	Inductance Pins 2 & 3	Inductance Range
42IF100		13:1	355 μH	0.6 μH	315 μH	+80% −20%
42IF110		35:1	298 μH	0.4 μH	287 μH	+85% −20%
42IF300		10:1	339 μH	0.4 μH	318 μH	+35% −20%

TABLE A2-2 AM Radio IF Transformer Without Internal Capacitor, Nominally Used at ~455 kHz

Part Number	Impedance Ratio	Turns Ratio	Inductance Pins 1 & 3	Inductance Pins 1 & 2	Inductance Pins 2 & 3	Inductance Range
42IF104	50K:500	22:1	627 µH	136 µH	180 µH	+85% −15%
42IF106	20K:5K	6:1	635 µH	65 µH	293 µH	+70% −20%

TABLE A2-3 FM Radio IF Transformer Without Internal Capacitor, Nominally Used at ~10.7 MHz

Part Number	Impedance Ratio	Turns Ratio	Inductance Pins 1 & 3	Inductance Pins 1 & 2	Inductance Pins 2 & 3	Inductance Range
42IF124	15K:500	14:1	2.6 µH	0.3 µH	0.3 µH	+50% −25%

IF Transformer with Internal Capacitor Removed

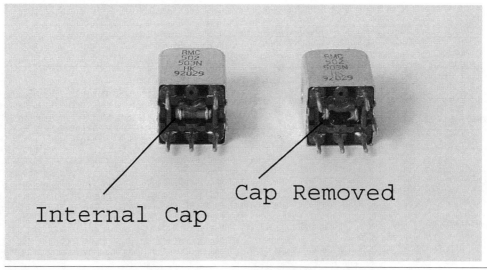

FIGURE A2-2 Example IF transformer with internal capacitor removed.

TABLE A2-4   AM Radio 455-kHz IF Transformers with Internal Capacitor Removed

Part Number	Impedance Ratio	Turns Ratio	Inductance Pins 1 & 3	Inductance Pins 1 & 2	Inductance Pins 2 & 3	Inductance Range
42IF101	60K:600	22:1	692 µH	208 µH	141 µH	+40% −12%
42IF102	30K:500	22:1	727 µH	95 µH	295 µH	+60% −20%
42IF103	20K:6K	6:1	576 µH	58 µH	268 µH	+50% −10%
42IF302	30K:500	22:1	682 µH	80 µH	296 µH	+30% −25%
42IF303	20K:5K	7:1	652 µH	56 µH	328 µH	+35% −25%

TABLE A2-5   FM Radio 10.7-MHz IF Transformers with Internal Capacitor Removed

Part Number	Impedance Ratio	Turns Ratio	Inductance Pins 1 & 3	Inductance Pins 1 & 2	Inductance Pins 2 & 3	Inductance Range
42IF122	15K:300	14:1	2.8 µH	0.3 µH	<0.1 µH	+70% −30%
42IF123	25K:4K	7:1	2.5 µH	0.7 µH	<0.1 µH	+80% −20%
42IF129	15K:100	18:1	6.1 µH	~0.1 µH	2.3 µH	+20% −35%

Note: All values of impedance ratio and turns ratio were referenced from Mouser Electronics Catalog No. 610, May–July 2002.

Appendix 3

Short Alignment Procedure for Superheterodyne Radios

IF Transformer Alignment

With a modulated 455-kHz signal, place the signal generator's output via a wire near the antenna coil or loop antenna. Set the variable capacitor to maximum capacitance or fully closed. With the earphone or loudspeaker, listen for the modulating signal and adjust each of the IF transformers for maximum volume.

Oscillator Coil and Trimmer Capacitor Adjustments

With the variable capacitor's trimmer capacitors for the RF and Oscillator sections set to half capacitance, provide a modulated 540-kHZ source (or 537-kHz source from Chapter 4) by coupling a wire near the antenna coil or loop antenna. Set the variable capacitor to an almost fully closed position with about 1 to 3 degrees of arc from the fully closed position. Adjust the oscillator coil for maximum volume. Then if possible, slide the coil of the ferrite bar antenna to adjust for maximum volume. If an RF transformer is used, adjust the RF transformer for maximum volume.

Now, set the test generator to 1600 kHz (or use the third harmonic of 537 kHz = 1611 kHz) with modulation. Readjust the variable capacitor to near minimum capacitance, or about 1 to 5 degrees of arc from the fully open position. Adjust the oscillator trimmer capacitor for maximum volume. Generally, the RF trimmer capacitor will not be set to an optimum for the low and high end of the AM band unless there is a back and forth adjustment between the oscillator coil, oscillator trimmer capacitor, and RF coil or antenna coil, etc. Therefore, as a compromise, one can adjust the RF trimmer capacitor for maximum signal strength as desired by the operator. That is, the RF trimmer capacitor can be adjusted for maximum signal at the mid band, the low end of the band, or the high end of the band. No generator is necessary, just tune to a station in the desired portion of the AM band and adjust the RF trimmer capacitor for maximum volume.

Index

Made in the USA
Columbia, SC
30 December 2024

50840271R00272